D1270269

Polyelectrolytes for Water and Wastewater Treatment

Editor

William L. K. Schwoyer

The Permutit Company
Research and Development Center
Monmouth Junction, New Jersey

CRC Press, Inc.
Boca Raton, Florida

Library of Congress Cataloging in Publication Data

Polyelectrolytes for water and wastewater treatment.

Bibliography: p.
Includes index.
1. Water — Purification — Flocculation. 2. Sewage
— Purification — Flocculation. 3. Polyelectrolytes.
I. Schwoyer, William L. K.
TD455.P64   628.1′622   80-15346
ISBN 0-8493-5439-0

Direct all inquiries to CRC Press, Inc., 2000 N.W. 24th Street, Boca Raton, Florida, 33431.

© 1981 by CRC Press, Inc.

International Standard Book Number 0-8493-5439-0

Library of Congress Card Number 80-15346
Printed in the United States

FOREWORD

The co-authors of this book are employed in the application of the particular aspect of polyelectrolyte research and development, manufacture or use about which they have written. This book reflects their experiences in, and philosophies concerning the use of, polyelectrolytes for water and wastewater treatment. Application work at this level frequently requires interpreting the more abstract, theoretical, and laboratory work in terms of operating full-scale equipment. This book should be useful for consulting engineers engaged in water and wastewater plant design, engineers responsible for plant operation, and as an auxiliary test for graduate and advanced undergraduate level environmental engineering students.

There is a certain amount of overlap concerning various aspects of chemical conditioning and colloid chemistry in certain chapters. The authors' lines of reasoning pertaining to these characteristics have been retained in order to afford the reader a clear understanding of how these considerations are applied to the particular piece of equipment or unit operation described.

The book also is a reflection of the state of knowledge of the various aspects of polyelectrolyte manufacture and use. It becomes evident while reading through the book that much more is known about the chemistry, and hence the control of polymer synthesis and manufacture, than is known about the waste streams to which they are applied. The variations in the levels of knowledge from facet to facet of the overall purview of polyelectrolytes and their uses reflects an economic fact of life. More effort is spent in income producing than in nonincome producing areas of our economy.

THE EDITOR

William Lawrence Kenneth Schwoyer has been employed by The Permutit Company, Inc. for the past eight years and is currently Development Manager, Liquid-Solids Separations, at their Princeton, New Jersey Research and Development Center. Prior to his employment with The Permutit Company, he was in charge of explosives physical testing and quality control for the Trojan-U.S. Powder Division of Commercial Solvents Corporation. He holds a B.S. degree in Chemistry from the Engineering College of Lehigh University. His publications include 14 U.S. Patents, in addition to numerous papers and articles pertaining to liquid-solids separation and sludge dewatering, in particular. Professional affiliations include American Institute of Chemical Engineers, American Chemical Society, and the American Institute of Mining Engineers.

CONTRIBUTORS

Ralph J. Chamberlain
Project Leader
Water Treating Chemicals Department
American Cyanamid Company
Stamford, Connecticut

Nathan M. Levine, Ph.D.
Manager,
Mining Applications Research
Celanese Plastics and Specialties
 Company
Jeffersontown, Kentucky

Lionel B. Luttinger, Ph.D.
Senior Chemist
The Permutit Company
Princeton, New Jersey

John G. Penniman, Jr.
President
Paper Chemistry Consulting
 Laboratory
Carmel, New York

Richard M. Schlauch
Research Associate
The Permutit Company
Research and Development Center
Monmouth Junction, New Jersey

Norman Vorchheimer, Ph.D.
Supervisor
Research Department
Betz Laboratories, Inc.
Trevose, Pennsylvania

John G. Walzer
District Engineer
Black Clawson Company
Middletown, Ohio

TABLE OF CONTENTS

Chapter 1

SYNTHETIC POLYELECTROLYTES

N. Vorchheimer

TABLE OF CONTENTS

I. INTRODUCTION

The invention and use of synthetic polyelectrolytes has grown rapidly in the last ten years. This can be attributed both to increasing awareness and governmental pressures for environmental purposes, and to increasing awareness of the economic benefits derivable from the use of these versatile molecules. The great bulk of the activity has naturally come from industrial laboratories, and it will be the author's intent to focus on the commercially important polymers.

Many people use the terms polyelectrolyte and water-soluble polymer synonymously, with some justification. Strictly speaking, a polyelectrolyte is a polymer having ionizable groups, usually one or more per repeat unit, while there are many water-soluble polymers that have no ionizable groups, such as polyvinylalcohol, unhydrolyzed acrylamide homopolymers, or polyethyleneoxides. However, the great majority of the nonionic polyacrylamides contain a minor percentage of carboxyl groups resulting from hydrolysis of amides, so that the number of water-soluble polymers with unionizable groups is small compared to the multitude of true polyelecholytes. In addition, there is a class of potentially important polyelectrolytes for water treatment which are not quite water-soluble, and there are water-insoluble plastics such as Surlyn which contain ionizable groups. For the most part, however, the term polyelectrolyte refers to water-soluble polymers.

Because of the extensiveness of the literature, and the existence of numerous review articles, the author will make no attempt to make this a comprehensive chapter, and apologizes in advance to those inventors and researchers of important materials whose work may have been overlooked. As previously mentioned, this chapter will concentrate on the polyelectrolytes that have achieved significance specifically for water treatment, by virtue of their ability to perform a task at a price which makes their use compelling. This will be referred to as cost-performance. In this way, the author hopes to achieve some measure of dissimilarity from those who have previously surveyed the field. A working knowledge of organic and polymer chemistry on the part of the reader has been assumed also.

The term ''synthetic'' will mean man-made, starting from small molecules. Thus, that large class of starch-based polyelectrolytes used for water treatment will only be mentioned in passing. It is also not the intent of this article to contend with the vast literature of polypeptides, except where they may find utility in water treatment. The extensive use of polyalkylene-oxides for other than water and waste treatment is also beyond the scope of this work.

Discussion will also include those polymers which may not now have commercial importance in the water treatment field, but which have some potential in this area.

The subject of molecular weight is a recurring thorn in the side of polymer chemists who must deal with a nontechnical audience on a daily basis. The patent and trade literature is replete with references to molecular weights for particular polymer systems. Rarely is mention made of the methods used for determining molecular weights, which except for monodisperse polymers, will give widely different answers. Also, intrinsic viscosity is frequently used to calculate molecular weights using the Mark-Houwink equation,[1] when the constants used are open to considerable question, because differences in polymerization conditions and monomers used may lead to polymers having different structures than those used to determine the constants.

In this chapter, the chief distinctions made with regard to molecular weight are between the high-molecular-weight vinyl addition polymers and the low- to medium-molecular-weight condensation or addition polymers. With the former type, molecular weights tend to be in the millions, while in the latter, molecular weights span the range from about 500 up to several hundred thousand. In general, since the polymers in question are always used in solution, the important effects of molecular weight on bulk properties are less a consideration. These differences will only be important for activity, where the effects can be dramatic, and for the physical form of the polymers.

II. PHYSICAL FORMS OF SYNTHETIC POLYELECTROLYTES

The physical state of polyelectrolytes has always been a vitally important area in the marketing and sale of these polymers for water and waste treatment. It has become even more important in recent years, when the distinctions, no longer clear-cut, have become a subject of intense competition. Also, much information on method of manufacture can be gained from the physical form of the polyelectrolyte.

Historically, synthetic polyelectrolytes used as flocculants were solids if they were high molecular weight (addition polymers) and liquids (actually concentrated solutions) if they were low-to-medium molecular weight (condensation polymers). The advent of inverse emulsion technology has resulted in the loss of this distinction, which also does not hold for the low- to medium-molecular-weight anionic deposit-control agents (*vide infra*), which are vinyl addition polymers available in both solution and dry form.

A. Solids

The existence of high-molecular-weight polyelectrolytes in solid form can be attributed to one major factor, economics. It would be desirable if all polymers used for water treatment were already in solution, easily dilutable for the application, not requiring elaborate dissolution or inversion equipment (for water-in-oil emulsion polymers) or precautions. However, the viscosity of high-molecular-weight polyelectrolyte solutions in water is severely concentration dependent — for most of these polymers, solutions having concentrations above 1% have viscosities so high that they are virtually unpumpable, and thus, unusable. Conversely, to prepare and ship polymer solutions that are 99% or more water is prohibitively expensive, both from the standpoint of freight charges and the necessity of an extra manufacturing step, since economics again dictate that polymers must be prepared at considerably higher concentrations.

There are at least three types of solid, high-molecular-weight polyelectrolytes: granules, flakes, and beads. All these result from differences in processing of polymers prepared in aqueous solution, or in an aqueous phase of a suspension.

I. Granular

The synthesis of high-molecular-weight polyelectrolytes in concentrated aqueous solution results in a rigid, tough, rubbery gel, comparable in consistency to a soft gum eraser. The isolation of polymer from this gel-like state can be accomplished in a number of ways. Use of a precipitation process, in which the water is extracted from the finely ground gel particles by a second solvent (in which the polymer is insoluble, but completely miscible with the water), results in solid polymer in granular form.[2] A number of techniques have been used for the grinding of the gel particles, which can be difficult because of their rubbery, but nevertheless tacky, state. These include various cutting devices for reducing the particle size.[3] Once ground and extracted, the particles can be dried and ground even further, if desired.

The second solvents (polymer nonsolvent) used most frequently are methanol and acetone, although many other solvents are suitable. Polymers manufactured using these techniques usually contain small amounts of water and the second solvent. The presence of small amounts of the second solvent may be beneficial for stabilizing the polymers against loss of molecular weight, when subsequently redissolved at low concentrations for use.[4]

In another granular polymer process the rubbery gel is cut into finely divided form and dried, usually in a dry gas stream, in the absence of a second solvent.[10] Salts have been added to the polymerization in this technique to reduce the tackiness of the particles and make them more free flowing.

A third method which results in polymer in finely divided granular form is precipitation polymerization. It has been believed that this method does not give high-molecular-weight polymers,[20] but recent technology has refuted this belief.[5,8] In this method, the monomer(s) are dissolved in water and another water-soluble solvent. The polymer once formed is insoluble in the dual solvent, and under the proper conditions precipitates in finely divided granular form. The second solvent, such as acetone,[5] t-butanol,[6] and combinations thereof,[5] can comprise up to 70% of the total solvent. The addition of salts[7] or polyvinylalcohols[8] are claimed to assist in the separation of the precipitated polymers in easily recoverable, nonsticky granules.

A precipitation process has been used in which the precipitating medium consists solely of an aqueous salt solution.[9] In this process, the precipitated polymer can be used directly as a stable suspension, or the polymer can be isolated by coagulation of the suspension.

Monomer concentrations of up to 50% of the total for precipitation polymerizations have been claimed. This process avoids the costly isolation processes required for solution polymerization.

2. Flake Form

If, instead of treating the rubbery gel by any of the processes outlined above, it is dried directly, typically with a drum drier, polymer in flake form results.[12] For processing purposes, polymerization at lower concentrations is desirable, in order to reduce the toughness of the gel. In this process, high temperatures are generally required to drive off the aqueous solvent, so that although the process is more economical than the solvent-extraction methods, degradation of the polymers is more severe. The addition of inorganic salts to the polymerization medium appears to minimize the subsequent degradation upon drying.[13]

3. Bead Form

The technology for preparing polyelectrolytes in bead form has been known for at least 17 years.[14] At one time, a number of companies marketed polymers in this form, but a combination of lack of adequate molecular-weight control and poor solubility resulted in the withdrawal of these types from the marketplace. Recently, improvements in both areas have resulted in a re-introduction of this type to the water-treatment market.[15]

Whether a system yields solid polymer in bead form or a stable polymer/water-in-oil emulsion (*vide infra*) is determined by the size of the particle of the discontinuous phase (in this case, the polymer/water phase). In both processes, the monomer or monomers in the aqueous phase are emulsified or suspended in a suitable nonsolvent, using surfactants, suspending agents, or combinations thereof. The water can first be suspended or emulsified followed by addition of the monomer(s).[16] The important parameters in determining the particle size for bead polymerization are stirring speed and agitator type, concentration and type of surfactant, kettle geometry, and solvent.[16]

The polymerization of water-soluble monomers in bead form has been accomplished both via a suspension process, where the droplets are coated by solid or liquid suspension stabilizers or agglomeration inhibitors, (which maintain the beads as discrete particles,[14]) or by an emulsion process, where a surface-active emulsifying agent is required.[16] Typical of suspending agents used in the former process are silanized inorganic hydroxy-oxides, such as silanized silicas, treated clays,[14] and organic polymers,[17] all of which have in common a low hydrophile-lipophile balance (HLB),[18] promoting water-in-oil suspensions. Surfactants used in the latter processes have been claimed to be primarily of the type to form water-in-oil emulsions (low HLB), for example, sorbitan mono-oleates,[16] and oil-in-water emulsions, such as the nonionic ethoxylated nonyl phenols.[19] In some cases, combinations of surfactants and suspending agents are used.[16] The process is nevertheless of the emulsion type.

Typical ratios of oil-to-aqueous phase (including monomers) in polymerizations yielding polymer beads range from 1-1 to 3-1, while surfactant or suspending agent concentrations ase usually 2% or less, based on the total recipe. It will be seen that these ratios differ markedly from those required for the inverse emulsion polymerization.

B. Liquids

The advantage of supplying a water treatment chemical in liquid form is primarily its ease of use. Whereas the solid products must be carefully dispersed and stirred before complete dissolution takes place, the dilution of a concentrated solution is relatively simple; even the water-in-oil emulsion polymer technology has advanced so that dilution has become simplified.

Nevertheless, there are problems associated with all forms of polyelectrolytes. In the case of solids, the primary problems are insoluble fractions (sometimes called "fish-eyes"), rate of dissolution, "slimes" (caused by spillage and subsequent moisture absorption), and difficulties associated with solution makeup. With solutions, the primary problem is one of stability: hydrolytic, bacterial (and fungal), and chemical (such as cross-linking). The problems associated with emulsions have been sensitivity to moisture, ease of inversion, and stability (tendency to separate).

Polyelectrolytes in solution form are generally the lower-molecular-weight species, and the products are known up to about 50% concentration. The author is aware of one class of polyelectrolyte where a relatively high-molecular-weight polymer is available commercially in aqueous solution of under 10% concentration. These are the aminomethylated polyacrylamides (*vide infra*).

There are, of course, many products which are available in both solution and solid form. These are predominantly the low-molecular-weight, anionic, deposit control agents made by free-radical addition polymerization, where the polymer melting points or Tgs are well above room temperature, and removing moisture is easier.

The technology associated with the preparation and use of polyelectrolytes in water-in-oil emulsion form is complex and extensive. Briefly, the commercial appearance of high-molecular-weight polyacrylamides of all charge types in inverse emulsion form is a recent development, although the basic technology goes back to the Vanderhoff patent.[21] There are two commercially available forms of these emulsions: those requiring a separate activator (surfactant) for inversion into aqueous solution, and the self-inverting emulsions, where the inverting surfactant is already present in the emulsion.[22]

The basic steps of preparing a polyelectrolyte in a water-in-oil emulsion form are first, dissolution of the monomer or monomers in the aqueous phase. This is added, usually with rapid agitation, to the hydrocarbon phase which contains the emulsifying agent. The polymerization is then initiated.[23]

The continuous oil phase usually consists of a hydrocarbon liquid. Environmental considerations dictate that these are preferably the nontoxic aliphatic type.[24] The recent technology indicates that the hydrocarbon phase constitutes about 25 to 35% of the total emulsion, while the aqueous phase consists of roughly equal parts of monomer(s) and water.[25] The ratio of monomer/water phase to oil phase is thus just the reverse of the ratios optimum for suspension polymerization, discussed above.

The emulsifying agent concentrations appear to be optimal in the 2 to 10% range (by weight, based on total emulsion), considerably higher than that used for suspension polymerization. Suitable emulsifying agents are of the low HLB type, similar to the emulsifying agents used most commonly in suspension polymerization.[21]

The inverting surfactants, conversely, are of the high HLB, water-soluble type.[22] As mentioned, they can either be added to the water, into which the polymer is to be dissolved, in which case the surfactant is commonly called an activator, or the surfactant can be added directly to the emulsion. In this case the emulsion is self-inverting, i.e., it will invert directly upon dilution in the water. In this latter case, great care must be exercised to avoid disruption of the water-in-oil emulsion prematurely. Moisture must also be carefully excluded from the self-inverting emulsions to avoid the formation of lumpy areas where the presence of the water has caused an inversion. The inverting surfactants are used at the 1 to 10% level, based on polymer weight.

It is also possible to add the already-formed, finely divided polymer to the water-in-oil emulsion,[22,26,28] but this is judged to be impractical. The addition of oil-soluble polymers, such as polyisobutylene and polybutadiene, in small amounts (0.3 to 0.75%, based on total emulsion) has been used to stabilize the emulsions and to minimize sediment formation.[27]

III. INITIATION

While it is beyond the scope of this chapter to discuss free-radical initiation processes, to which the reader is referred elsewhere,[29] there are some special techniques used which bear mentioning here.

In order to prepare polyelectrolytes having molecular weights in the millions, very low initiator concentrations are required. Furthermore, conversion of monomer to polymer must be as close to quantitative as possible, both for economics of manufacture and for toxicological reasons. In addition, in order to maximize molecular weight and to minimize cross-linking reactions, starting temperatures are kept as low as possible. Low starting temperatures are also important in those solution processes using high monomer concentrations, where the polymerization is essentially adiabatic, resulting in high final temperatures. The extremely high viscosities attained make stirring and cooling virtually impossible. The combined effects of the above considerations tend to result in rapid depletion of initiators and consequently, incomplete polymerization. This problem has been solved by the use of dual initiator systems, consisting in one case of a redox couple for low temperature initiation and an azo initiator for the thermal generation of radicals at the higher final temperatures.[30] Another method uses a redox couple where the oxidizing component also decomposes thermally to form radicals, such as the persulfate salts. In the latter case, an excess of oxidant over reducing agent is used.[31] In a third method, a combination of radiation for the low temperature initiation and an azo compound or peroxide for the thermal initiation is used.[34]

For those processes where these techniques are not used, residual monomers have been removed by chemical means, such as reaction with sulfite,[99] or with ammonia or amines.[104]

It should also be mentioned that at the high solution concentrations used, an increase in the propagation rate and molecular weight (known as the Trommsdorff effect[32]) is usually observed as the viscosity of the medium increases significantly.

IV. ANIONIC POLYELECTROLYTES

Today all of the commercially important anionic polyelectrolytes are made by free-radical addition polymerization. By virtue of their different uses, they can be conveniently divided into two classes: low ($\leqslant$100,000) and high molecular weight ($\geqslant$ 1,000,000). It is recognized that these are arbitrary distinctions, but they are meaningful boundaries when utility is considered.

A. Carboxyl-Containing Anionic Polyelectrolytes

Carboxyl-containing anionic polyelectrolytes are by far the most important class of anionics, both in the high- and low-molecular-weight range. High-molecular-weight carboxyl-containing anionic polyelectrolytes are used as flocculants for water and wastewater treatment. With few exceptions, such as separation of red-mud from dissolved alumina,[33] these polymers do not have high charge densities, i.e., the carboxyl content constitutes a minor part of the molecule.

The most common of these anionic polymers are based on acrylamide and are generally prepared in one of two ways. The first, polymerization/hydrolysis of acrylamide, is shown below (Equation 1).

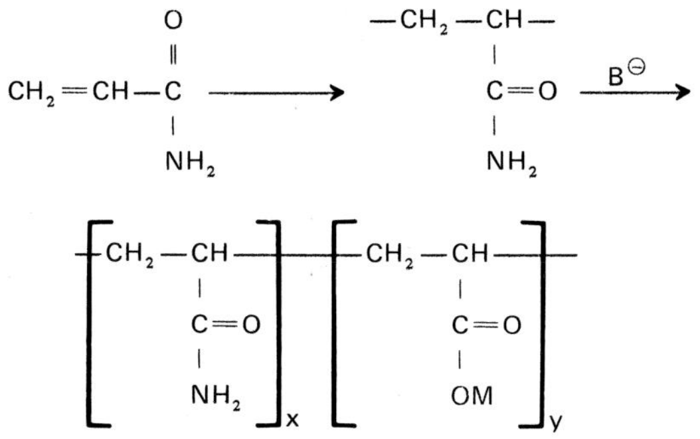

$$
CH_2{=}CH{-}\overset{\displaystyle O}{\overset{\displaystyle \|}{C}}\underset{\displaystyle NH_2}{|} \longrightarrow \quad \begin{matrix} -CH_2-CH- \\ | \\ C{=}O \\ | \\ NH_2 \end{matrix} \xrightarrow{\ B^{\ominus}\ }
$$

$$
\left[\begin{matrix} CH_2-CH \\ | \\ C{=}O \\ | \\ NH_2 \end{matrix}\right]_x \left[\begin{matrix} CH_2-CH \\ | \\ C{=}O \\ | \\ OM \end{matrix}\right]_y \tag{1}
$$

where M = metal salt, and x is usually >y.

The reaction is shown as a two-step process, although it is likely that most of the commercial polymers currently available are prepared by simultaneous polymerization and hydrolysis. Specifically, a basic material such as sodium hydroxide[35] or sodium carbonate[12] is added to the acrylamide to give the desired level of hydrolysis, followed by polymerization. The hydrolysis is assisted by the heat of polymerization, and can be readily controlled by the amount of base added and by the reaction conditions.[36]

The hydrolysis of acrylamide polymers in emulsion form presents particular difficulties because of the danger of destabilization of the emulsion by the basic hydrolysis agent. One solution utilizes alkali-stable surfactants, such as ethoxylated oleyl alcohol, added to the already formed polymer emulsion prior to addition of the hydrolysis agent.[38] While hydrolysis of ester or nitrile groups could theoretically give similar results it is this author's judgment that all of the commercially available materials are based on acrylamide monomers.

A commonly used alternative method involves copolymerization of acrylamide with acrylic acid or acrylate salts (Equation 2).[37] The polymerization techniques are similar to those for polymerization/hydrolysis, but one could speculate whether polymers prepared by the two methods differ in structure.

$$
\left.\begin{array}{c}
\text{O}\\
\parallel\\
CH_2{=}CHC{-}NH_2\\
\\
\text{O}\\
\parallel\\
CH_2{=}CH{-}C{-}OR
\end{array}\right\}
\longrightarrow
\left[\begin{array}{c}
CH_2{-}CH\\
\mid\\
C{=}O\\
\mid\\
NH_2
\end{array}\right]_x
\left[\begin{array}{c}
CH_2{-}CH\\
\mid\\
C{=}O\\
\mid\\
OR
\end{array}\right]_y
$$

(2)

R = H, or metal salt.

Based on published reactivity ratios,[39] we would expect a roughly random copolymerization of the two monomers. This is a crude extrapolation of the published data on isothermal copolymerizations to the more nearly adiabatic processes in current use. The only factor mitigating against a random hydrolysis process might be possible neighboring group effects which would tend to "cluster" the carboxyl groups along the chain (Equation 3).

$$
\begin{array}{ccc}
\diagup CH_2 \diagdown \diagup CH_2 \diagdown \diagup & \diagup CH_2 \diagdown \diagup CH_2 \diagdown \diagup\\
CH \qquad CH & CH \qquad CH\\
\mid \qquad \mid & \mid \qquad \mid\\
C{=}O \qquad C{=}O \xrightarrow{\ OH^{\ominus}\ } & C{=}O \qquad C{=}O\\
\mid \qquad \mid & \mid \qquad \mid\\
OR \qquad NH_2 & OR \qquad O_{\ominus}
\end{array}
$$

(3)

In any case, both types of polymers prepared at comparable molecular weights appear to function similarly in their application.

Although other monomers could be copolymerized to give carboxyl-containing polymers, such as maleic anhydride (followed by hydrolysis), methacrylic acid, or itaconic acid, it is believed that all of the high-molecular-weight copolymerized anionics currently available are based on acrylic acid. These other monomers take on significance in the low-molecular-weight species.

In contrast to the high-molecular-weight, carboxyl-containing polyelectrolytes, where a low charge density is the rule, most of the low-molecular-weight polymers are made by homopolymerization of carboxyl-containing monomers, or by extensive hydrolysis of other groups to the carboxyl stage. These polymers are used in the general area of deposit control, which encompasses a number of processes, such as: dispersion of already-formed precipitates having scale-forming tendencies; inhibition of crystallization of potential deposits; change of crystal habit of precipitating species; and flocculation of precipitates into easily removed form. These materials are used in applications where the formation of scale onto a surface would have detrimental effects. Typical would be heat-transfer surfaces in boilers and cooling towers, and scale formation in the paper-making process. The scale-forming chemicals generally result from ions already present in the aqueous systems combining with other components to form insoluble salts. These other components could be ions resulting from other water-treating chemicals, or minor impurities arising from the manufacturing process, such as resin acids. Other depositing species can originate as a result of concentration cycles, such as occur in cooling towers.

The most common low-molecular-weight, carboxyl-containing polymers used for water treatment are polyacrylic (I) and polymethacrylic acids (II) or their salts (Equation 4). The former polymers are used primarily for deposit control in cooling tower water and paper mill processes, while the latter are most prevalent in boiler systems.

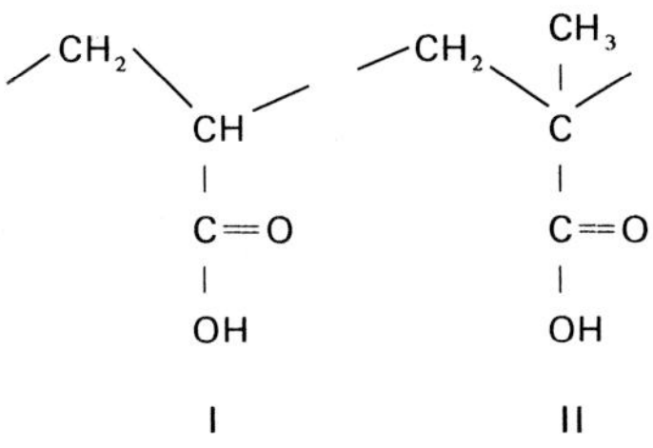

$$\underset{I}{\underset{\displaystyle \overset{\displaystyle CH_2}{\big\backslash} \overset{\displaystyle CH}{\underset{\displaystyle \big| }{}} \overset{\displaystyle C=O}{\underset{\displaystyle \big| }{}} OH}{} \qquad \underset{II}{\underset{\displaystyle \overset{\displaystyle CH_2}{\big\backslash} \overset{\displaystyle CH_3}{\underset{\displaystyle C}{\big| }} \overset{\displaystyle C=O}{\underset{\displaystyle \big| }{}} OH}{}} \qquad\qquad (4)$$

While these polymers have been made over a broad range of molecular weights, it appears that the most useful for deposit control have molecular weights under about 10,000, and in some cases the molecular weights can be under 1000.[40] In order to prepare polymers in these molecular weight ranges, a number of techniques can be used. High initiator concentrations, high concentrations of chain transfer agents, and combinations thereof appear to be the most prevalent. Milchem, Inc., has a series of patents[41] for preparing polymers having molecular weights under 1000 by an interesting process utilizing a composition containing up to 5% initiator concentrations, up to 15% chain transfer agents, and up to 90% acrylic acid. The chain transfer agents are mercaptans such as 2-mercaptoethanol and thioglycolic acid, and the reaction is carried out in aqueous solution. Low concentrations of a catalytic ion such as ferric, ferrous, or cupric can be added to speed the initiation. The interesting feature of this polymerization is the rapid addition of monomer, initiator, and chain transfer agent sequentially and in increments to the solvent heated to the reaction temperatures. The rapid contact of all the reagents results in a very rapid, exothermic reaction, leading to the desired low molecular weight. The rapid reaction with all the polymer components present probably also results in a relatively narrow molecular weight distribution.

With polymers in the low-molecular-weight range, the end group concentration becomes correspondingly higher, and may be of significance in affecting activity.[42] In a typical polyacrylic acid where high initiator concentrations are coupled with chain transfer agents, two typical end group structures might be as shown below (Equation 5).

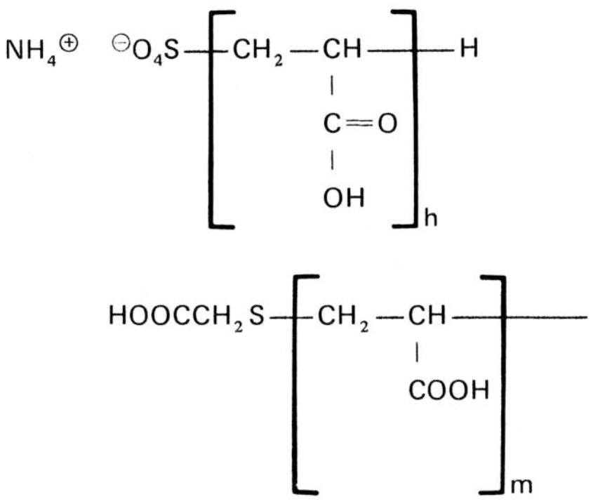

$$NH_4^{\oplus}\ \ {}^{\ominus}O_4S - \left[CH_2 - \underset{\underset{\displaystyle OH}{\underset{\displaystyle |}{\underset{\displaystyle C=O}{|}}}}{CH} \right]_h - H$$

$$HOOCCH_2S - \left[CH_2 - \underset{\underset{\displaystyle COOH}{|}}{CH} \right]_m - \qquad\qquad (5)$$

In this example, the initiator was ammonium persulfate (thermal initiation) and the chain transfer agent was thioglycolic acid. The value of n or m could be as low as ten for commercially available polymers.

Other carboxyl-containing deposit control agents which are of commercial importance are low-molecular-weight polyacrylamides having the majority of the amide groups hydrolyzed (Equation 1, where $y > x$),[43] acrylonitrile copolymers (Equation 6),[44] copolymers of acrylic acid with 2-hydroxypropyl acrylate (Equation 7),[45] and hydrolyzed homo- and copolymers of maleic anhydride (Equations 8 and 9).

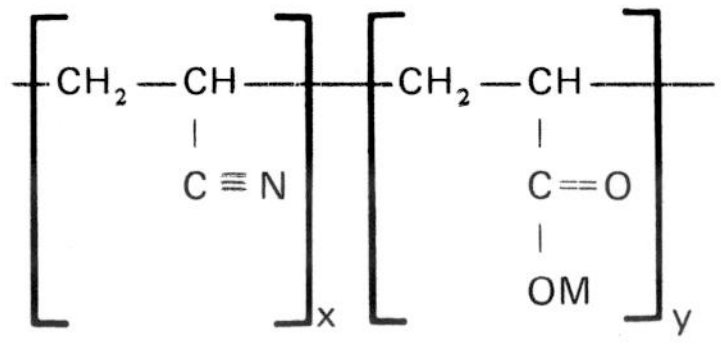

$$\left[\begin{array}{c} CH_2-CH- \\ | \\ C\equiv N \end{array}\right]_x \left[\begin{array}{c} CH_2-CH- \\ | \\ C=O \\ | \\ OM \end{array}\right]_y \tag{6}$$

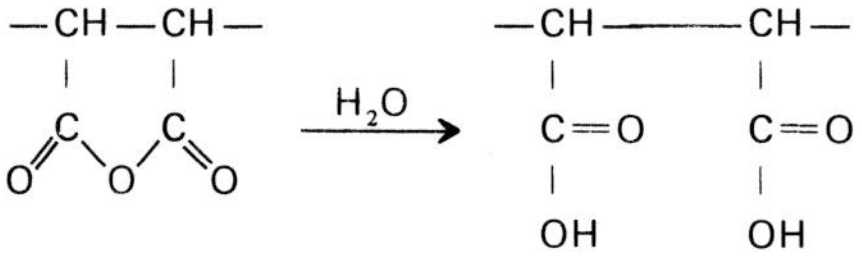

where M = H or metal ion

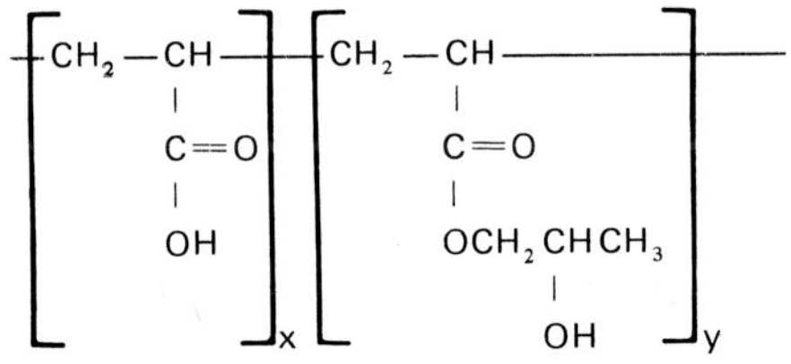

$$\left[\begin{array}{c} CH_2-CH- \\ | \\ C=O \\ | \\ OH \end{array}\right]_x \left[\begin{array}{c} CH_2-CH- \\ | \\ C=O \\ | \\ OCH_2CHCH_3 \\ | \\ OH \end{array}\right]_y \tag{7}$$

$$\begin{array}{c} -CH-CH- \\ \diagdown C \diagup O \diagdown C \diagup \\ O \qquad\qquad O \end{array} \xrightarrow{H_2O} \begin{array}{c} -CH-CH- \\ | \qquad | \\ C=O \quad C=O \\ | \qquad | \\ OH \qquad OH \end{array} \tag{8}$$

$$-CH_2-CH-CH-CH- \longrightarrow -CH_2-CH-CH-CH- \tag{9}$$

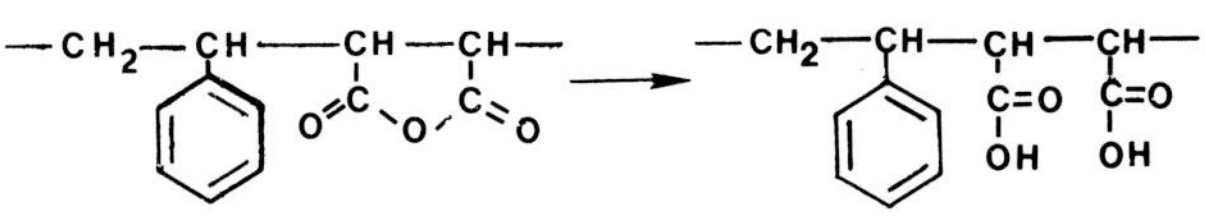

An important feature of these latter polymers is the frequency of the carboxyl groups. In the hydrolyzed polymaleic anhydride, the "charge density," i.e., the number of potential charges per unit weight is twice that of the polyacrylic acids. This may be of importance in certain deposit control applications where adsorption onto the crystal or nucleating site is a kinetically important mechanism.[55]

Although at one time it was believed that homopolymerization of maleic anhydride was impossible, there have been many reported techniques for this polymerization since 1961,[46] including the use of peroxides in aromatic solvents,[47] electrolytic polymerization,[48] and gamma or UV irradiation.[46] Polymaleic anhydride is now an article of commerce.

Commercially important copolymers of maleic anhydride include those with styrene (*vide infra*), diisobutylene (III),[49] and 1-octadecene (IV) (Equation 10).[50] It should be noted that these polymers are some of the newest polymers for water treatment, and they have promise to become increasingly important in the future.

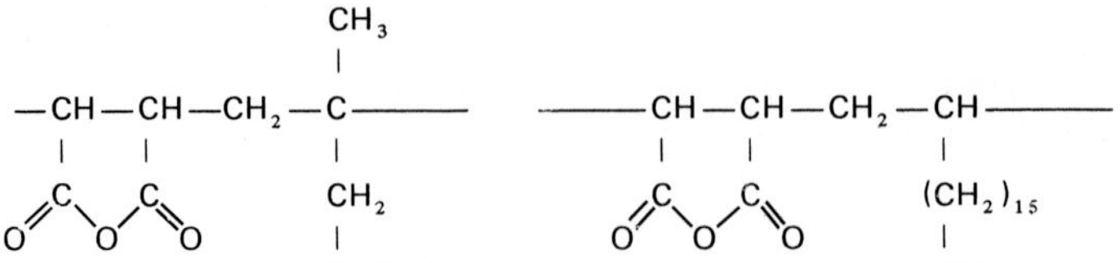

$$\begin{array}{cc} \begin{array}{c} CH_3 \\ | \\ -CH-CH-CH_2-C- \\ | \quad | \qquad\qquad | \\ \diagdown C \diagup O \diagdown C \diagup \quad CH_2 \\ O \qquad O \qquad | \\ \qquad\qquad C(CH_3)_3 \end{array} & \begin{array}{c} -CH-CH-CH_2-CH- \\ | \quad | \qquad\qquad | \\ \diagdown C \diagup O \diagdown C \diagup \quad (CH_2)_{15} \\ O \qquad O \qquad | \\ \qquad\qquad CH_3 \end{array} \\ \text{III} & \text{IV} \end{array} \tag{10}$$

Many of the copolymers mentioned in Equation 10 appear to function in applications where polyacrylic acid or hydrolyzed polyacrylamides are ineffective. It appears that the combination of the hydrophilic carboxyl group with a hydrophobic group leads to unique activity for many depositing species. This hydrophilic-hydrophobic combination has become of interest in the area of flocculants as well.

B. Sulfur-Containing Anionic Polyelectrolytes

There are two other classes of commercially important anionic polyelectrolytes, those containing sulfur atoms and those containing phosphorous. The water-soluble, sulfur-containing polyelectrolytes are comprised primarily of polymeric sulfonic acids, RSO_3H, although species such as polyvinyl sulfuric acid, made from polyvinyl alcohol and chlorosulfonic acid (Equation 11),[51] are known.

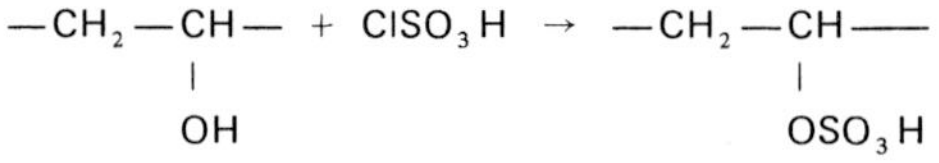

$$-CH_2-CH- + ClSO_3H \rightarrow -CH_2-CH-- $$
$$\qquad\quad | \qquad\qquad\qquad\qquad\qquad | $$
$$\qquad\quad OH \qquad\qquad\qquad\qquad\quad OSO_3H \qquad\qquad (11)$$

The sulfonic acid polyelectrolytes have been made both by polymerization and copolymerization of sulfonic acid-containing monomers, and by direct sulfonation of the appropriate polymer.

The simplest monomers which can produce sulfonic acid-containing polyelectrolytes are vinylsulfonic acid and its salts (V). The monomer and its polymers have been known for many years.[56]

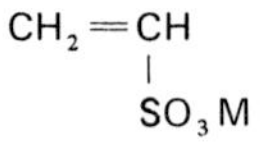

$$CH_2=CH$$
$$\qquad\quad |$$
$$\qquad\quad SO_3M$$

V

Sodium polyvinyl sulfonate (VI)[57] and copolymers with methacrylic acid (VII)[58] have been reported as useful deposit-control agents, but it is believed that there currently are no commercially available polymers for water treatment which utilize vinylsulfonic acid.

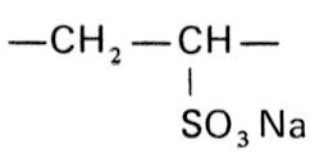

$$-CH_2-CH- \qquad\qquad \left[\begin{array}{c} CH_3 \\ | \\ CH_2-C \\ | \\ C=O \\ | \\ OH \end{array}\right]_x \left[\begin{array}{c} \\ CH_2-CH- \\ | \\ SO_3M \\ \\ \end{array}\right]_y$$
$$\qquad\quad |$$
$$\qquad\quad SO_3Na$$

VI 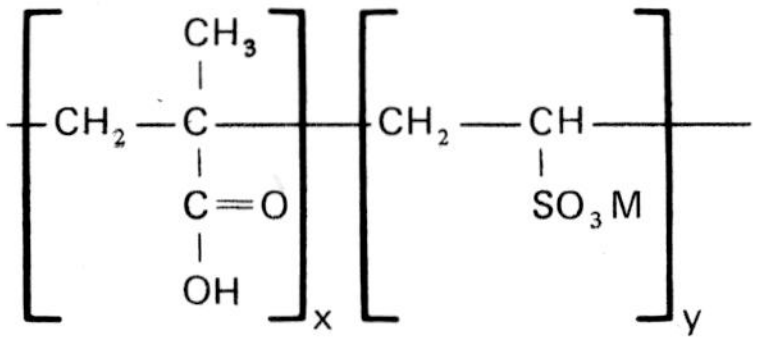   VII

Another sulfonic acid, which reportedly gives useful low-molecular-weight deposit-control agents when copolymerized with fumaric acid, is allyl sulfonic acid (VIII) (Equation 12).[59] This polymer is also believed not to be of current commercial interest.

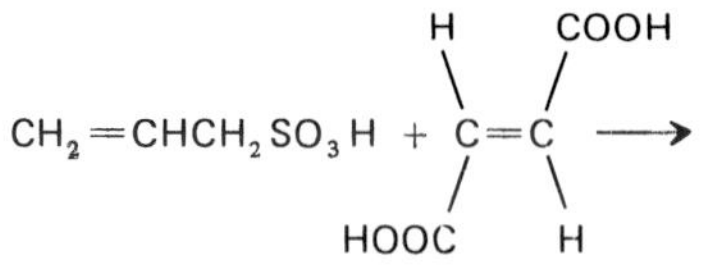

$$CH_2\!=\!CHCH_2SO_3H \ + \ \underset{HOOC}{\overset{H}{\diagup}}C\!=\!C\underset{H}{\overset{COOH}{\diagdown}} \longrightarrow$$

VIII

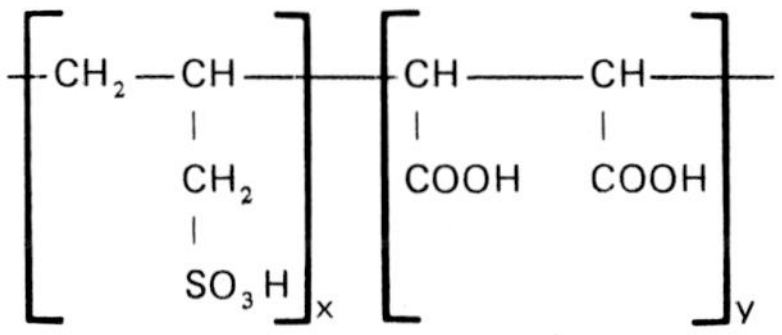

$$\left[\begin{array}{c} \text{-CH}_2\text{--CH---} \\ | \\ \text{CH}_2 \\ | \\ \text{SO}_3\text{H} \end{array}\right]_x \left[\begin{array}{cc} \text{CH---} & \text{CH--} \\ | & | \\ \text{COOH} & \text{COOH} \end{array}\right]_y \qquad (12)$$

2-Acrylamido-2-methyl propane sulfonic acid (AMPS) is a monomer offered by Lubrizol Corporation that has been used to prepare both high-molecular-weight anionic flocculants, by homopolymerization or copolymerization with acrylamide or sodium acrylate,[52] and low-molecular-weight deposit-control agents, primarily as copolymers with acrylic acid.[53] It is believed that AMPS polymers have significant potential in the area of deposit control.

This monomer is prepared from isobutylene, acetyl sulfonic acid, and acrylonitrile (Equation 13).[54]

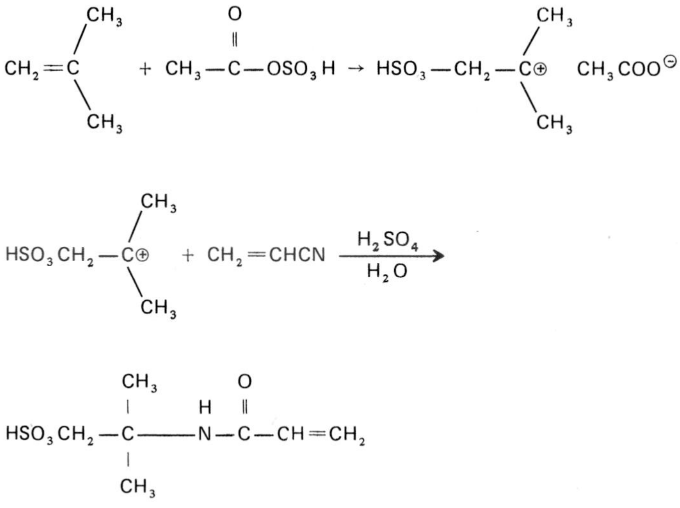

$$CH_2\!=\!\underset{CH_3}{\overset{CH_3}{C}} \ + \ CH_3\!-\!\overset{O}{\underset{\|}{C}}\!-\!OSO_3H \ \rightarrow \ HSO_3\!-\!CH_2\!-\!\underset{CH_3}{\overset{CH_3}{C\oplus}} \quad CH_3COO^{\ominus}$$

$$HSO_3CH_2\!-\!\underset{CH_3}{\overset{CH_3}{C\oplus}} \ + \ CH_2\!=\!CHCN \ \xrightarrow[\text{H}_2\text{O}]{\text{H}_2\text{SO}_4}$$

$$HSO_3CH_2\!-\!\underset{CH_3}{\overset{CH_3}{\underset{|}{C}}}\!-\!-\!-\!\overset{H}{N}\!-\!\overset{O}{\underset{\|}{C}}\!-\!CH\!=\!CH_2$$

AMPS   (13)

Sulfonic acid-containing acrylates, such as 2-sulfoethylmethacrylate (IX) and 3-sulfopropylacrylate (X) have been used to prepare polyelectrolytes for water treatment, but here again there is some question of their commercial utility. Homopolymers of IX,[60] and copolymers of X with acrylamide[61] have been reported.

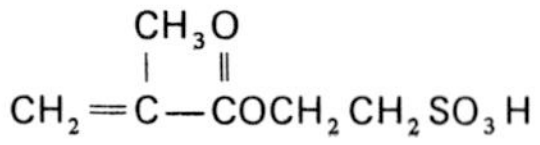

$$CH_2\!=\!\overset{CH_3}{\underset{|}{C}}\!-\!\overset{O}{\underset{\|}{C}}OCH_2CH_2SO_3H \qquad\qquad CH_2\!=\!CH\overset{O}{\underset{\|}{C}}\!-\!OCH_2CH_2CH_2SO_3H$$

IX 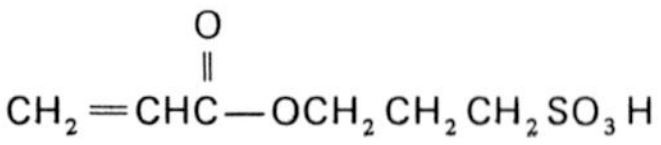   X

Polymeric derivatives of styrene sulfonic acid have significant commercial utility. Although polymers and copolymers made from monomeric styrene sulfonic acid (XI) are known (Equation 14), most of the currently available polymers are believed to be prepared by sulfonation of the already existing polymer (Equation 15). It is likely that sulfonation occurs primarily at the para position.

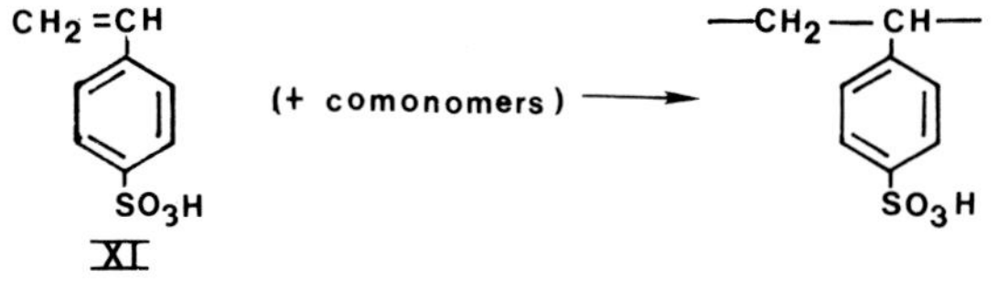

(14)

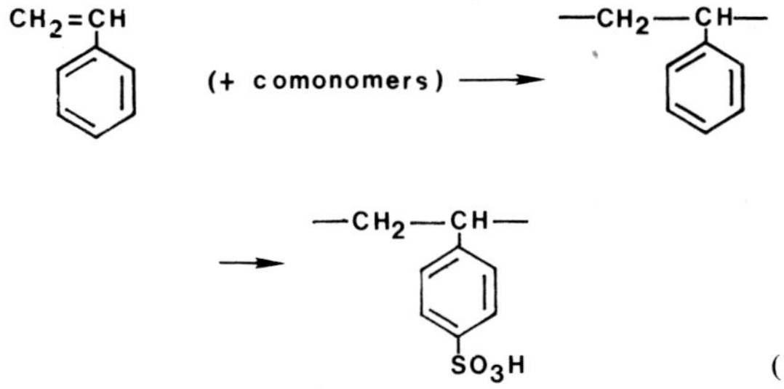

(15)

A number of sulfonation techniques have been used. Sulfur trioxide has been used, either alone,[64] in adducts with ethers or ketones,[63] or with complexating agents.

The sulfur trioxide-trialkylphosphate complex, which is reported to affect the sulfonation without the introduction of cross-links, appears to be a particularly useful method.[66] Other sulfonation techniques utilize sulfuric acid or chlorosulfonic acid.[63]

Although high-molecular-weight polystyrene sulfonic acid flocculants are known,[91] the majority of the commercially important polystyrene sulfonic acids used for water treatment are deposit control agents, and include polystyrene sulfonic acids of varying molecular weight,[67] and low-molecular-weight hydrolyzed sulfonated styrene-maleic anhydride copolymers (Equation 16).[64,68]

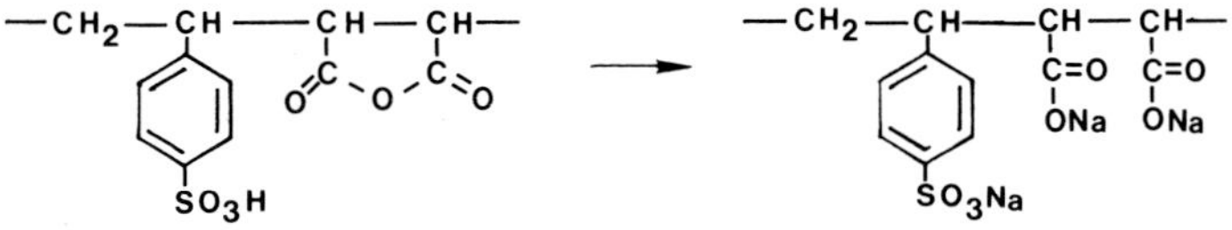

(16)

The repeat unit is as shown here because of the expected tendency of styrene and maleic anhydride to copolymerize in an alternating fashion.[69] This structure would, of course, only be applicable to copolymers having a 1 to 1 M ratio of monomers, or to other ratios of monomers polymerized to low conversion. Normally, commercial processes strive for maximum conversion to polymer.

C. Phosphorus-Containing Polyelectrolytes

Anionic polyelectrolytes containing phosphorous are among the most recent developments in the area of water treatment. Consequently, there is the least amount of published information available. From what is known, the primary interest in phosphorous-containing polyelectrolytes appears to be in the areas of deposit control and chelation, and the primary functionality is in the form of phosphonic acids. It also does not appear that there are currently any polymeric phosphonates beyond the developmental stage.

Many of the polymeric phosphonic acids reported are prepared using similar chemistry to that used for the preparation of the well-known phosphonic acids used for deposit control and corrosion inhibition, e.g., nitrilo-tris-methylenephosphonic acid, prepared by the Mannich reaction of an amine with formaldehyde and phosphorous acid (Equation 17).

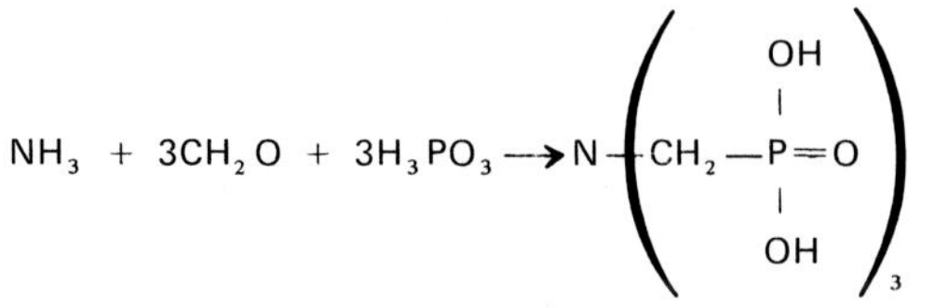

$$NH_3 \ + \ 3CH_2O \ + \ 3H_3PO_3 \longrightarrow N \left(CH_2 - \overset{\overset{\displaystyle OH}{|}}{\underset{\underset{\displaystyle OH}{|}}{P}} = O \right)_3 \qquad (17)$$

The reaction has been extended to prepare polyphosphonic acids by the reaction of polyamines with formaldehyde and phosphorous acid. Many of these reactions are similar to those used to prepare the low- to medium-molecular-weight catonic condensation polymers, discussed in Section V.B.

In one of the simpler examples, a polyether terminated with amino groups is phosphomethylated (Equation 18).[70]

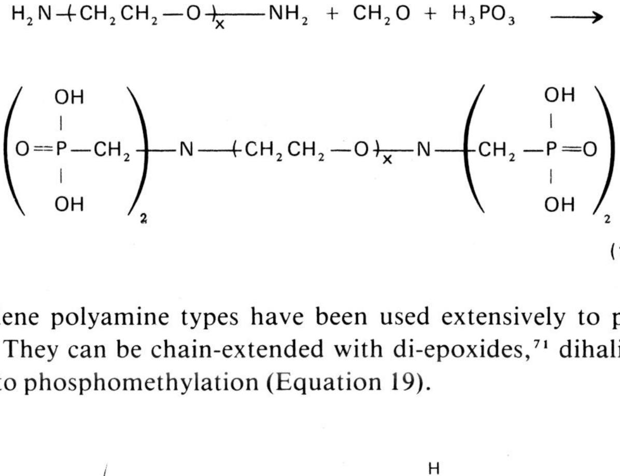

$$H_2N(CH_2CH_2-O)_x-NH_2 \ + \ CH_2O \ + \ H_3PO_3 \longrightarrow$$

$$\left(O = \overset{\overset{\displaystyle OH}{|}}{\underset{\underset{\displaystyle OH}{|}}{P}} - CH_2 \right)_2 N - (CH_2CH_2-O)_x - N \left(CH_2 - \overset{\overset{\displaystyle OH}{|}}{\underset{\underset{\displaystyle OH}{|}}{P}} = O \right)_2$$

$$(18)$$

The polyethylene polyamine types have been used extensively to prepare polyphosphonic acids. They can be chain-extended with di-epoxides,[71] dihalides, or epihalohydrins,[72] prior to phosphomethylation (Equation 19).

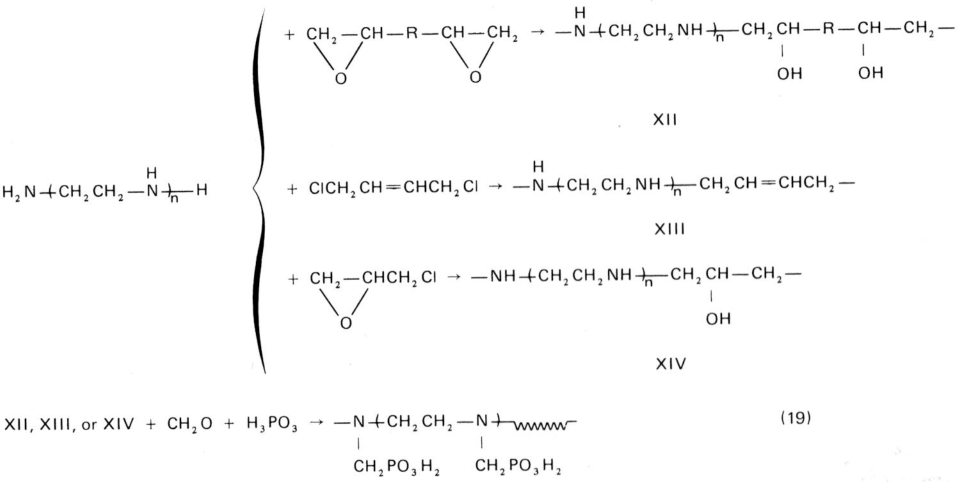

$$H_2N(CH_2CH_2-\overset{\overset{\displaystyle H}{|}}{N})_n-H \begin{cases} + \ CH_2-CH-R-CH-CH_2 \rightarrow -\overset{\overset{\displaystyle H}{|}}{N}(CH_2CH_2NH)_n-CH_2\underset{\underset{\displaystyle OH}{|}}{CH}-R-\underset{\underset{\displaystyle OH}{|}}{CH}-CH_2- \\[1.2em] \qquad\qquad \text{XII} \\[1em] + \ ClCH_2CH=CHCH_2Cl \rightarrow -\overset{\overset{\displaystyle H}{|}}{N}(CH_2CH_2NH)_n-CH_2CH=CHCH_2- \\[1.2em] \qquad\qquad \text{XIII} \\[1em] + \ CH_2-CHCH_2Cl \rightarrow -NH(CH_2CH_2NH)_n-CH_2\underset{\underset{\displaystyle OH}{|}}{CH}-CH_2- \\[1.2em] \qquad\qquad \text{XIV} \end{cases}$$

$$\text{XII, XIII, or XIV} + CH_2O + H_3PO_3 \rightarrow -\underset{\underset{\displaystyle CH_2PO_3H_2}{|}}{N}(CH_2CH_2-\underset{\underset{\displaystyle CH_2PO_3H_2}{|}}{N})\wwww \qquad (19)$$

In the above equation, n can be one or more, or a mixture. These polyethylene polyamines are products of the reaction of ammonia with ethylenedichloride, discussed in Section V.B.

In another variation, phosphomethylated amines are chain-extended with dihalides or epihalohydrins to give quaternized polyphosphonic acids (Equation 20).[73]

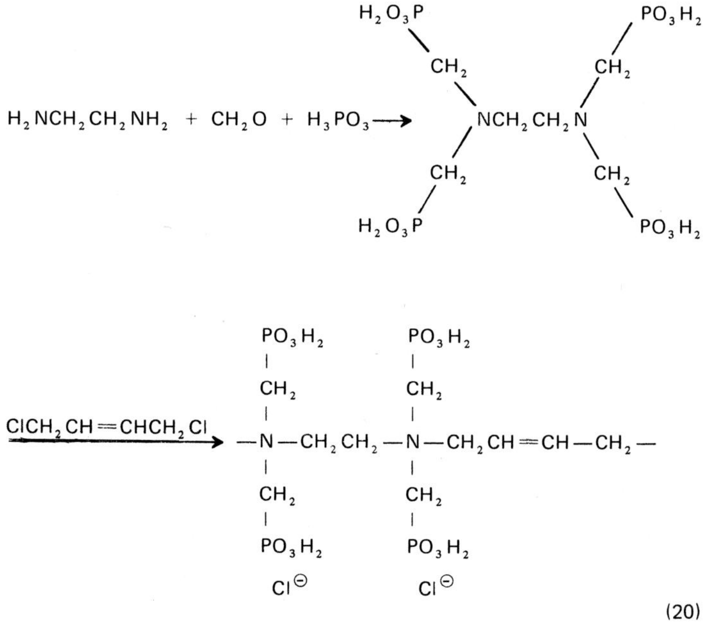

$$H_2NCH_2CH_2NH_2 \; + \; CH_2O \; + \; H_3PO_3 \longrightarrow$$

$$\begin{array}{ccc}
H_2O_3P & & PO_3H_2 \\
\diagdown & & \diagup \\
CH_2 & & CH_2 \\
\diagdown & & \diagup \\
& NCH_2CH_2N & \\
\diagup & & \diagdown \\
CH_2 & & CH_2 \\
\diagup & & \diagdown \\
H_2O_3P & & PO_3H_2
\end{array}$$

$$\xrightarrow{ClCH_2CH=CHCH_2Cl}$$

$$\begin{array}{ccc}
PO_3H_2 & & PO_3H_2 \\
| & & | \\
CH_2 & & CH_2 \\
| & & | \\
-N-CH_2CH_2- & N-CH_2CH=CH-CH_2- \\
| & & | \\
CH_2 & & CH_2 \\
| & & | \\
PO_3H_2 & & PO_3H_2 \\
Cl^{\ominus} & & Cl^{\ominus}
\end{array}$$

$$(20)$$

If converted to the salt form, or used at high pH, these polymers would be polyampholytes (containing both positive and negative charges), where the negative charges would predominate (Equation 21).

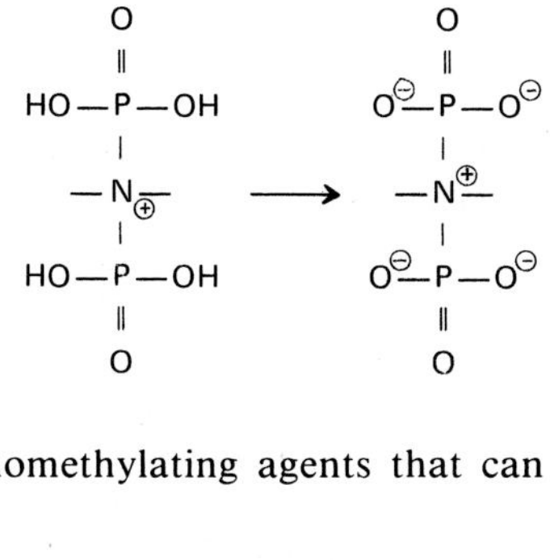

$$\begin{array}{ccc}
O & & O \\
\| & & \| \\
HO-P-OH & & O^{\ominus}-P-O^{\ominus} \\
| & & | \\
-N_{\oplus}- & \longrightarrow & -N^{\oplus}- \\
| & & | \\
HO-P-OH & & O^{\ominus}-P-O^{\ominus} \\
\| & & \| \\
O & & O
\end{array} \qquad (21)$$

Other phosphomethylating agents that can be used are chloromethylenephosphonic acid[74]

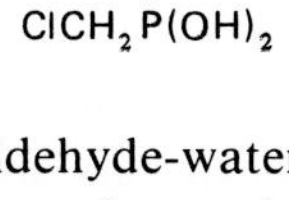

$$\begin{array}{c}
O \\
\| \\
ClCH_2P(OH)_2
\end{array}$$

and phosphorous trichloride-formaldehyde-water.[75] Other polyphosphonic acids have been prepared by reaction of a hydrocarbon polymer (e.g., polyethylene) by oxidative chlorophosphonylation,[76] followed by esterification or hydrolysis (Equation 22).

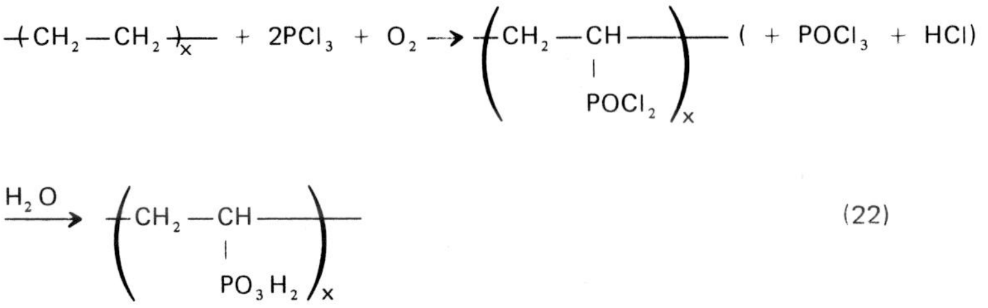

$$+CH_2-CH_2\frac{}{}_x + 2PCl_3 + O_2 \rightarrow \left(CH_2-CH \underset{POCl_2}{\overset{|}{}}\right)_x (+ POCl_3 + HCl)$$

$$\xrightarrow{H_2O} \left(CH_2-CH \underset{PO_3H_2}{\overset{|}{}}\right) \tag{22}$$

An interesting reaction (to prepare polymeric ester chain condensates of ethane-1-hydroxy-1, 1-diphosphonic acid) utilizes the reaction of phosphorous acid and acetic anhydride in a large excess of acetic anhydride (Equation 23).[77]

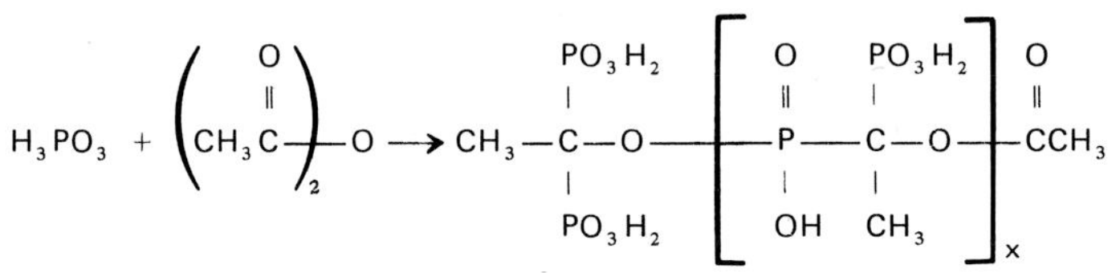

$$H_3PO_3 + \left(CH_3C\overset{O}{\overset{\|}{}}\right)_2 O \rightarrow CH_3-\underset{PO_3H_2}{\overset{PO_3H_2}{\underset{|}{\overset{|}{C}}}}-O \left[\underset{OH}{\overset{O}{\overset{\|}{}}}P\underset{CH_3}{\overset{PO_3H_2}{\underset{|}{\overset{|}{C}}}}-O\right]_x \overset{O}{\overset{\|}{}}CCH_3$$

$$\tag{23}$$

where x = 1 to ~16. These water-soluble polyphosphonic acids are used as sequestrants and detergent builders.[78]

V. CATIONIC POLYELECTROLYTES

The chemistry of cationic polyelectrolytes is more complex than that of the anionic species. Practically all of the important anionic polymers can be prepared by relatively simple free-radical polymerization, and the most common reactions on existing polymers are hydrolyses, sulfonations, or phosphomethylations. The chemistry of synthesizing cationic polyelectrolytes, however, involves free-radical addition polymerization, condensation reactions, epoxide addition reactions, and a variety of reactions on existing polymer backbones, including Hofman degradations, Mannich reactions, and nucleophilic displacements.

Another feature distinguishing the two major types of polyelectrolytes is function. High-molecular-weight anionic polyelectrolytes are used generally as flocculants, while the low-molecular-weight anionics are dispersants, just the opposite. In the cationic series, all molecular weight ranges function as flocculants. This, of course, is primarily because the majority of particulate matter found in nature bears negative charges.

Furthermore, although free-radical initiation can be used to prepare all molecular weights of the anionic polymers, this polymerization has found its greatest utility on the cationic side only for the high-molecular-weight flocculants. The great majority of the commercially important low-molecular-weight cationics are manufactured by condensation polymerization. There are, of course, a few notable exceptions to these generalizations.

A. Chain-Reaction Polymers[109]

Most of the high-molecular-weight cationic polyelectrolytes available today are made by free-radical copolymerization of acrylamide with minor amounts (10 mol % or less) of the cationic monomer. The ratio of acrylamide to the cationic monomer is dictated by economics and activity; for most of the applications in water treatment, 10 mol % is sufficient. Furthermore, most of the cationic monomers available are of sufficient molecular weight that 10 mol % is equivalent in most cases to greater than

30 wt %. Since the cationic monomer is also a considerably more expensive monomer than acrylamide (as of April 1, 1978, acrylamide was roughly 50 ¢/lb, while the cationic monomers cost more than $1.00/lb), economics favor utilizing lower amounts of the cationic monomer.

As the application of polymers into such areas as waste treatment increases because of increasing environmental pressures, the variety of useful cationic polyelectrolytes will increase, and polymers containing higher cationic content will become more prevalent.

Because of solubility, pH, stability of charge, and polymerization behavior, practically all of the cationic monomers used for water treatment polymers are quaternized aminoesters or amino amides, although the parent tertiary aminoalkylesters[178] and tertiary aminoalkyl amides[154] are referred to extensively in the patent literature, very often in conjunction with the quaternized monomers. There have been commercial cationic flocculants which consisted of copolymers of acrylamide with unquaternized tertiary aminoalkyl esters, but their use has not been extensive.

Some of the more common quaternized monomers are shown below:

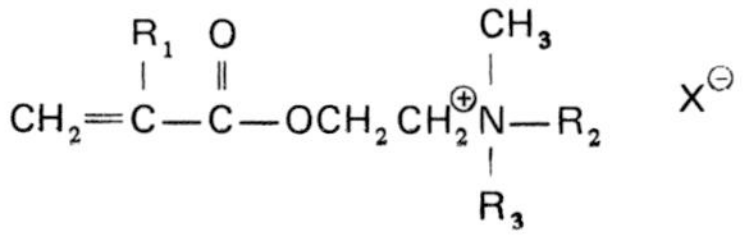

$$CH_2{=}C(R_1){-}\overset{O}{\overset{\|}{C}}{-}OCH_2CH_2\overset{\oplus}{\underset{R_3}{\overset{CH_3}{N}}}{-}R_2 \quad X^{\ominus}$$

XV

where R_1 = H or CH_3; R_2, R_3 = CH_3 or CH_3CH_2; and X^- = Cl^- or $CH_3SO_4^-$

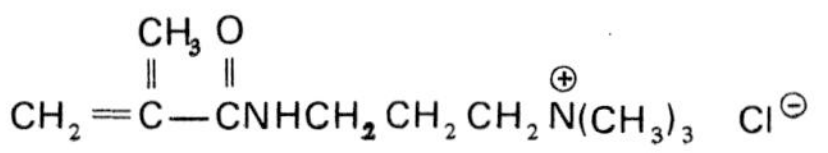

$$CH_2{=}C(\overset{CH_3}{\overset{\|}{}}){-}\overset{O}{\overset{\|}{C}}NHCH_2CH_2CH_2\overset{\oplus}{N}(CH_3)_3 \quad Cl^{\ominus}$$

XVI

$$CH_2{=}CH\overset{O}{\overset{\|}{C}}NH\overset{CH_3}{\underset{CH_3}{\overset{\|}{C}}}{-}CH_2CH_2\overset{\oplus}{N}(CH_3)_3 \quad Cl^{\ominus}$$

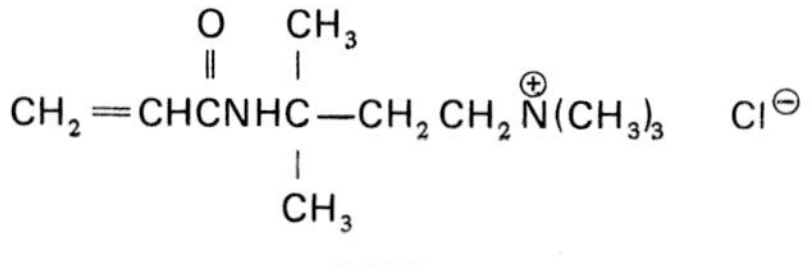

XVII

The esters are generally prepared from the tertiary amino alcohol by an esterification or trans-esterification reaction, followed by quaternization (Equation 24).

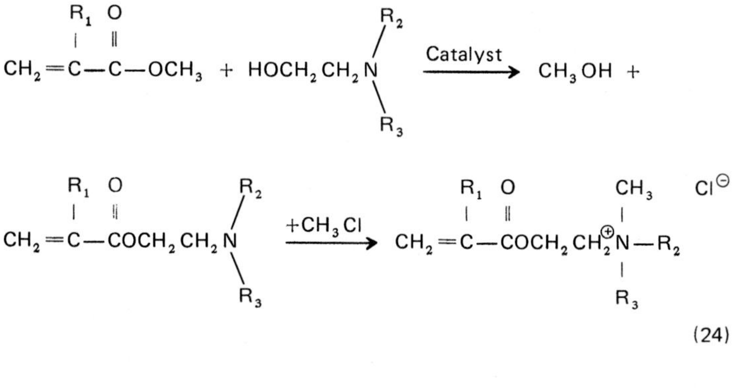

$$CH_2{=}C(R_1){-}\overset{O}{\overset{\|}{C}}{-}OCH_3 + HOCH_2CH_2N\overset{R_2}{\underset{R_3}{}} \xrightarrow{Catalyst} CH_3OH +$$

$$CH_2{=}C(R_1){-}\overset{O}{\overset{\|}{C}}OCH_2CH_2N\overset{R_2}{\underset{R_3}{}} \xrightarrow{+CH_3Cl} CH_2{=}C(R_1){-}\overset{O}{\overset{\|}{C}}OCH_2CH_2\overset{\oplus}{\underset{R_3}{\overset{CH_3}{N}}}{-}R_2 \quad Cl^{\ominus}$$

(24)

The amide XVI can be prepared by addition of the amine to methacrylic acid or its ester, followed by decomposition of the resultant β-amino-propionamide (Equation 25)[88]. The amine can then be suitably quaternized.

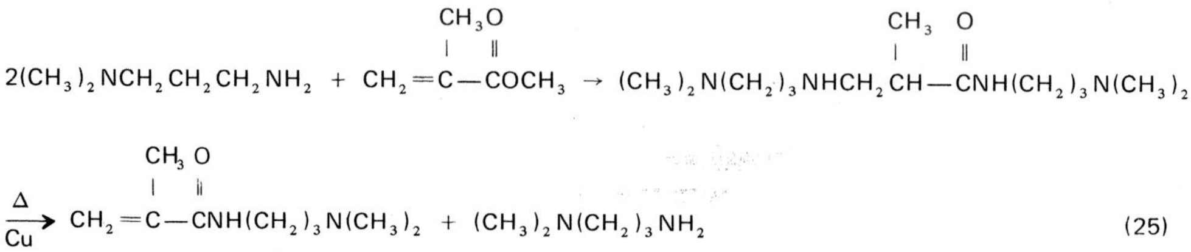

$$2(CH_3)_2NCH_2CH_2CH_2NH_2 \ + \ CH_2{=}\overset{\overset{\displaystyle CH_3}{|}}{C}{-}\overset{\overset{\displaystyle O}{\|}}{C}OCH_3 \ \rightarrow \ (CH_3)_2N(CH_2)_3NHCH_2\overset{\overset{\displaystyle CH_3}{|}}{C}H{-}\overset{\overset{\displaystyle O}{\|}}{C}NH(CH_2)_3N(CH_3)_2$$

$$\xrightarrow[\text{Cu}]{\Delta} \ CH_2{=}\overset{\overset{\displaystyle CH_3}{|}}{C}{-}\overset{\overset{\displaystyle O}{\|}}{C}NH(CH_2)_3N(CH_3)_2 \ + \ (CH_3)_2N(CH_2)_3NH_2 \tag{25}$$

Structure XVII, 3-acrylamido-3-methylbutyl trimethylammonium chloride, is apparently made by addition of dimethylamine to isoprene under basic conditions, followed by a Ritter reaction of the product with acrylonitrile, and finally quaternization (Equation 26).[79]

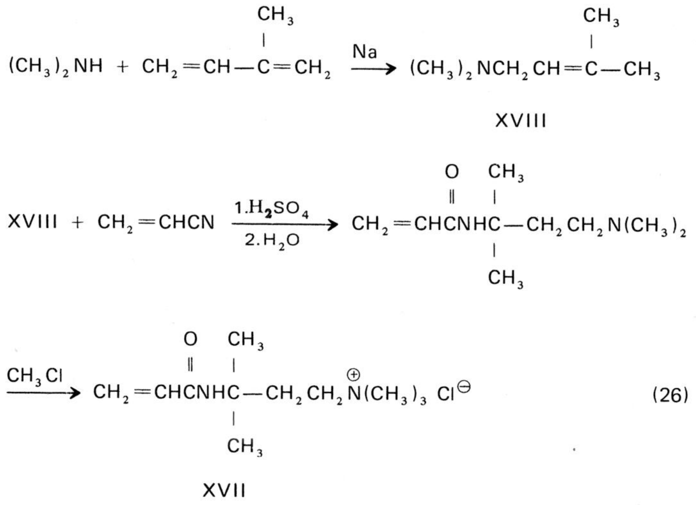

$$(CH_3)_2NH \ + \ CH_2{=}CH{-}\overset{\overset{\displaystyle CH_3}{|}}{C}{=}CH_2 \ \xrightarrow{\text{Na}} \ (CH_3)_2NCH_2CH{=}\overset{\overset{\displaystyle CH_3}{|}}{C}{-}CH_3$$

$$\text{XVIII}$$

$$\text{XVIII} \ + \ CH_2{=}CHCN \ \xrightarrow[\text{2.H}_2\text{O}]{\text{1.H}_2\text{SO}_4} \ CH_2{=}CHC\overset{\overset{\displaystyle O}{\|}}{}NH\overset{\overset{\displaystyle CH_3}{|}}{\underset{\underset{\displaystyle CH_3}{|}}{C}}{-}CH_2CH_2N(CH_3)_2$$

$$\xrightarrow{CH_3Cl} \ CH_2{=}CHC\overset{\overset{\displaystyle O}{\|}}{}NH\overset{\overset{\displaystyle CH_3}{|}}{\underset{\underset{\displaystyle CH_3}{|}}{C}}{-}CH_2CH_2\overset{\oplus}{N}(CH_3)_3 \ \ Cl^{\ominus} \tag{26}$$

$$\text{XVII}$$

An hydroxyl-containing quaternized monomer, which is probably not currently used in water treatment polymers, is made by reacting a tertiary amine with an epihalohydrin in an aprotic solvent. The product is then reacted with an unsaturated acid.[94] The preparation of 2-hydroxy-3-acryloxypropyltrimethylammonium chloride is illustrated in Equation 27. Both homopolymers and copolymers with acrylamide are known.

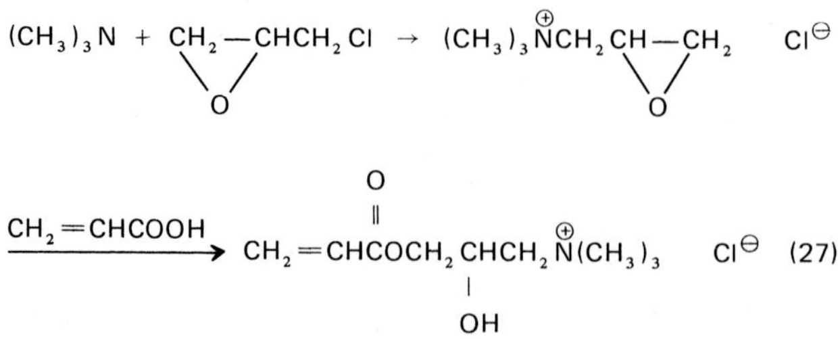

$$(CH_3)_3N \ + \ CH_2{-}CHCH_2Cl \ \rightarrow \ (CH_3)_3\overset{\oplus}{N}CH_2CH{-}CH_2 \ \ Cl^{\ominus}$$

$$\xrightarrow{CH_2{=}CHCOOH} \ CH_2{=}CHC\overset{\overset{\displaystyle O}{\|}}{}OCH_2\overset{\underset{\underset{\displaystyle OH}{|}}{}}{C}HCH_2\overset{\oplus}{N}(CH_3)_3 \ \ Cl^{\ominus} \tag{27}$$

The diallyldialkylammonium halides represent the one commercially important exception to the ester and amide cationic monomers. These are prepared from allylchloride and the corresponding amine (Equation 28).[80]

$$2CH_2{=}CHCH_2Cl \ + \ (CH_3)_2NH \xrightarrow{\text{base}}$$

$$(CH_2{=}CHCH_2)_2\overset{\oplus}{N}(CH_3)_2 \quad Cl^{\ominus} \tag{28}$$

Because of the difunctionality of this type of monomer, one would normally expect free-radical polymerization to produce a branched, and eventually cross-linked, insoluble structure, in analogy to the use of such cross-linking monomers as divinylbenzene. The actual observation of soluble, apparently linear, polymers has been explained by an alternating intramolecular-intermolecular mechanism (Equation 29).[81]

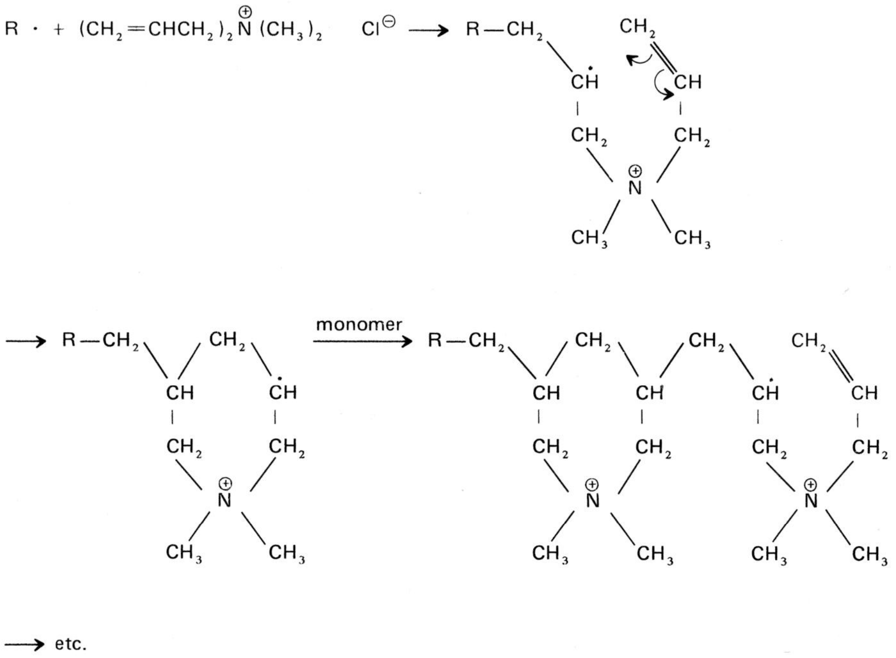

$$R\cdot \ + \ (CH_2{=}CHCH_2)_2\overset{\oplus}{N}(CH_3)_2 \quad Cl^{\ominus} \longrightarrow$$

$$\xrightarrow{\quad} \ R{-}CH_2 \cdots CH_2 \xrightarrow{\text{monomer}} R{-}CH_2 \cdots \tag{29}$$

$$\longrightarrow \ \text{etc.}$$

There has been considerable debate in the recent literature concerning the size of the rings formed during the polymerization (Equation 30).[82]

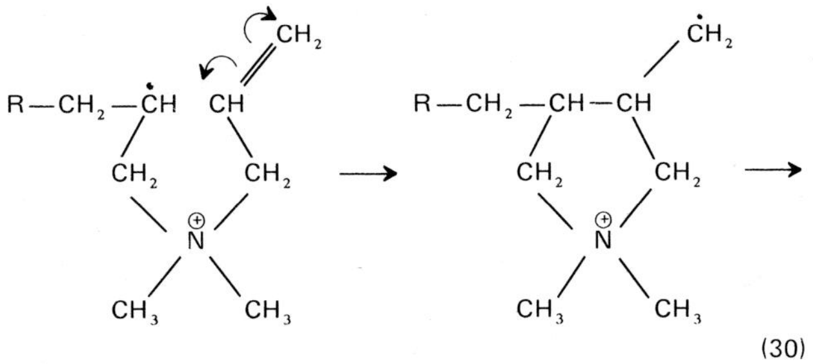

$$\tag{30}$$

This class of monomers has been used to prepare both low- to medium-molecular-weight polymers by homopolymerization,[83] and high-molecular-weight polymers by copolymerization with acrylamide.[84]

There are two other routes to quaternized monomers and polymers which have been extensively studied, but whose products have seen limited utility for water treatment, probably because of economics. Polyvinylbenzyltrimethylammonium chloride homo-

and copolymers can be prepared either by homo- or copolymerization of the monomer[21] or by reaction of vinylbenzyl chloride polymers with trimethylamine (Equation 31).[95] The vinylbenzyl chloride polymers can correspondingly be prepared by polymerization of vinylbenzyl chloride,[96] by chloromethylation of polystyrene,[97] or by chlorination of polymethylstyrenes (Equation 32).[98] In the latter method, mixtures of isomers are initially polymerized.

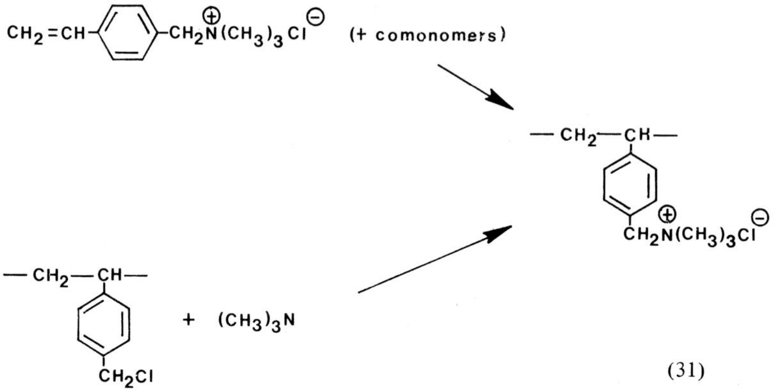

(31)

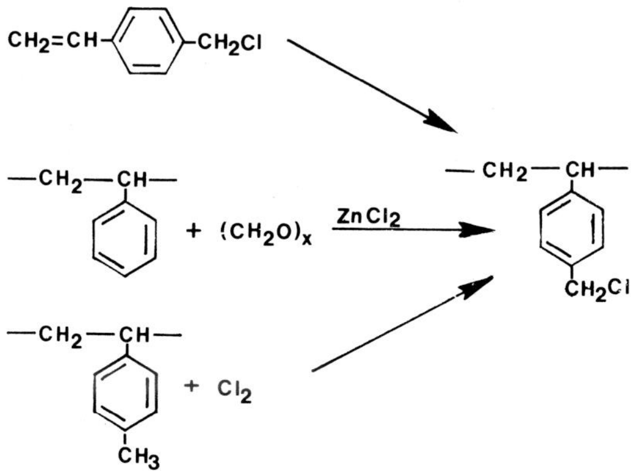

(32)

Derivatives of pyridine have also been used extensively to prepare polyelectrolytes, and are among some of the earliest cationic polyelectrolytes known.[100] The monomers used most frequently are the quaternized 2-vinyl- (XIX), 4-vinyl- (XX), and 2-methyl-5-vinyl pyridines (XXI).

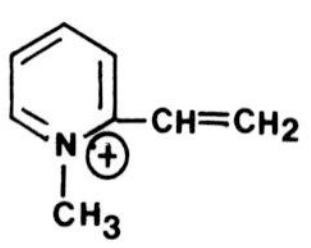

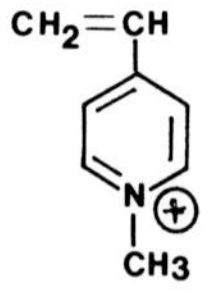

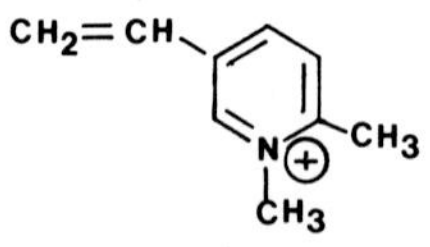

These monomers have been both homo- and copolymerized to give water-soluble polymers.[101,102] The corresponding unquaternized monomers have also been used to prepare cationic polymers.[102] Quaternization of the already formed polymers has also been reported.[103]

There are at least three types of non-quaternized, cationic, high-molecular-weight addition polymers of current commercial importance. They are polyamines, and are prepared by reactions performed on existing polymer backbones. A high-molecular-weight polyvinylimidazoline is prepared by reaction of polyacrylonitrile with ethylenediamine, using thioacetamide as a catalyst (Equation 33).[8] The polymer is recovered as the acid salt.

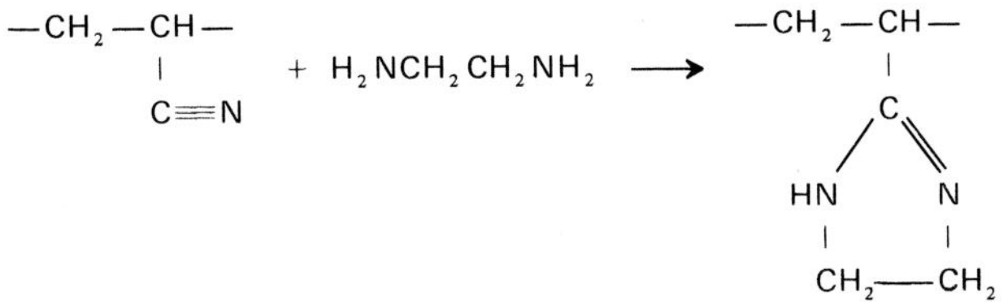

$$-CH_2-CH- \atop | \atop C\equiv N \quad + H_2NCH_2CH_2NH_2 \longrightarrow$$

$$-CH_2-CH- \atop | \atop C \atop HN\diagdown N \atop | \quad | \atop CH_2-CH_2$$

(33)

There has been a great deal of activity in the literature and on a commercial scale with the Mannich reaction of polyacrylamide with formaldehyde and dimethylamine to form the cationic aminomethylated polyacrylamides (Equation 34)[86,105]

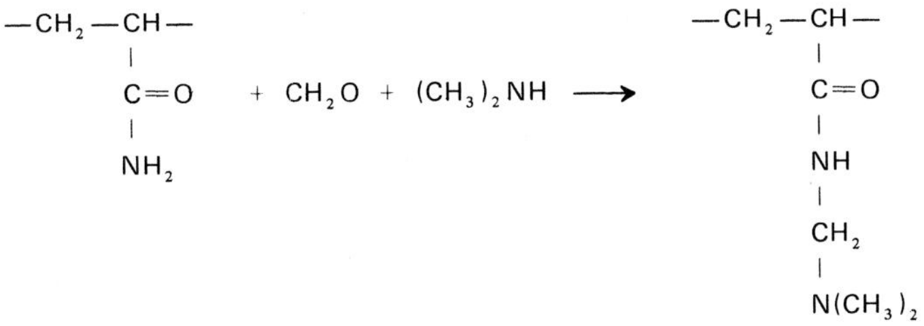

$$-CH_2-CH- \atop | \atop C=O \atop | \atop NH_2 \quad + CH_2O + (CH_3)_2NH \longrightarrow \quad -CH_2-CH- \atop | \atop C=O \atop | \atop NH \atop | \atop CH_2 \atop | \atop N(CH_3)_2$$

(34)

The resulting polyamine can be subsequently quaternized with any of the usual quaternizing agents.[87] The primary concerns with this system are to maximize conversion of the amide groups, to avoid cross-linking reactions, which can occur by reaction of the intermediate hydroxymethylated amide (Equation 35), and to minimize residual formaldehyde and dimethylamine, both highly objectionable because of toxicity and odor. Residual formaldehyde can also lead to instability by cross-linking reactions with unconverted amide groups.

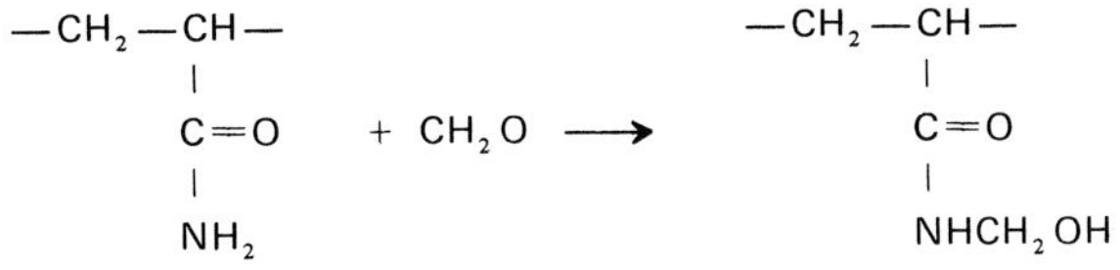

$$-CH_2-CH- \atop | \atop C=O \atop | \atop NH_2 \quad + CH_2O \longrightarrow \quad -CH_2-CH- \atop | \atop C=O \atop | \atop NHCH_2OH$$

XXII

(35)

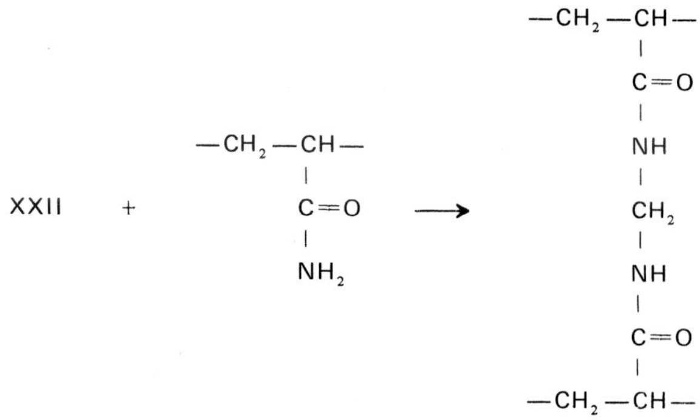

$$XXII \quad + \quad
\begin{array}{c}
-CH_2-CH- \\
| \\
C=O \\
| \\
NH_2
\end{array}
\quad \longrightarrow \quad
\begin{array}{c}
-CH_2-CH- \\
| \\
C=O \\
| \\
NH \\
| \\
CH_2 \\
| \\
NH \\
| \\
C=O \\
| \\
-CH_2-CH-
\end{array}$$

(35 continued)

A third method of creating a commercially important cationic polyamine involves reaction of polyacrylic acid homo- or copolymers with ethylenimine (Equation 36).[88] The polymer is utilized in the salt form.

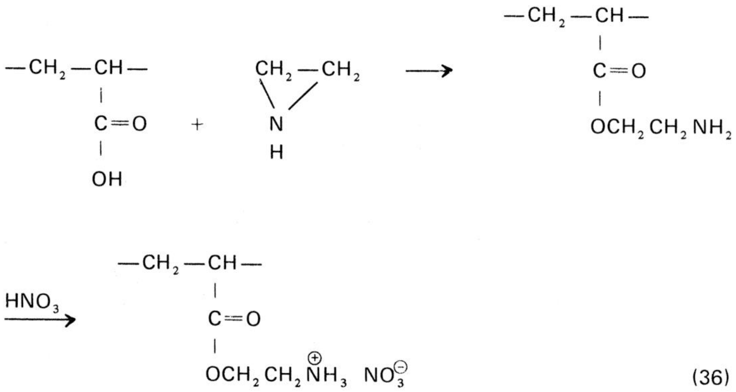

$$
\begin{array}{c}
-CH_2-CH- \\
| \\
C=O \\
| \\
OH
\end{array}
\quad + \quad
\begin{array}{c}
CH_2-CH_2 \\
\diagdown \quad \diagup \\
N \\
| \\
H
\end{array}
\quad \longrightarrow \quad
\begin{array}{c}
-CH_2-CH- \\
| \\
C=O \\
| \\
OCH_2CH_2NH_2
\end{array}
$$

$$
\xrightarrow{HNO_3} \quad
\begin{array}{c}
-CH_2-CH- \\
| \\
C=O \\
| \\
OCH_2CH_2\overset{\oplus}{N}H_3 \quad NO_3^{\ominus}
\end{array}
$$

(36)

Other polyamines reported in the literature, but which apparently have not yet achieved significance, at least in water treatment, include polyvinylamine, which classically has been made by polymerization of N-vinylphthalimide or N-vinylsuccinimide, followed by hydrolysis (Equation 37).[89]

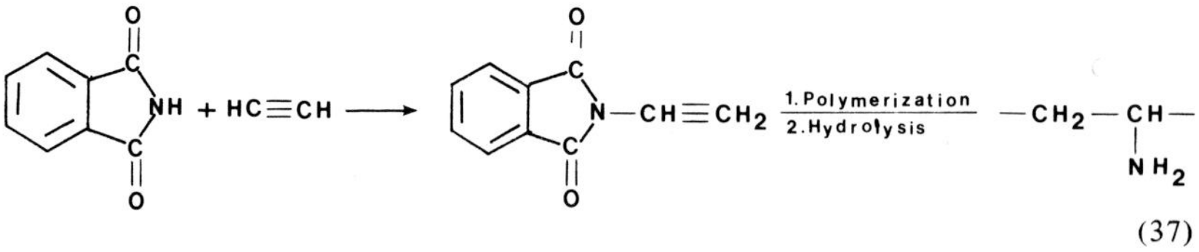

$$
\text{(phthalimide)} \; NH + HC\equiv CH \longrightarrow \text{(N-vinylphthalimide)} \; N-CH\equiv CH_2 \xrightarrow[\text{2.Hydrolysis}]{\text{1.Polymerization}}
\begin{array}{c}
-CH_2-CH- \\
| \\
NH_2
\end{array}
$$

(37)

Recently, an improved, apparently more economical route to polyvinylamine from N-vinylacetamide has been developed (Equation 38).[90] This polymer was subsequently used as a base for nonabsorbable dyes.

$$
\underset{O}{\overset{O}{\underset{\|}{CH_3CH}}} + 2\, CH_3\overset{O}{\overset{\|}{C}}NH_2 \xrightarrow{H^{\oplus}} CH_3CH\left(NH\overset{O}{\overset{\|}{C}}CH_3\right)_2 \xrightarrow{\Delta} CH_2=CHNH\overset{O}{\overset{\|}{C}}CH_3 + CH_3COOH
$$

XXIII

(38)

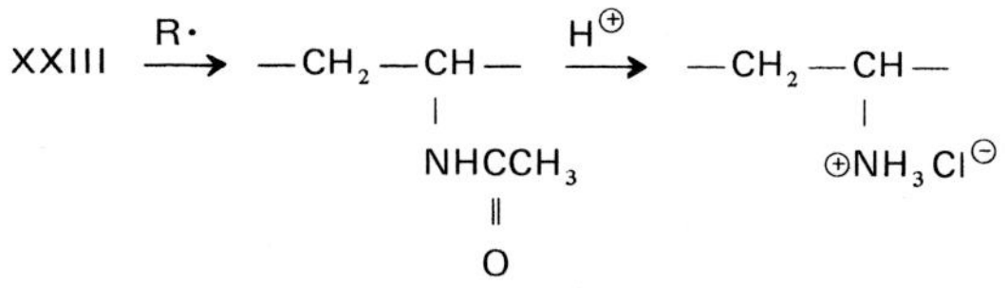

$$XXIII \xrightarrow{R\cdot} \begin{array}{c} -CH_2-CH- \\ | \\ NHCCH_3 \\ \| \\ O \end{array} \xrightarrow{H^{\oplus}} \begin{array}{c} -CH_2-CH- \\ | \\ {}^{\oplus}NH_3\,Cl^{\ominus} \end{array}$$

(38 continued)

The Hofmann reaction on polyacrylamide has also been used to introduce amino groups (Equation 39).[86] Conversions up to about 90% have been reported, even with very high-molecular-weight polyacrylamide.[92]

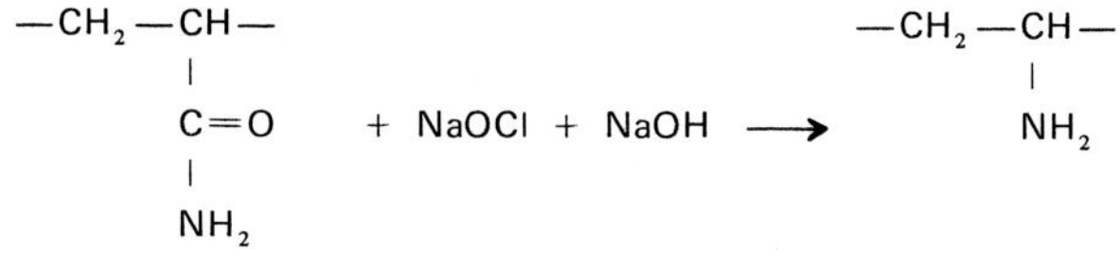

$$\begin{array}{c} -CH_2-CH- \\ | \\ C=O \\ | \\ NH_2 \end{array} + NaOCl + NaOH \longrightarrow \begin{array}{c} -CH_2-CH- \\ | \\ NH_2 \end{array}$$

(39)

Polymeric carboxylic acid groups have been converted to amines via the Schmidt reaction (Equation 40).[93]

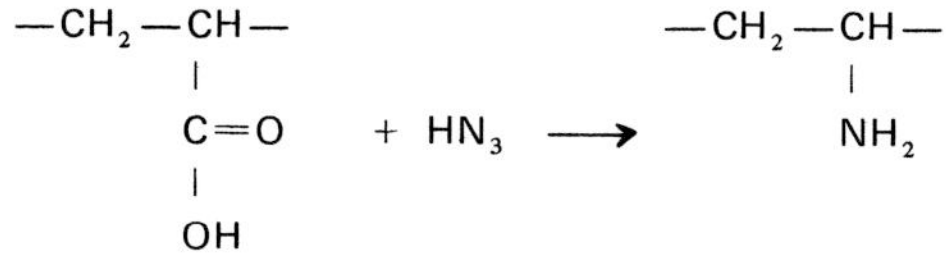

$$\begin{array}{c} -CH_2-CH- \\ | \\ C=O \\ | \\ OH \end{array} + HN_3 \longrightarrow \begin{array}{c} -CH_2-CH- \\ | \\ NH_2 \end{array}$$

(40)

Since conversions of only 50 to 60% were reported, this actually results in a polyampholyte (see Section VII) (Equation 41), having both anionic and cationic charges. The structure shown in Equation 41 does not imply that the carboxyl and amine groups alternate in a regular fashion.

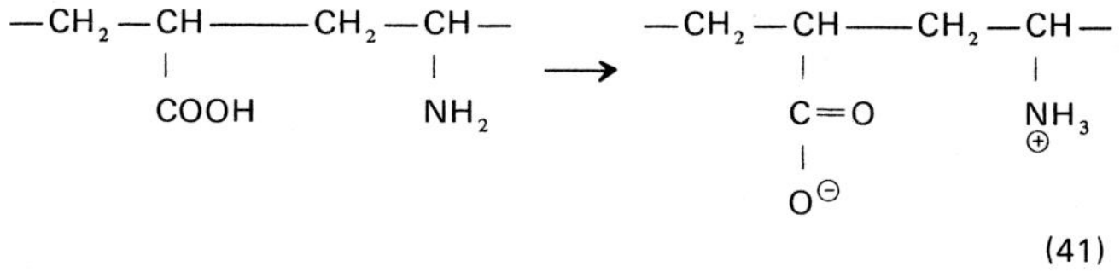

$$\begin{array}{cc} -CH_2-CH\!\!-\!\!-\!\!-CH_2-CH- \\ | \quad\quad | \\ COOH \quad\quad NH_2 \end{array} \longrightarrow \begin{array}{cc} -CH_2-CH\!\!-\!\!-\!\!-CH_2-CH- \\ | \quad\quad | \\ C=O \quad\quad NH_3 \\ | \quad\quad {}^{\oplus} \\ O^{\ominus} \end{array}$$

(41)

Other reactions which have converted existing polymers to cationic species include the reaction of a tertiary amine-primary amine with polyacrylonitrile,[155] polyacrylamide,[156] or a styrene-maleic anhydride copolymer[157] (Equation 42). The last polymer mentioned has also been quaternized.

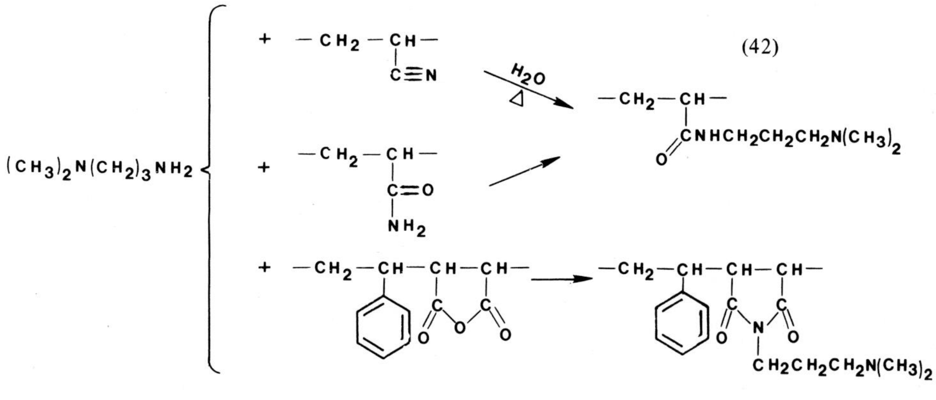

(42)

Cationic polymers were obtained when adducts of amines with epichlorohydrin were reacted with polyacrylic acids[158] or polyacrylamides[159] (Equation 43).

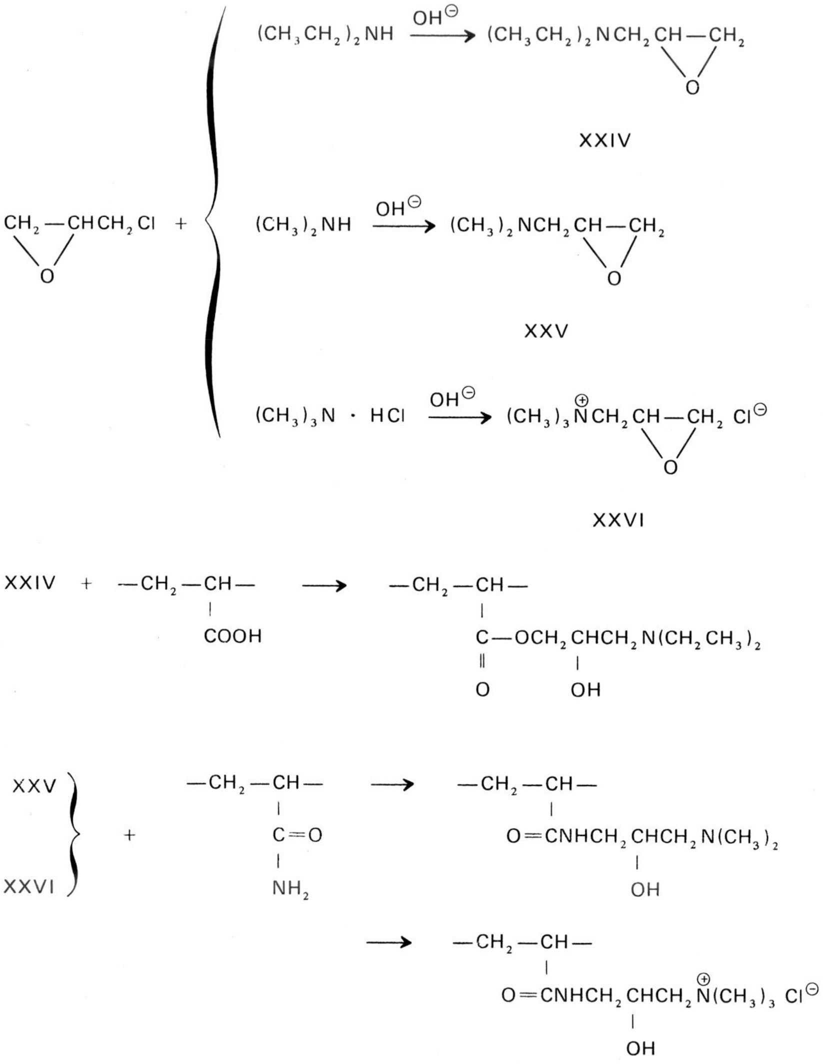

$$CH_2-CHCH_2Cl \ + \ \begin{cases} (CH_3CH_2)_2NH \xrightarrow{OH^\ominus} (CH_3CH_2)_2NCH_2CH-CH_2 \\ \qquad\qquad\qquad\qquad\qquad\qquad\qquad XXIV \\ (CH_3)_2NH \xrightarrow{OH^\ominus} (CH_3)_2NCH_2CH-CH_2 \\ \qquad\qquad\qquad\qquad\qquad\qquad XXV \\ (CH_3)_3N \cdot HCl \xrightarrow{OH^\ominus} (CH_3)_3\overset{\oplus}{N}CH_2CH-CH_2 \ Cl^\ominus \\ \qquad\qquad\qquad\qquad\qquad\qquad\qquad XXVI \end{cases}$$

XXIV + —CH$_2$—CH— $\longrightarrow$ —CH$_2$—CH—
| |
COOH   C—OCH$_2$CHCH$_2$N(CH$_2$CH$_3$)$_2$
                                    ‖ |
                                    O   OH

$$XXV \atop XXVI \Big\} \ + \ \begin{matrix} -CH_2-CH- \\ | \\ C=O \\ | \\ NH_2 \end{matrix} \ \longrightarrow \ \begin{matrix} -CH_2-CH- \\ | \\ O=CNHCH_2CHCH_2N(CH_3)_2 \\ | \\ OH \end{matrix}$$

$$\longrightarrow \ \begin{matrix} -CH_2-CH- \\ | \\ O=CNHCH_2CHCH_2\overset{\oplus}{N}(CH_3)_3 \ Cl^\ominus \\ | \\ OH \end{matrix}$$

(43)

Also, cationic polymers resulted when dimethylaminoethyl chloride was reacted with polyvinylalcohol (Equation 44),[160] when a polybutadiene polymerized primarily via a 1,2 mode was reacted with HBr under free-radical conditions, followed by reaction with a tertiary amine (Equation 45),[161] and when polymerized 2-chloromethylbutadiene was reacted with a tertiary amine (Equation 46).[162]

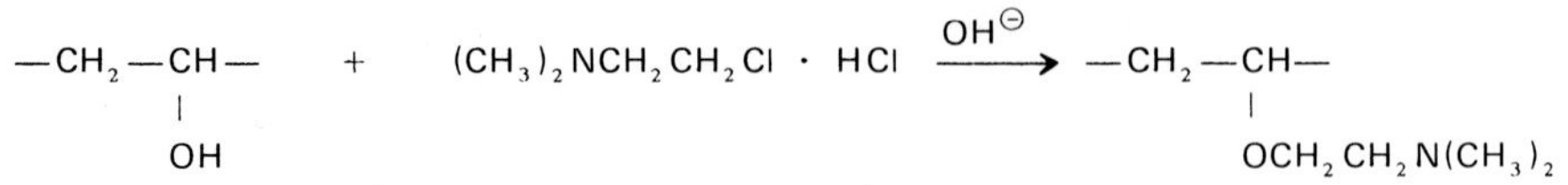

—CH$_2$—CH— + (CH$_3$)$_2$NCH$_2$CH$_2$Cl $\cdot$ HCl $\xrightarrow{OH^\ominus}$ —CH$_2$—CH—
| |
OH   OCH$_2$CH$_2$N(CH$_3$)$_2$

(44)

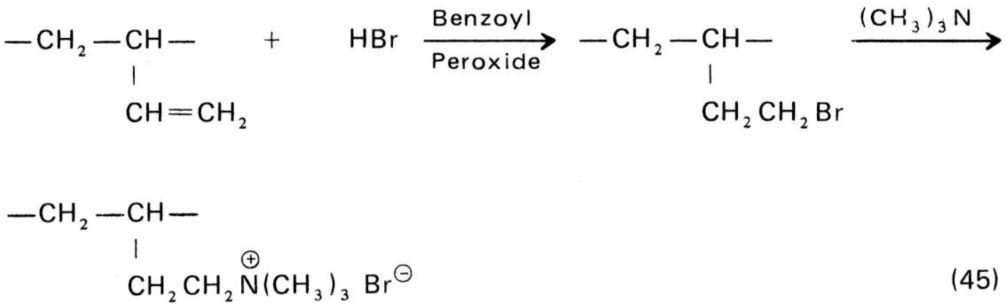

$$-CH_2-CH- \quad + \quad HBr \xrightarrow[\text{Peroxide}]{\text{Benzoyl}} \quad -CH_2-CH- \xrightarrow{(CH_3)_3N}$$
$$\quad\quad | \quad\quad\quad\quad\quad\quad\quad\quad\quad\quad\quad\quad\quad\quad\quad | $$
$$\quad CH=CH_2 \quad\quad\quad\quad\quad\quad\quad\quad\quad\quad CH_2CH_2Br$$

$$-CH_2-CH-$$
$$\quad\quad |$$
$$\quad CH_2CH_2\overset{\oplus}{N}(CH_3)_3 \ Br^{\ominus} \tag{45}$$

$$-CH_2-C=CH-CH_2- \quad + \quad (CH_3)_3N \longrightarrow -CH_2-C=CH-CH_2-$$
$$\quad\quad\quad | \quad\quad\quad\quad\quad\quad\quad\quad\quad\quad\quad\quad\quad\quad\quad\quad\quad\quad | $$
$$\quad\quad CH_2Cl \quad\quad\quad\quad\quad\quad\quad\quad\quad\quad\quad\quad CH_2\overset{\oplus}{N}(CH_3)_3 \ Cl^{\ominus} \tag{46}$$

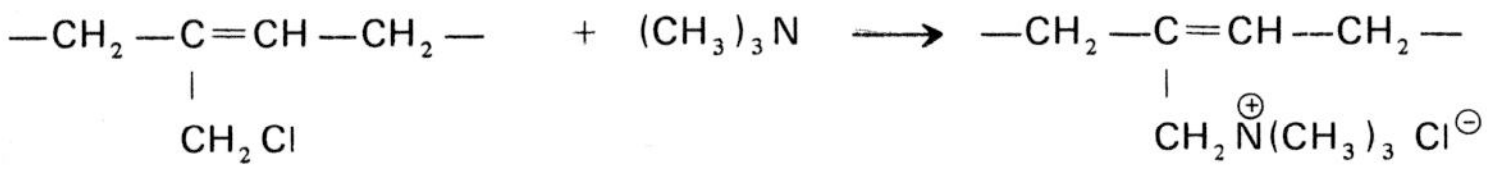

Another type of chain reaction polymerization has been used to prepare cationic polyelectrolytes. In this method, epichlorohydrin is polymerized under anhydrous conditions through the epoxide group to give a polyether having pendant chloromethyl groups, which are subsequently quaternized with a tertiary amine, usually trimethylamine (Equation 47).

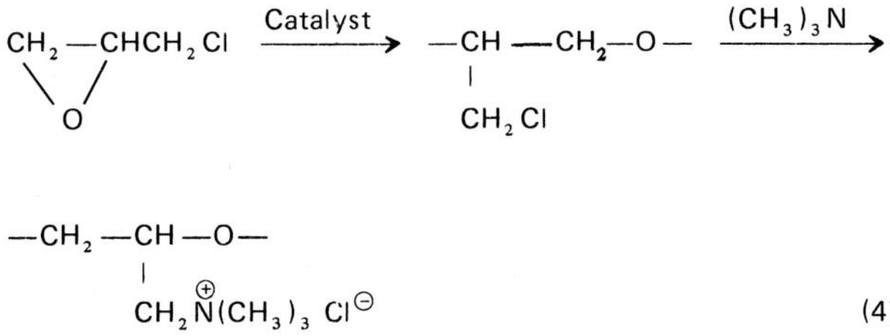

$$CH_2-CHCH_2Cl \xrightarrow{\text{Catalyst}} -CH-CH_2-O- \xrightarrow{(CH_3)_3N}$$
$$\backslash/ | $$
$$O CH_2Cl$$

$$-CH_2-CH-O-$$
$$\quad\quad\quad |$$
$$\quad CH_2\overset{\oplus}{N}(CH_3)_3 \ Cl^{\ominus} \tag{47}$$

Both acidic[106] and basic[107] catalysts can be used to polymerize the epichlorohydrin. Effective products can be made which have only sufficient quaternization to render the polymer water soluble,[108] but complete quaternization appears to be more desirable.[106] Polyepichlorohydrin has also been reacted with ethylene diamine to give a cationic polyelectrolyte (Equation 48), which could be further chain extended with epichlorohydrin or ethylene dichloride (see Section V.B.).[163]

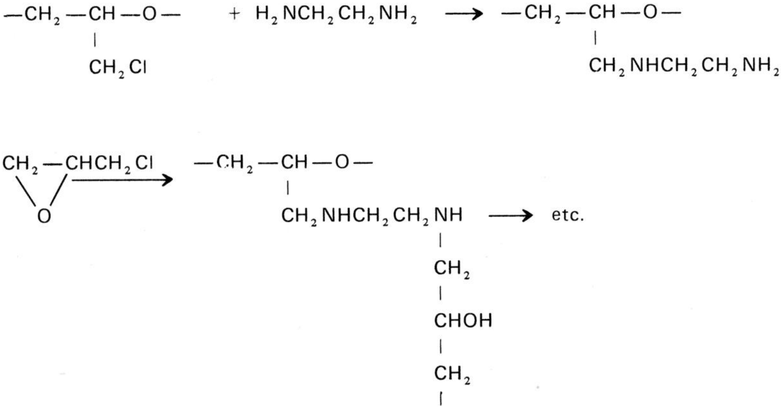

$$-CH_2-CH-O- \quad + \quad H_2NCH_2CH_2NH_2 \longrightarrow -CH_2-CH-O-$$
$$\quad\quad\quad | \quad\quad\quad\quad\quad\quad\quad\quad\quad\quad\quad\quad\quad\quad\quad\quad\quad\quad | $$
$$\quad CH_2Cl \quad\quad\quad\quad\quad\quad\quad\quad\quad\quad\quad\quad\quad CH_2NHCH_2CH_2NH_2$$

$$CH_2-CHCH_2Cl \qquad -CH_2-CH-O-$$
$$\backslash/ \xrightarrow{} | $$
$$O CH_2NHCH_2CH_2NH \longrightarrow \text{etc.} \tag{48}$$
$$ | $$
$$ CH_2$$
$$ | $$
$$ CHOH$$
$$ | $$
$$ CH_2$$
$$ | $$

B. Step-Reaction Polymerization

This author considers the area of production of low- to medium-molecular-weight cationic polyeletrolytes via condensation polymerization to be the most competitive and challenging field of polyelectrolyte chemistry, and one that has produced the most imaginative inventions in the field. To produce a high charge density cationic polymer at low cost with optimum molecular weight, when molecular weight is inherently limited by the chemistry of the reactions involved, is a formidable task indeed. The profusion of patent activity and variety of products available is a tribute to the creativity of the chemists working in this area. The diversity of products in light of the limited number of suitable raw materials again attests to the inventiveness of the researchers.

At the outset, a few definitions and principles are in order. The definitions of Billmeyer[109] can be used to distinguish the step-reaction, or condensation polymers to be discussed here. Although some of the polymers fit the classical definition of condensation polymerization, involving loss of a small molecule at each reaction step, most of the polymers are formed without loss of small molecules. In every other respect, however, the polymerizations meet the criteria of step-reaction polymers.

In this section, polymerization reactions involving amines predominate. In these reactions, ammonia is tetrafunctional, since it can react completely to form quaternary ammonium salts. By the same token, a primary amine is trifunctional, a secondary amine bifunctional, and a tertiary amine is monofunctional. As we shall see, these are important considerations for the synthesis of cationic polyelectrolytes formed by step-reaction processes.

1. Epihalohydrin Types

In this class of cationic polymers, epichlorohydrin (1-chloro-2,3-epoxypropane) (XXVII), because of cost, is used almost exclusively. In reactions with amines it is bifunctional, and will therefore form polymers with ammonia, primary and secondary amines. It will not react with tertiary amines to form polymers, but has been used to synthesize two molecules which could be useful for the introduction of cationic groups into polymers having suitable hydroxyl or amine functionality, such as polysaccharides or polyvinylalcohols (Equation 49).[110]

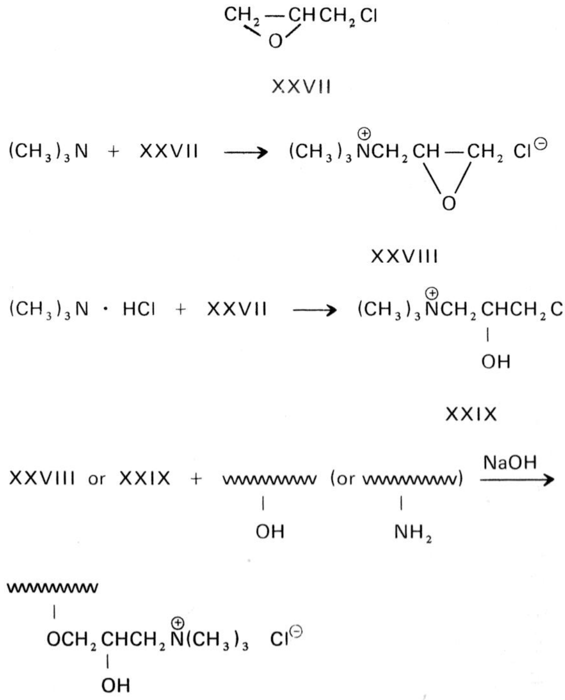

$$CH_2—CHCH_2Cl$$
$$\diagdown O \diagup$$

XXVII

$$(CH_3)_3N \ + \ XXVII \ \longrightarrow \ (CH_3)_3\overset{\oplus}{N}CH_2CH—CH_2 \ \ Cl^{\ominus}$$
$$\diagdown O \diagup$$

XXVIII

$$(CH_3)_3N \cdot HCl \ + \ XXVII \ \longrightarrow \ (CH_3)_3\overset{\oplus}{N}CH_2CHCH_2Cl \ \ Cl^{\ominus}$$
$$| \\ OH$$

XXIX

$$XXVIII \ or \ XXIX \ + \ \text{wwwww} \ (or \ \text{wwwwww}) \ \xrightarrow{\ NaOH\ }$$
$$| \qquad\qquad\qquad | \\ OH \qquad\qquad\quad NH_2$$

$$\text{wwwwww} \\ | \\ OCH_2CHCH_2\overset{\oplus}{N}(CH_3)_3 \ \ Cl^{\ominus} \\ | \\ OH$$

(49)

The reaction of epichlorohydrin with secondary amines produces linear polyquaternary ammonium salts (Equation 50). Polymers such as this in which the charged atoms are in the backbone are called ionenes.[111]

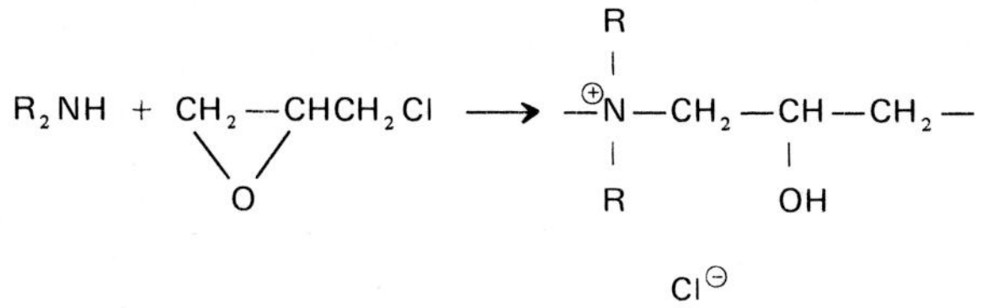

$$R_2NH + CH_2\underset{\diagdown\!O\!\diagup}{-}CHCH_2Cl \longrightarrow \overset{\overset{R}{|}}{\underset{\underset{R}{|}}{\overset{\oplus}{-}N}}-CH_2-\underset{\underset{OH}{|}}{CH}-CH_2-$$

$$Cl^{\ominus}$$

(50)

This elegant reaction has been the subject of much study and, when R = CH$_3$, a great deal of commercial activity[112]. A number of water-treatment companies offer polyelectrolytes that are believed to be of this type. The polymerization usually proceeds in two stages: low temperature addition of the amine to the halohydrin, or vice versa (Equation 51),[113] followed by polymerization at higher temperatures (Equation 52). Unless the first reaction is carried out at temperatures below about 15°C, some polymerization occurs during the first stage. It is also not significant in terms of product which of the two adducts forms in the first part of the reaction; indeed, both species may be present.

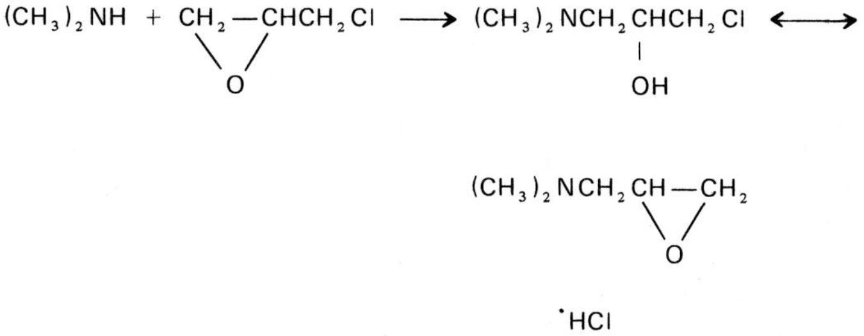

$$(CH_3)_2NH + CH_2\underset{\diagdown\!O\!\diagup}{-}CHCH_2Cl \longrightarrow (CH_3)_2N\underset{\underset{OH}{|}}{CH_2CHCH_2}Cl \longleftrightarrow$$

$$(CH_3)_2NCH_2CH\underset{\diagdown\!O\!\diagup}{-}CH_2$$

$$\cdot HCl$$

(51)

$$\left.\begin{array}{c} \overset{\cdot\,HCl}{(CH_3)_2NCH_2CH\underset{\diagdown\!O\!\diagup}{-}CH_2} \\[3mm] (CH_3)_2NCH_2\underset{\underset{OH}{|}}{CHCH_2}Cl \end{array}\right\} \longrightarrow -\overset{\overset{CH_3}{|}}{\underset{\underset{CH_3}{|}}{N_{\oplus}}}-CH_2\underset{\underset{OH}{|}}{CHCH_2}- $$

$$Cl^{\ominus}$$

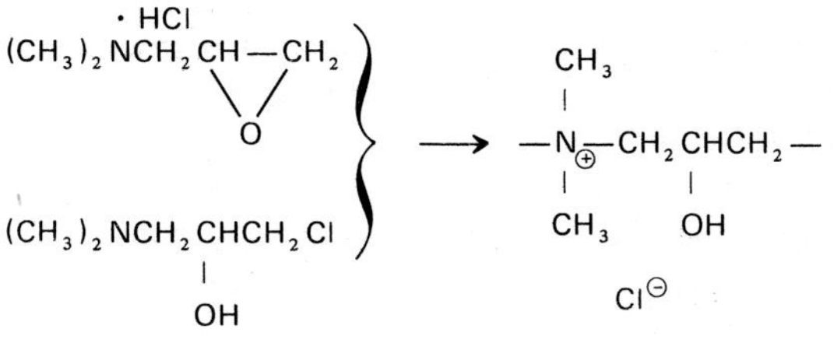

(52)

In order to achieve highest molecular weights for this type of polymerization, both reactants must be present in as close to stoichiometric amounts as possible. Nevertheless, even under the most ideal of conditions, number-average molecular weights for this polymer are usually 10,000 or under. Since other ionenes formed by condensation reactions have been made at considerably higher molecular weights, failure to achieve higher molecular weights may be due to side reactions.[114]

Higher molecular weights for this type of polymer have been achieved by terpolymerization with small amounts of amines having functionality greater than 2.[115] In these cases, the higher molecular weights are achieved at the expense of linearity (Equation 53).

$$(CH_3)_2NH + RNH_2 + CH_2-CHCH_2Cl \underset{O}{\overset{}{\longrightarrow}} \left[\begin{array}{c} CH_3 \\ | \\ -N{\oplus}-CH_2CHCH_2- \\ | \quad\quad | \\ CH_3 \quad OH \end{array}\right]\left[\begin{array}{c} R \\ | \\ -N-CH_2CH-CH_2- \\ | \\ OH \end{array}\right]$$

$$\underset{\substack{CH_2-CHCH_2Cl \\ \diagdown\diagup \\ O}}{\longrightarrow} \left[\begin{array}{c} CH_3 \\ | \\ -N{\oplus}-CH_2CHCH_2- \\ | \quad\quad | \\ CH_3 \quad OH \end{array}\right]\left[\begin{array}{c} R \\ | \\ -N{\oplus}-CH_2CH-CH_2- \\ | \quad\quad | \\ CH_2 \quad OH \\ CHOH \\ CH_2 \\ | \end{array}\right] \quad (53)$$

where moles CH_2-$CHCH_2Cl \cong$ total moles of amines.

The molecular weights and extent of branching of these types of polymers can be controlled by a number of variables: type of polyfunctional amine, ratio of polyfunctional amine to secondary amine, concentration, ratio of epichlorohydrin to amines, and method of addition of reagents. A trifunctional primary amine will give the lowest extent of branching, and will conversely require the greatest extent of reaction to achieve a maximum molecular weight. More complex amines, such as the hexafunctional ethylenediamine, will give more extensive branching, and a higher molecular weight at the same reaction extent as the simpler amines. The more complex amines must therefore be used at lower ratios. While an optimum combination of properties for certain uses (i.e., effect of extent of branching on activity) appears to be achieved when the polyfunctional amine is used at less than 5 mol% of the total amines,[116] useful polyelectrolytes have been prepared where the polyfunctional amine is present at up to 33 mol%.[117]

Extent of reaction can also be controlled by the stoichiometry and mode of addition of the epichlorohydrin. One technique is to react a roughly equimolar amount of epichlorohydrin with the amines until no further increase in viscosity occurs. The epichlorohydrin can also be used to "titrate" the amine functionalities. In this case, a major portion of the epichlorohydrin is added and reacted, followed by the addition of incremental portions, which are then allowed to react (illustrated with ammonia as the polyfunctional amine in Equation 54). These increments are added in gradually smaller amounts, since as the reaction (and extent of branching) proceeds, the polymer can approach the gel point of "infinite" molecular weight, where solubility and utility are lost. The incremental portions can be allowed to react fully, until no further viscosity increase is observed, or the reaction can be "short-stopped" by the addition of acid. Some techniques favor the addition of an inorganic base to speed the reaction, although there is some evidence that this may affect the stability of the product.[117]

The concentration of reactants can also be used to control the extent of reaction. At higher concentrations, reaction is more rapid and branching reactions are more frequent than at lower concentrations, resulting in a more rapid buildup in viscosity and approach to the gel point. A reaction product which is close to the gel point can be diluted to retard further reaction, followed by addition of more epichlorohydrin to increase molecular weight (and branching). This can be done in successive steps until a highly viscous, completely water-soluble, ungelled product at low concentrations is obtained.

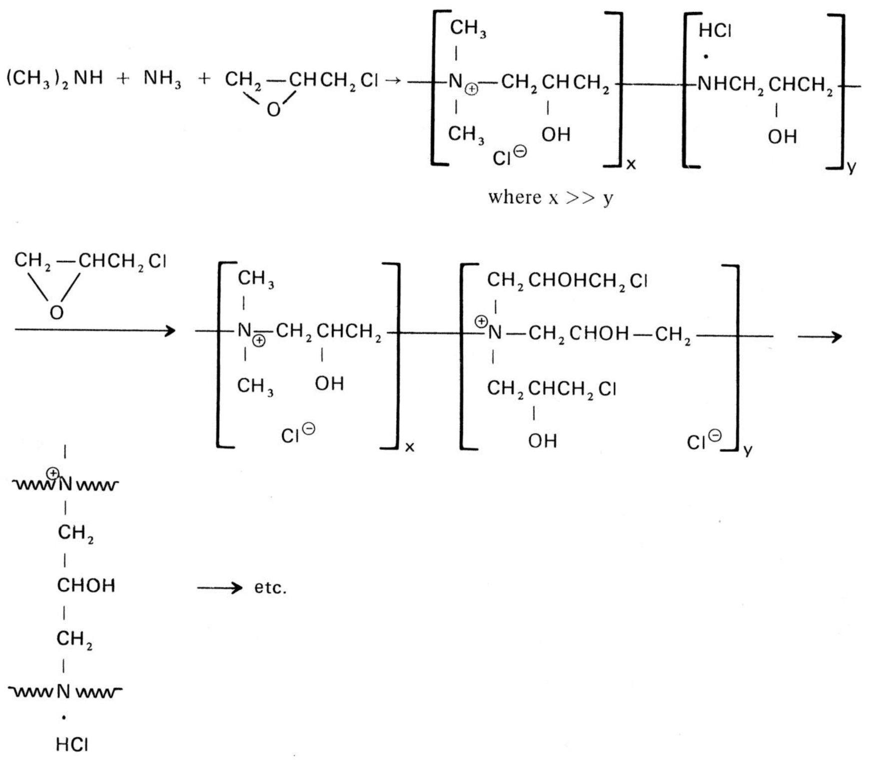

$$(CH_3)_2NH \; + \; NH_3 \; + \; CH_2\!-\!CHCH_2Cl \rightarrow \left[\begin{array}{c} CH_3 \\ | \\ N_{\oplus}\!-\!CH_2CHCH_2 \\ | \quad\quad | \\ CH_3 \quad OH \\ Cl^{\ominus} \end{array} \right]_x \left[\begin{array}{c} HCl \\ \cdot \\ NHCH_2CHCH_2 \\ | \\ OH \end{array} \right]_y$$

where x >> y

$$CH_2\!-\!CHCH_2Cl$$

$$\xrightarrow{\quad} \left[\begin{array}{c} CH_3 \\ | \\ N_{\oplus}\!-\!CH_2CHCH_2 \\ | \quad\quad | \\ CH_3 \quad OH \\ Cl^{\ominus} \end{array} \right]_x \left[\begin{array}{c} CH_2CHOHCH_2Cl \\ | \\ {}_{\oplus}N\!-\!CH_2CHOH\!-\!CH_2 \\ | \\ CH_2CHCH_2Cl \\ | \\ OH \quad\quad Cl^{\ominus} \end{array} \right]_y \longrightarrow \quad (54)$$

$$\begin{array}{c} | \\ \text{wwww}\overset{\oplus}{N}\text{ www} \\ | \\ CH_2 \\ | \\ CHOH \quad \longrightarrow \quad etc. \\ | \\ CH_2 \\ | \\ \text{www N www} \\ \cdot \\ HCl \end{array}$$

In all of the above examples, the product consists primarily of a polyquaternary ammonium salt, as evidenced by the inertness of the product to chlorinated waters.[115]

Many useful polyelectrolytes have been prepared with polyfunctional amines and epichlorohydrin alone. The same reaction mechanisms can be written for these polymerizations, and in all cases a complex, branched reaction product, containing mixtures of quaternary ammonium salts and amines, is obtained. All of the techniques for controlling molecular weights and extent of branching discussed above have been used with these polymers.

The simplest members of the series, ammonia-epichlorohydrin polymers (Equation 55) have been made using an excess of ammonia over epichlorohydrin.[118] The products are probably of low molecular weight and only lightly branched. It is believed that these polymers ase not of great current commercial interest.

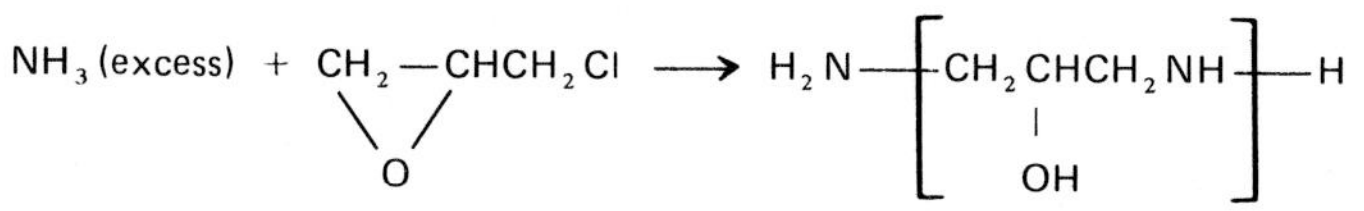

$$NH_3 \text{ (excess)} \; + \; CH_2\!-\!CHCH_2Cl \; \longrightarrow \; H_2N\!-\!\left[\begin{array}{c} CH_2CHCH_2NH \\ | \\ OH \end{array} \right]\!-\!H \quad\quad (55)$$

Epichlorohydrin polymers with monomethylamine, however, are of recent commercial importance (Equation 56).[116] For this system, both the method of introduction of the total amount of reactants into the reaction vessel, followed by molecular weight buildup under alkaline conditions,[119] and the epichlorohydrin incremental addition method to increase molecular weight[120] have been reported. In the preferred latter technique, additional amine is added when the desired viscosity is reached to prevent further molecular weight increase by reaction with remaining epichlorohydrin functionalities. This product has also been quaternized to give useful products.[121]

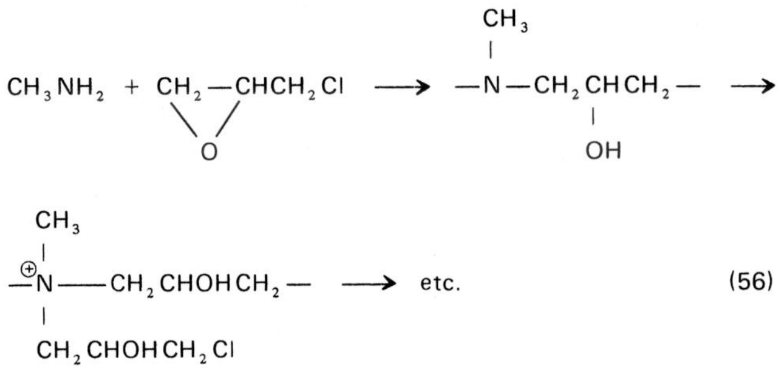

$$CH_3NH_2 \ + \ CH_2\underset{\diagdown \diagup}{\overset{}{-}}CHCH_2Cl \ \longrightarrow \ \overset{\overset{\displaystyle CH_3}{|}}{-N}-CH_2\underset{|}{\overset{}{C}H}CH_2- \ \longrightarrow$$

with O in the epoxide and OH below.

$$\overset{\overset{\displaystyle CH_3}{|}}{\underset{\underset{\displaystyle CH_2CHOHCH_2Cl}{|}}{\overset{\oplus}{-}N}}-CH_2CHOHCH_2- \ \longrightarrow \ etc. \qquad (56)$$

Mixtures of methylamine with polyfunctional amines have also been used.[123]

In one variation of the use of methylamine with epichlorohydrin,[124] 2 mol of epichlorohydrin are reacted with 1 mol of methylamine to give roughly the dichloride (XXX) (branching reactions involving the tertiary amine undoubtedly take place — Equation 58), which is then reacted with N,N,N',N'-tetramethylethylenediamine (Equation 57). The product is substantially quaternized.

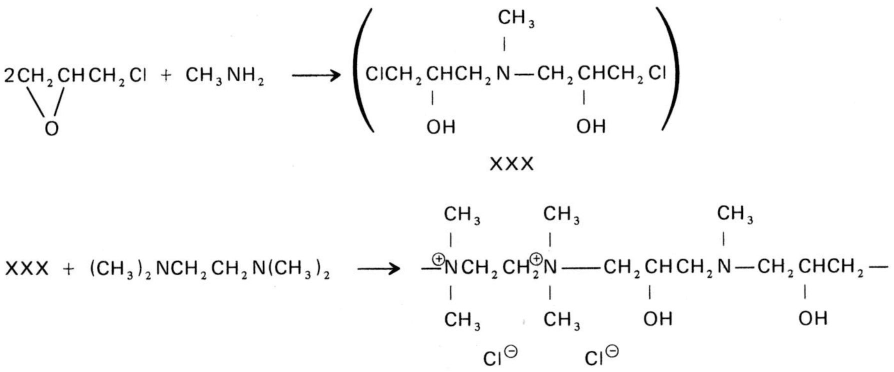

$$2CH_2\underset{\diagdown \diagup}{\overset{}{-}}CHCH_2Cl \ + \ CH_3NH_2 \ \longrightarrow \ \left(ClCH_2\underset{|}{\overset{}{C}HCH_2}\overset{\overset{\displaystyle CH_3}{|}}{N}-CH_2\underset{|}{\overset{}{C}HCH_2}Cl \right)$$

with OH under each CHCH$_2$ group.

XXX

$$XXX \ + \ (CH_3)_2NCH_2CH_2N(CH_3)_2 \ \longrightarrow \ \overset{\oplus}{N}CH_2CH_2\overset{\oplus}{N}--CH_2CHCH_2N-CH_2CHCH_2-$$

with CH$_3$ and CH$_3$ above, CH$_3$, CH$_3$, OH, OH below, and $Cl^{\ominus}$ $Cl^{\ominus}$ below. (57)

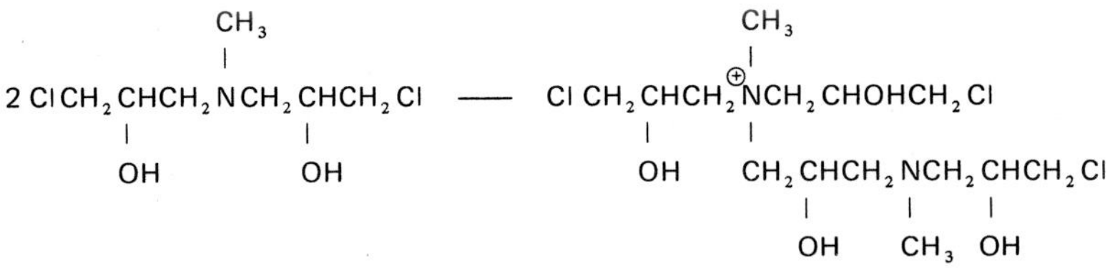

$$2 ClCH_2\underset{|}{\overset{}{C}HCH_2}\overset{\overset{\displaystyle CH_3}{|}}{N}CH_2\underset{|}{\overset{}{C}HCH_2}Cl \ \overline{} \ ClCH_2CHCH_2\overset{\overset{\displaystyle CH_3}{|}}{\overset{\oplus}{N}}CH_2CHOHCH_2Cl$$

with OH OH below left, and below right: OH, CH$_2$CHCH$_2$NCH$_2$CHCH$_2$Cl, with OH CH$_3$ OH below. (58)

Useful products, using techniques similar to those described above, have been prepared from N,N-dimethyl-1,3-propane diamine and epichlorohydrin.[122] An idealized product is shown in Equation 59. In this case, after the desired viscosity is reached, further reaction is prevented by addition of acid, which converts any reactive amines to relatively unreactive salts. Furthermore, the reaction does not require addition of inorganic bases to proceed smoothly.

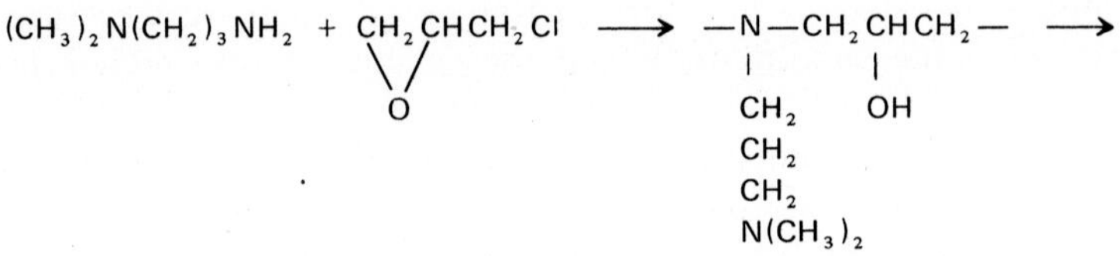

$$(CH_3)_2N(CH_2)_3NH_2 \ + \ CH_2\underset{\diagdown \diagup}{\overset{}{-}}CHCH_2Cl \ \longrightarrow \ -N-CH_2CHCH_2- \ \longrightarrow$$

with below: CH$_2$ OH, CH$_2$, CH$_2$, N(CH$_3$)$_2$.

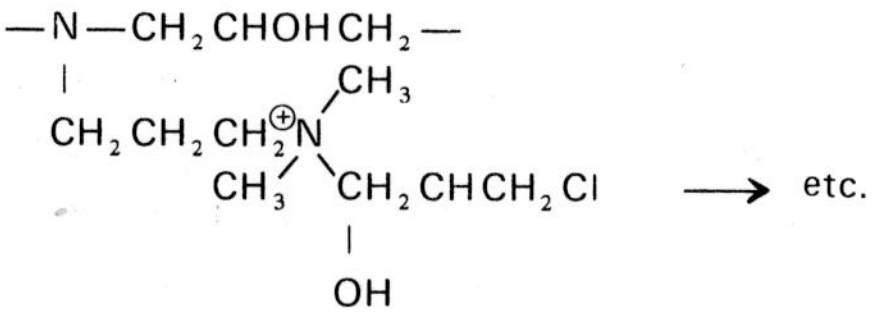

$$-N-CH_2CHOHCH_2-$$

(structure for Equation 59)

$$\text{(59)}$$

The patent literature is replete with references to the cationic polymers prepared by the reaction of polyalkylene polyamines with epichlorohydrin. The most common polyalkylene polyamines utilized are derived from ammonia and ethylene dichloride (XXXI) (see Section V.B.2 for a discussion of the synthesis of these amines),[125] although others, where the alkyl chain connecting the nitrogens is longer, have been used.[126]

$$H_2N+CH_2CH_2NH+_xH$$

XXXI

Equation 60 illustrates the use of diethylenetriamine for polymer formation.

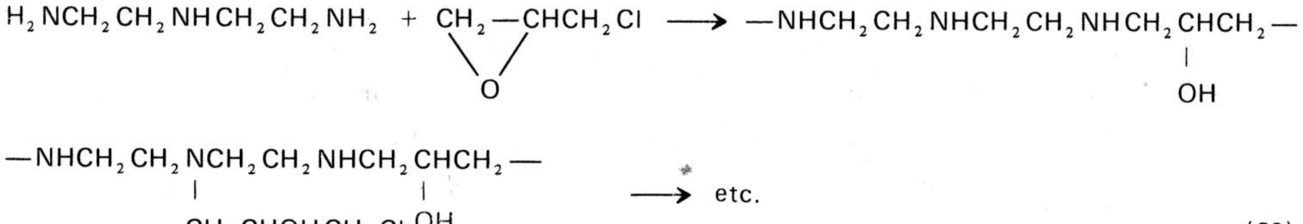

$$H_2NCH_2CH_2NHCH_2CH_2NH_2 \; + \; CH_2-CHCH_2Cl \longrightarrow -NHCH_2CH_2NHCH_2CH_2NHCH_2CHCH_2-$$

(structure continued)

$$\text{(60)}$$

One type of polymer used extensively in water treatment is prepared using two types of condensation reactions. In the first, a polyamide is formed by reaction of adipic acid with a polyalkylene polyamine, with the polyamidation reaction probably proceeding primarily through the primary amino groups. The low-molecular-weight condensate is then reacted with epichlorohydrin to achieve higher molecular weights (Equation 61).[127] These polymers have also been quaternized to give useful, stable products.[128]

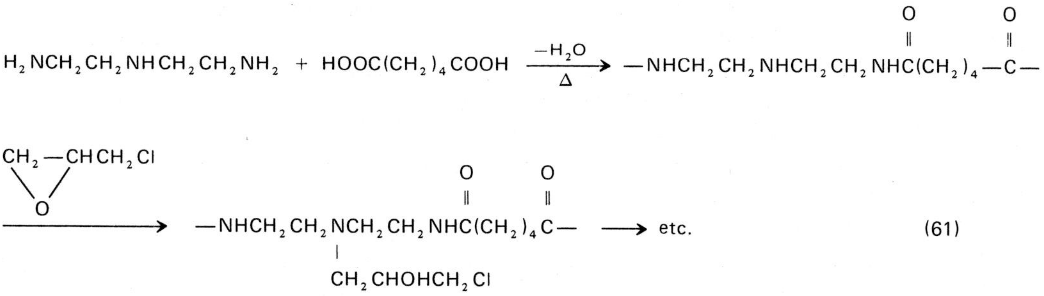

$$H_2NCH_2CH_2NHCH_2CH_2NH_2 \; + \; HOOC(CH_2)_4COOH \xrightarrow[\Delta]{-H_2O}$$

(structure for Equation 61)

$$\text{(61)}$$

The utility of the reactions discussed above is obvious. There are numerous other examples of epichlorohydrin-amine polymerizations, with a great many variations being possible. The number of possible combinations of readily available amines and epichlorohydrin is staggering. When epoxides, halides, quaternizing agents, and other reagents (e.g., aldehydes) are also used, the list becomes nearly endless. It is the author's opinion that this is still an area where a fertile imagination can invent new and useful molecules.

2. Dihaloalkane Types

A second class of cationic condensation polyelectrolyte is prepared by the nucleophilic displacement reaction of amines with dihaloalkanes or their derivatives. As with epichlorohydrin-based cationic polyelectrolytes, the number of possible products is large, and the field can only be surveyed here in a cursory manner.

When ditertiary amines or secondary amines are reacted with dihaloalkanes, ionenes result. In the classic studies of Rembaum and associates[111] ditertiary amines were reacted with unsubstituted dihaloalkanes (Equation 62), and polymer formation as a function of the length of the alkyl chain was studied.

$$(CH_3)_2N(CH_2)_xN(CH_3)_2 \ + \ X\text{---}(CH_2)_y\text{---}X \ \longrightarrow \ \overset{\overset{\displaystyle CH_3}{|}}{\underset{\underset{\displaystyle CH_3}{|}}{\overset{\oplus}{N}}}\text{---}(CH_2)_x\text{---}\underset{Cl^{\ominus}}{\overset{\overset{\displaystyle CH_3}{|}}{\underset{\underset{\displaystyle CH_3}{|}}{\overset{\oplus}{N}}}}\text{---}(CH_2)_y\text{---} \qquad (62)$$

Although molecular weights approaching 100,000 were obtained for the polymer where $x = 3, y = 4$,[129] and it was shown that increasing charge density led to increasing flocculation activity,[130] at the present time the cost of the starting materials has limited the commercial utility of these ionenes. In a variation of this reaction, a high-molecular-weight ionene was prepared from 3,3-dimethylaminopropylchloride (Equation 63).[131] Marvel and co-workers in previous attempts in the 1930s had only succeeded in synthesizing low-molecular-weight polymers.[132] Although this polymer could be expected to show good activity because of the high charge density and molecular weight, it appears to have little current commercial interest for water treatment because of cost.

$$(CH_3)_2NCH_2CH_2CH_2Cl \ \longrightarrow \ \underset{\underset{\underset{\displaystyle Cl^{\ominus}}{}}{\underset{\displaystyle CH_3}{|}}}{\overset{\overset{\displaystyle CH_3}{|}}{\overset{\oplus}{N}}}\text{---}CH_2CH_2CH_2\text{---} \qquad (63)$$

Ionenes have also been prepared from the reaction of dihaloalkenes, such as 1,4-dichloro-2-butene, with dimethylamine (Equation 64),[133] or with ditertiary amines. In the latter case, the ditertiary amine can be prepared by reaction of 2 mol of dimethylamine with one of epichlorohydrin (Equation 65).[134] In the former case, linear polymers result when dimethylamine is used alone, and branched polymers are formed when minor amounts of a polyfunctional amine are used in conjunction with dimethylamine.

$$(CH_3)_2NH \ + \ ClCH_2CH{=}CHCH_2Cl(trans) \ + \ NaOH \ \longrightarrow$$

$$\underset{\underset{\underset{\displaystyle Cl^{\ominus}}{}}{\underset{\displaystyle CH_3}{|}}}{\overset{\overset{\displaystyle CH_3}{|}}{\overset{\oplus}{N}}}\text{---}CH_2CH{=}CH\text{---}CH_2\text{---} \qquad (64)$$

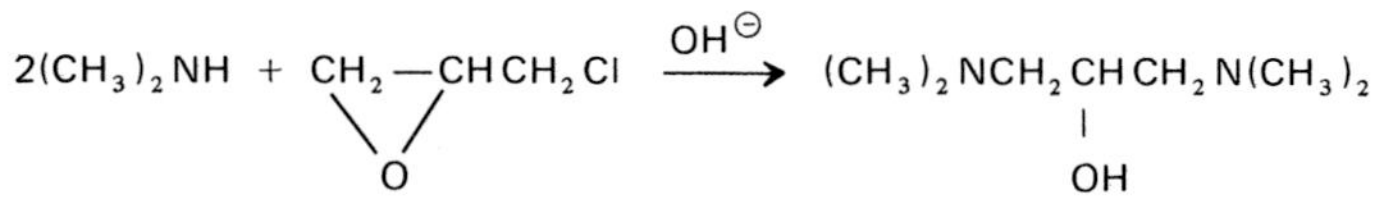

$$2(CH_3)_2NH + CH_2\!-\!CHCH_2Cl \xrightarrow{OH^{\ominus}} (CH_3)_2NCH_2CHCH_2N(CH_3)_2$$

$$\underset{O}{\diagdown\diagup} \qquad\qquad \underset{OH}{|}$$

XXXII

(64 continued)

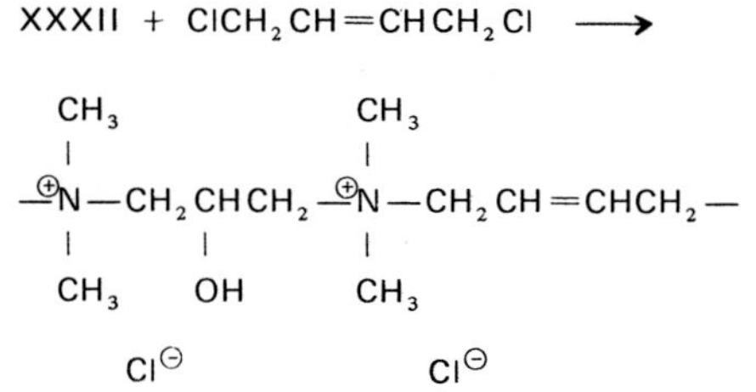

$$XXXII + ClCH_2CH\!=\!CHCH_2Cl \longrightarrow$$

$$\overset{CH_3}{\underset{|}{}} \qquad\qquad \overset{CH_3}{\underset{|}{}}$$

$$\overset{\oplus}{-}N\!-\!CH_2CHCH_2\!-\!\overset{\oplus}{N}\!-\!CH_2CH\!=\!CHCH_2\!-$$

$$\underset{CH_3}{|} \quad \underset{OH}{|} \quad \underset{CH_3}{|}$$

$$\qquad Cl^{\ominus} \qquad\qquad Cl^{\ominus} \qquad\qquad (65)$$

Ionenes prepared by the reaction of 1,4-dichloro-2-butene with the di- (linear polymer) or tritertiary amines (branched polymer) produced by the reaction of dimethylamine/formaldehyde with substituted or unsubstituted phenols, respectively, have been reported (Equation 66).[140]

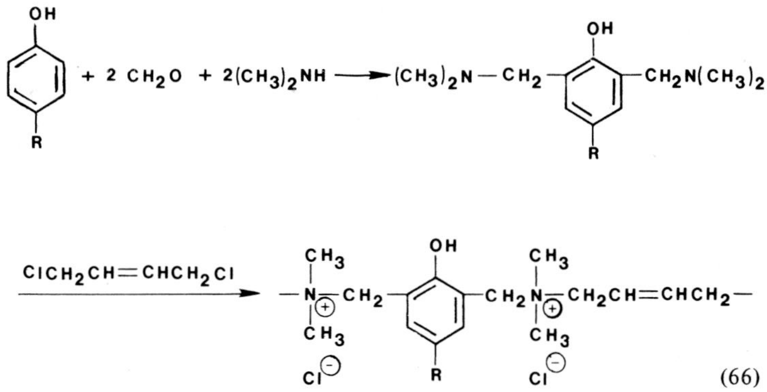

$$(66)$$

The adduct of dimethylamine and epichlorohydrin (XXXII) has also been reacted with 2,2′-dichlorodiethylether to give an ionene (XXXIII) (Equation 67).[136]

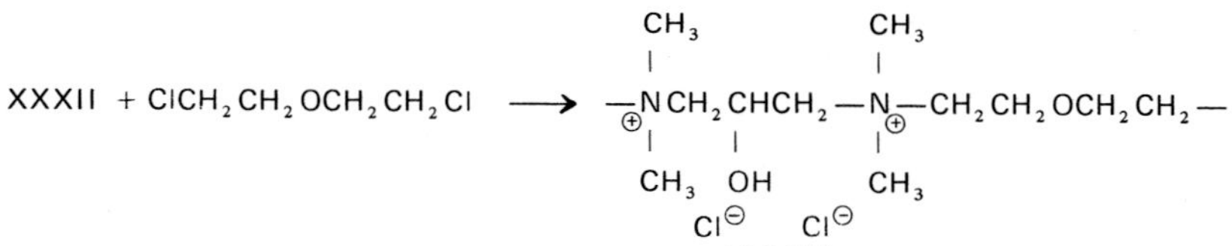

$$XXXII + ClCH_2CH_2OCH_2CH_2Cl \longrightarrow$$

XXXIII

$$(67)$$

Many ionenes have been prepared using N,N,N′,N′-tetramethylethylene-diamine (TMEDA) as the ditertiary amine. These include reactions with dihaloalkanes,[135] however, the main commercial interest appears to be with the reaction of TMEDA with 2,2′-dichlorodiethylether (Equation 68).[138]

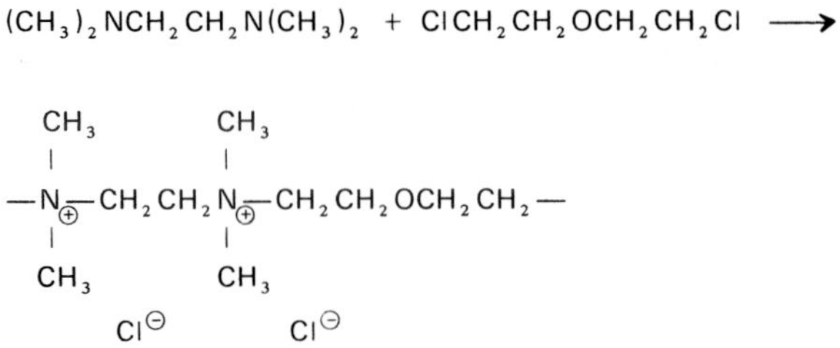

$$(CH_3)_2NCH_2CH_2N(CH_3)_2 \ + \ ClCH_2CH_2OCH_2CH_2Cl \ \longrightarrow$$

$$
\begin{array}{cc}
CH_3 & CH_3 \\
| & | \\
-N_\oplus-CH_2CH_2N_\oplus-CH_2CH_2OCH_2CH_2- \\
| & | \\
CH_3 & CH_3 \\
\quad Cl^\ominus & \quad Cl^\ominus
\end{array}
$$

(68)

The linear polymer (XXXIII)[137] has been chain extended with epichlorohydrin to give branched, higher-molecular-weight polyelectrolytes. In this case, epichlorohydrin reacts with the hydroxyl group to form the branch (Equation 69).

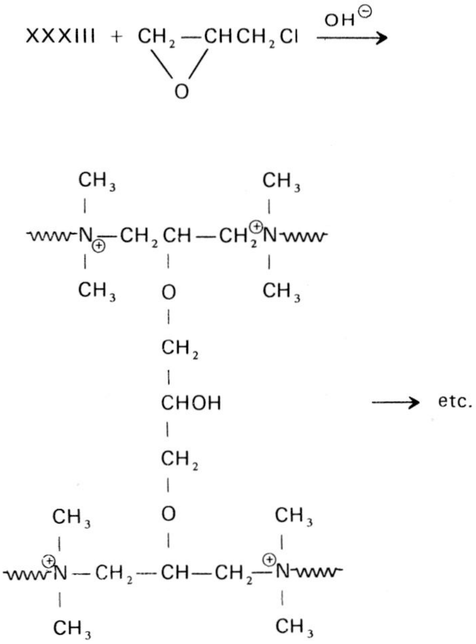

$$XXXIII \ + \ CH_2-CHCH_2Cl \ \xrightarrow{OH^\ominus}$$

$$
\begin{array}{ccc}
CH_3 & & CH_3 \\
| & & | \\
\text{www-}N_\oplus-CH_2CH-CH_2^\oplus N\text{-www} \\
| & | & | \\
CH_3 & O & CH_3 \\
& | \\
& CH_2 \\
& | \\
& CHOH & \longrightarrow \ \text{etc.} \\
& | \\
& CH_2 \\
& | \\
CH_3 & O & CH_3 \\
| & | & | \\
\text{www-}\overset{\oplus}{N}-CH_2-CH-CH_2-\overset{\oplus}{N}\text{-www} \\
| & & | \\
CH_3 & & CH_3
\end{array}
$$

(69)

Novel functional ionenes have been prepared by reacting diprimary amines, such as ethylenediamine, with four moles of a vinyl monomer, including acrylamide and many of the anionic and cationic monomers discussed above, to give the ditertiary amine via a Michael addition. The ditertiary amine is then reacted with the dihaloalkane, such as 2,2'-dichlorodiethylether (Equation 70).[139]

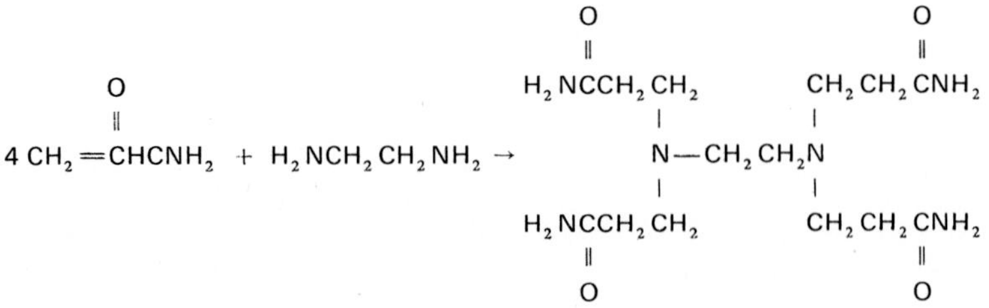

$$
4\ CH_2{=}CHC\overset{O}{\overset{\|}{N}}H_2 \ + \ H_2NCH_2CH_2NH_2 \ \rightarrow \
\begin{array}{cc}
\overset{O}{\overset{\|}{H_2NCCH_2}}CH_2 & \overset{O}{\overset{\|}{CH_2CH_2CNH_2}} \\
| & | \\
N-CH_2CH_2N \\
| & | \\
H_2NCCH_2CH_2 & CH_2CH_2CNH_2 \\
\overset{\|}{O} & \overset{\|}{O}
\end{array}
$$

XXXII

$$\text{XXXIV} + \text{ClCH}_2\text{CH}_2\text{OCH}_2\text{CH}_2\text{Cl} \longrightarrow$$

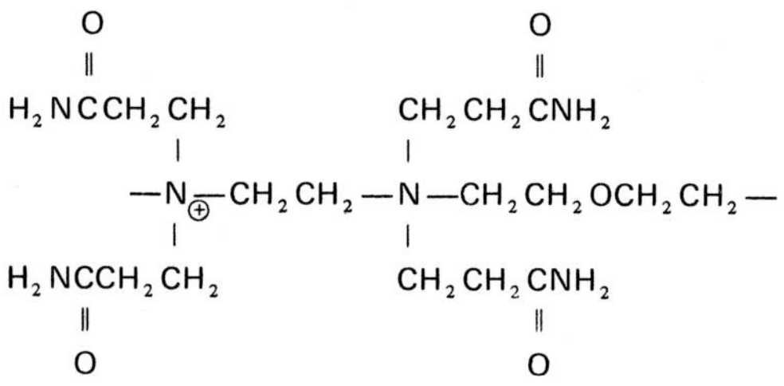

$$
\begin{array}{ccc}
\text{O} & & \text{O} \\
\| & & \| \\
\text{H}_2\text{NCCH}_2\text{CH}_2 & & \text{CH}_2\text{CH}_2\text{CNH}_2 \\
| & & | \\
-\text{N}_{\oplus}\text{CH}_2\text{CH}_2-\text{N}-\text{CH}_2\text{CH}_2\text{OCH}_2\text{CH}_2- \\
| & & | \\
\text{H}_2\text{NCCH}_2\text{CH}_2 & & \text{CH}_2\text{CH}_2\text{CNH}_2 \\
\| & & \| \\
\text{O} & & \text{O}
\end{array}
\qquad (70)
$$

Useful branched polyamines have been prepared from ammonia and ethylenedichloride (Equation 71). This reaction is a variation of the process for the manufacture of ethylenediamine and its higher homologues, the so-called "polyethyleneamines" (Equation 72).[141]

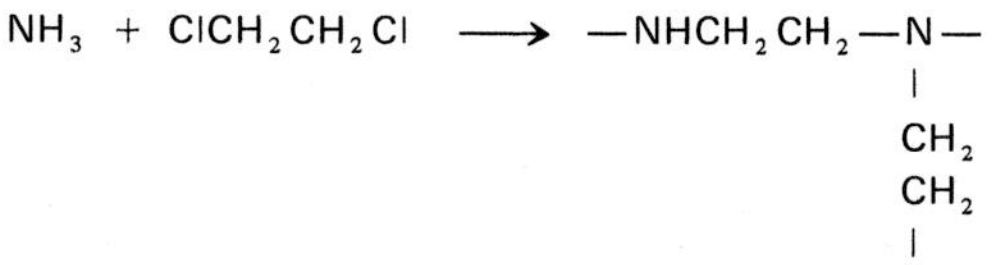

$$
\text{NH}_3 + \text{ClCH}_2\text{CH}_2\text{Cl} \longrightarrow
\begin{array}{c}
-\text{NHCH}_2\text{CH}_2-\text{N}- \\
| \\
\text{CH}_2 \\
\text{CH}_2 \\
|
\end{array}
\qquad (71)
$$

$$\text{NH}_3 + \text{ClCH}_2\text{CH}_2\text{Cl} \longrightarrow \text{H}_2\text{NCH}_2\text{CH}_2\text{NH}_2 + \text{H}_2\text{NCH}_2\text{CH}_2\text{NHCH}_2\text{CH}_2\text{NH}_2$$

$$+ \text{H}_2\text{NCH}_2\text{CH}_2\text{NHCH}_2\text{CH}_2\text{NHCH}_2\text{CH}_2\text{NH}_2 + \cdots$$

$$+ \text{H}_2\text{N}\!\!-\!\!(\text{CH}_2\text{CH}_2\text{NH})_x\!\!-\!\!\text{H} \qquad (72)$$

Although this reaction has been known for a long time, it is only relatively recently that technology for the preparation of water-soluble polyelectrolytes useful for water treatment has been developed.[142]

Other polyamines have been prepared from the reaction of ethylene dichloride with diethylenetriamine,[143] higher homologues of ethylenediamine, including complex mixtures,[144] and hexamethylenediamine and higher homologues (XXXV).[116] Epichlorohydrin has also been used to chain extend these systems.

$$\text{H}_2\text{N}\!\!-\!\!(\text{CH}_2\text{CH}_2\text{CH}_2\text{CH}_2\text{CH}_2\text{CH}_2\text{NH})_x\!\!-\!\!\text{H}$$

$$\text{XXXV}$$

Depending on extent of branching these polymers would be mixtures of amines and quaternary ammonium salts.

3. Aziridine Types

The controlled polymerization of ethylenimine has been known at least since 1939[145] and this polymer is probably one of the earliest known cationic polyamines used for water treatment.[146] The polymerization has been a frequent subject of study, and proceeds in bulk or in solution, generally with an acid catalyst. The product is a branched structure (Equation 73) and molecular weights up to 100,000 are known. At the higher molecular weights, the structure is highly branched and cross-linked although still water soluble.[147]

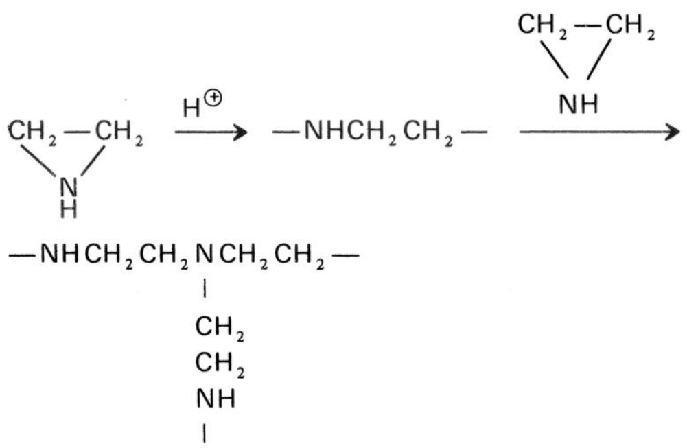

$$
\begin{array}{c}
\underset{\underset{H}{N}}{CH_2-CH_2} \xrightarrow{H^{\oplus}} -NHCH_2CH_2- \xrightarrow{\underset{\underset{NH}{\diagup}}{CH_2-CH_2}} \\[2em]
-NHCH_2CH_2NCH_2CH_2- \\
\mid \\
CH_2 \\
CH_2 \\
NH \\
\mid \qquad\qquad\qquad (73)
\end{array}
$$

Ethylenimine is prepared commercially either from 2-aminoethanol via the sulfate, or from 2-aminoethylchloride (Equation 74).[148]

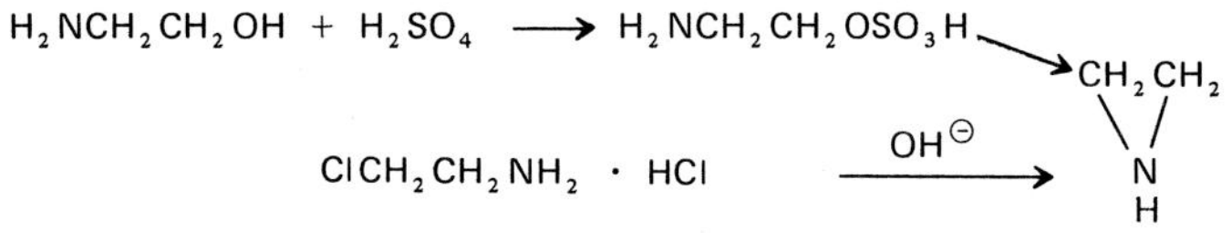

$$
H_2NCH_2CH_2OH + H_2SO_4 \longrightarrow H_2NCH_2CH_2OSO_3H
$$

$$
ClCH_2CH_2NH_2 \cdot HCl \xrightarrow{OH^{\ominus}} \underset{\underset{H}{N}}{CH_2CH_2}
$$

$$(74)$$

Although the monomer is highly toxic, with strong carcinogenic properties, polyethylenimine is relatively nontoxic.[148] The polymer is available in a number of molecular weight ranges. While a number of modifications of polyethylenimine are known, including copolymerization[149] and a quaternized form,[150] only polyethylenimine itself is believed to be of importance for water treatment.

4. Hydrophobic Cationic Polymers

A recent development in the area of cationic polyelectrolytes for water and waste treatment has been the "hydrophobic" polymers. These are polyelectrolytes, containing hydrophobic groups, which have reduced solubility in water, but which still are finding utility for water treatment, where some hydrophobicity may be of benefit. The impact of these polymers in the water treatment area has yet to be determined.

These polymers include regularly alternating copolymers of styrene and imidazolines (Equation 75),[151] cationic aminomethylated (see Section IVA) acrylamide-styrene copolymers or acrylamide-styrene-quaternized aminoacrylate terpolymers,[152] and aminomethylated acrylamide copolymers with other hydrophobic monomers, such as methyl methacrylate.[153] The latter polymers can also be quaternized.

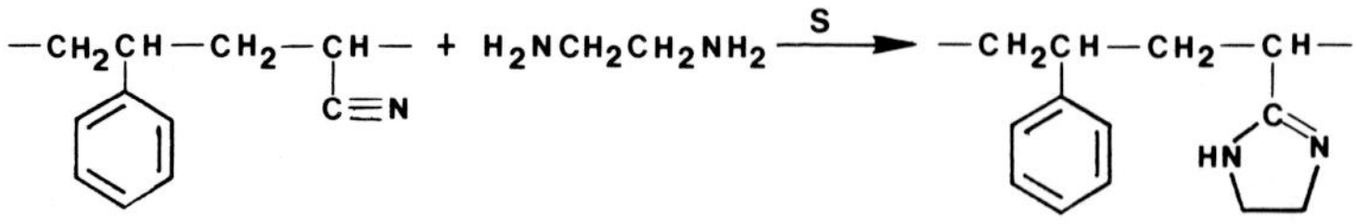

$$
-CH_2CH-CH_2-\underset{\underset{C\equiv N}{|}}{CH}- + H_2NCH_2CH_2NH_2 \xrightarrow{S} -CH_2CH-CH_2-CH-
$$

$$(75)$$

5. Miscellaneous Cationic Polyelectrolytes

A limited number of additional cationic polyelectrolytes deserve mention, not because of commercial utility, but because they demonstrate again the versatility possible with this area of polymer synthesis. A number of cationic sulfonium monomers, homo- and copolymers have been reported. These include the sulfonium acrylate esters (XXXVI),[164] the diallyl sulfonium monomers (XXXVII),[165] and the vinylbenzyl sulfonium monomers (XXXVIII).

$$CH_2=CH-\overset{\overset{O}{\|}}{C}-OCH_2\,CH_2\,\overset{\oplus}{\underset{}{S}}(CH_3)_2\,CH_3\,SO_4^{\ominus}$$

XXXVI

$$(CH_2=CHCH_2)_2\,\overset{\oplus}{\underset{}{S}}-CH_3 \quad CH_3\,SO_4^{\ominus}$$

XXXVII

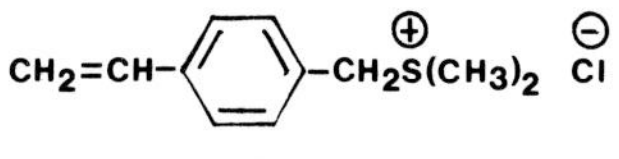

$$CH_2=CH-\langle\bigcirc\rangle-CH_2\overset{\oplus}{S}(CH_3)_2 \quad \overset{\ominus}{Cl}$$

__XXXVIII__

The vinylbenzylsulfonium polymers have been prepared both by polymerization of the sulfonium monomers,[166] and by reaction of vinylbenzylchloride polymers with dialkyl-sulfides.[167]

Interesting, branched polyaminoamides have been prepared by the reaction of ethylenediamine and higher homologues with methyl acrylate or methyl methacrylate (Equation 76).[168]

$$H_2NCH_2CH_2NH_2 \;+\; CH_2=CH\overset{\overset{O}{\|}}{C}OCH_3 \;\longrightarrow\; -NHCH_2CH_2NHCH_2CH_2\overset{\overset{O}{\|}}{C}-$$

$$\xrightarrow{\;CH_2=CH\overset{\overset{O}{\|}}{C}OCH_3\;}$$

$$-NHCH_2CH_2NCH_2CH_2\overset{\overset{O}{\|}}{C}-$$
$$\underset{\underset{\underset{-NCH_2CH_2NHCH_2CH_2\overset{\overset{O}{\|}}{C}-}{C=O}}{\underset{}{\overset{CH_2}{\underset{CH_2}{|}}}}{|}$$

$$\longrightarrow \text{ etc.} \tag{76}$$

Finally, cyclic amidine polymers were prepared by reacting cyanohydrins (e.g., acetone + HCN) with diamines such as ethylenediamine (Equation 77).[169]

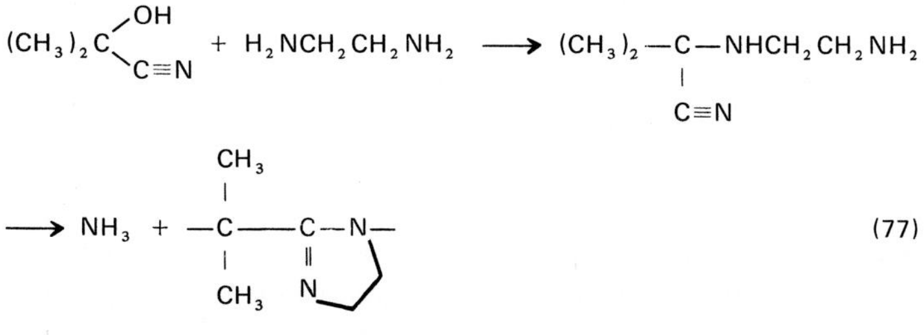

$$(CH_3)_2C\overset{OH}{\underset{C\equiv N}{\big\langle}} \;+\; H_2NCH_2CH_2NH_2 \;\longrightarrow\; (CH_3)_2-\underset{\underset{C\equiv N}{|}}{C}-NHCH_2CH_2NH_2$$

$$\longrightarrow NH_3 \;+\; -\underset{\underset{CH_3}{|}}{\overset{\overset{CH_3}{|}}{C}}-\underset{\underset{N}{\|}}{C}-N- \tag{77}$$

VI. NONIONIC POLYELECTROLYTES

The true nonionic "polyelectrolytes" used in water treatment are few in number indeed — comprising chiefly unhydrolyzed, high-molecular-weight polyacrylamide and the polyethylene oxides. Because the polymerizations of acrylamide intended to produce high molecular weight polymer are conducted under adiabatic conditions, in aqueous solution, conditions which favor hydrolysis (Equation 78), truly nonionic polyacrylamide is difficult to prepare, and most commercially available "nonionics" contain about 1 or 2% of hydrolyzed amide groups.

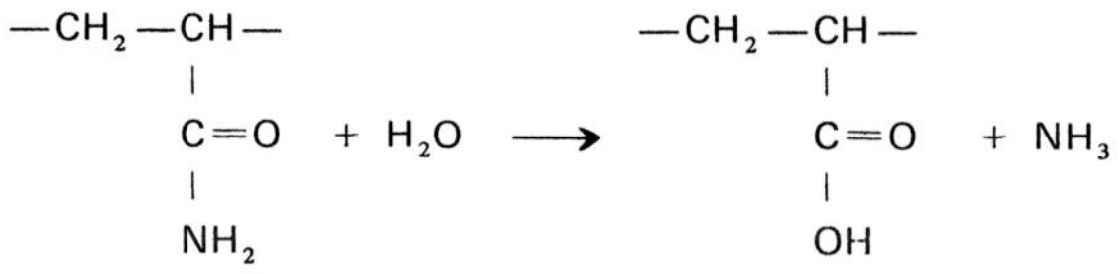

$$
\begin{array}{ccc}
-CH_2-CH- & & -CH_2-CH- \\
\quad | & & \quad | \\
C=O \;\; + H_2O \;\; \longrightarrow & & C=O \;\; + NH_3 \\
\quad | & & \quad | \\
NH_2 & & OH \qquad\qquad (78)
\end{array}
$$

Polyacrylamides having less than 1% hydrolysis have been prepared by a specific combination of concentration (30 to 60%), temperature (10 to 30°C), pH (4.7 to 5.9), initiator system (sodium bromate-sodium sulfite), and thickness of the reaction zone (at least one dimension being less than ¾ in.,[170] and by the addition of selected amines to the polymerization medium.[171] The polymerization of acrylamides in emulsion form may also reduce the extent of "self" hydrolysis, because of the reduced concentrations of water in these systems, and possibly the better heat transfer possible.

The other nonionic polymers which have seen significant use for water treatment are the high-molecular-weight poly (ethylene oxides) (Equation 79). Ethylene oxide polymers are available in two molecular weight ranges: liquids, waxes, and greases, with molecular weights up to about 20,000, and the "Polyox" series, having molecular weights of 100,000 to about five million.[172] Only the high-molecular-weight polymers appear to have shown utility in water treatment.

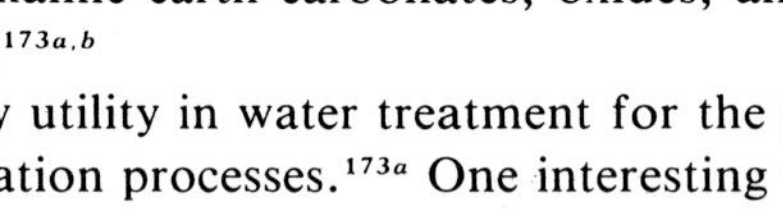

$$
\begin{array}{ccc}
CH_2-CH_2 & \longrightarrow & -CH_2CH_2O- \qquad\qquad (79) \\
\;\;\backslash\;/ & & \\
O & &
\end{array}
$$

Each of the series is manufactured by a different process. The low molecular weights are prepared by polymerization in the presence of Lewis acids or bases, while the high molecular weights require a heterogeous reaction system utilizing a variety of catalysts, such as the alkaline earth carbonates, oxides, and amides, and the aluminum alkyls and alkoxides.[173a,b]

The primary utility in water treatment for the poly (ethylene oxides) appears to be in mining flotation processes.[173a] One interesting use for the nonionic polymers is for the reduction of turbulent flow, the so-called drag reduction. This finds application in areas such as increased water flow through hoses for firefighting, and for increasing the speed of naval vessels.[174]

VII. AMPHOTERIC POLYELECTROLYTES

Polyelectrolytes having both negative and positive charges on the same chain are called amphoteric or polyampholytes. At the present time, it is believed that polyampholytes have little commercial utility for water treatment. It should be noted that the net charge on polymers of this type is strongly dependent upon the pH of the medium, since only quaternary ammonium salts have charges which are independent of pH.

Most of the polyampholytes associated with water treatment reported in the literature are cationic polyacrylamides having a minor extent of hydrolysis of amide groups, because of hydrolysis during polymerization, or where a separate hydrolysis step was carried out.[175] A number of polymers which are amphoteric by virtue of an incomplete cationization reaction on an anionic polymer have already been mentioned. One polyampholyte has been prepared by reaction of polymeric anhydrides with ammonia, followed by Hofmann degradation of the product (Equation 80).[176]

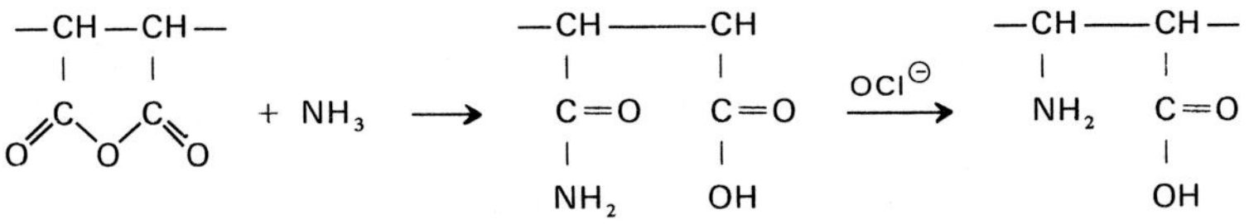

$$
\begin{array}{ccc}
\underset{\overset{|}{O}\overset{C}{\diagdown}\underset{O}{O}\overset{C}{\diagup}\overset{|}{O}}{-CH-CH-} + NH_3 \longrightarrow & \begin{array}{cc} -CH & CH \\ | & | \\ C=O & C=O \\ | & | \\ NH_2 & OH \end{array} \overset{OCl^{\ominus}}{\longrightarrow} & \begin{array}{cc} -CH & CH- \\ | & | \\ NH_2 & C=O \\ & | \\ & OH \end{array}
\end{array}
$$

(80)

One example of a polyampholyte prepared by copolymerization of an anionic monomer with a cationic monomer is shown in Structure XXXIX.[52]

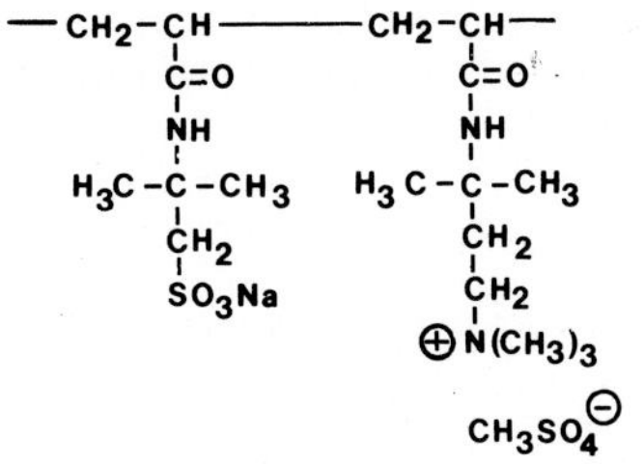

$$
\begin{array}{cc}
-CH_2-CH & CH_2-CH- \\
| & | \\
C=O & C=O \\
| & | \\
NH & NH \\
| & | \\
H_3C-C-CH_3 & H_3C-C-CH_3 \\
| & | \\
CH_2 & CH_2 \\
| & | \\
SO_3Na & CH_2 \\
& | \\
& \overset{\oplus}{N}(CH_3)_3 \\
& CH_3SO_4^{\ominus}
\end{array}
$$

__XXXIX__

Proteins are, of course, polyampholytes, and enzymatic treatment of proteins to give water-soluble polymers useful as coagulants has been reported.[177]

REFERENCES

1. **Cantow, M. J. R. and Johnson, J. F.,** Molecular-weight determination, in *Encyclopedia of Polymer Science and Technology,* Vol. 9, Bikales, N. M., Ed., Interscience, New York, 1968, 182.
2. **Hess, I. H. and Kurtz, K. K.,** Process for water-soluble polymer recovery, U.S. Patent 3,046,259, 1962.
3. **Terenzi, J. F.,** Particulation of polymer by extruding a solution thereof into a liquid stream of fluid, U.S. Patent 3,042,970, 1962.
4. **Sheats, G. F. and Linke, W. F.,** Process for stabilizing and storing of aqueous solutions of polyacrylamide, U.S. Patent 3,163,619, 1964.
5. **Monagle, D. J., Shyluk, W. P., and Smith, V. W., Jr.,** Preparation of acrylamide-type water-soluble polymers, U.S. Patent 3,509,113, 1970.
6. **Monagle, D. J. and Shyluk, W. P.,** Preparation of acrylamide-type water-soluble polymers, U.S. Patent 3,336,269, 1967.
7. **Monagle, D. J.,** Preparation of acrylamide-type water-soluble polymers, U.S. Patent 3,336,270, 1967.
8. **Hirate, E., Isaoka, S., and Sakai, S.,** Process for producing high molecular weight acrylamide water-soluble polymers by controlling the viscosity of the polymerization reaction medium with a water-miscible organic solvent, U.S. Patent 3,969,329, 1976.
9. **Volk, H. and Hamlin, P. J.,** Acrylic acid polymers, U.S. Patents 3,493,500, 1970; 3,658,772, 1972.

10. **Gershberg, D. B.,** Process for recovering acrylamide polymers from gels, U.S. Patent 3,714,136, 1973.
11. **Gershberg, D. B.,** Sodium sulfate coated polyacrylamide gel particles, U.S. Patent 3,766,120, 1973.
12. **Proffitt, A. C.,** Process for making hydrolyzed polyacrylamide, U.S. Patent 3,022,279, 1962.
13. **Kolodny, E. R.,** Polymeric composition and process for preparing the same, U.S. Patent 3,215,680, 1965.
14. **Friedrich, R. E., Wiley, R. M., and Garrett, W. L.,** Inverse suspension polymerization of water-soluble unsaturated monomers, U.S. Patent 2,982,749, 1961.
15. **Urick, J. M.,** Microbeads — a novel approach to retention aids and flocculation, *Pulp Pap.,* 51(8), 84, 1977.
16. **Morningstar-Paisley, Inc.,** Polymers of acrylamide and other monomers, British Patent 991,416, 1965.
17. **Hamann, H. C.,** Stabilized suspension polymerization, U.S. Patent 3,557,061, 1971.
18. **Griffin, W. C.,** Emulsions, in *Encyclopedia of Chemical Technology,* Vol. 8, 2nd ed., Standen, A., Ed., Interscience, New York, 1964, 117.
19. **Zimmermann, J. W. and Kühlkamp, A.,** Oil-in-water bead polymerization of water-soluble monomers, U.S. Patent 3,211,708, 1965.
20. **Thomas, W. M.,** Acrylamide polymers, in *Encyclopedia of Polymer Science and Technology,* Vol. 1, Bikales, N. M., Ed., Interscience, New York, 1964, 177.
21. **Vanderhoff, J. W. and Wiley, R. M.,** Water-in-oil emulsion polymerization process for polymerizing water-soluble monomers, U.S. Patent 3,284,393, 1966.
22. **Anderson, D. R.,** Process for rapidly dissolving water-soluble polymers, U.S. Patent 3,624,019, 1971.
23. **Anderson, D. R. and Frisque, A. J.,** Stable high solids water-in-oil emulsions of water soluble polymers, U.S. Patent 3,826,771, 1974.
24. **Goretta, L. A. and Sibert, F. J.,** Process for the preparation of water-in-oil emulsions of water-soluble vinyl carboxylic acid polymers and copolymers, U.S. Patent 4,070,321, 1978.
25. **Kane, J.,** High shear mixing of latex polymers, U.S. Patent 3,996,180, 1976.
26. **Venema, G. J.,** Process and apparatus for dissolving water-soluble polymers and gums in water involving inversion of water-in-oil emulsions, U.S. Patent 3,852,234, 1974.
27. **Slovinsky, M., Ryan, R. C., and Phillips, K. G.,** Stabilized water-in-oil emulsions utilizing minor amounts of oil-soluble polymers, U.S. Patent 3,915,920, 1975.
28. **Anderson, D. R. and Frisque, A. J.,** Rapid dissolving water-soluble polymers, U.S. Patent 3,734,873, 1973.
29. **Eastmond, G. C.,** Free radical polymerization, in *Encyclopedia of Polymer Science and Technology,* Vol. 7, Bikales, N. M., Ed., Interscience, New York, 1967, 361.
30. **Gill, E. A.,** Polymerization with a redox and azoisobutyronitrile catalyst, U.S. Patent 3,573,263, 1971.
31. **Henley, E. J. and Bell, R. C.,** Polymerization process for water-soluble polymers, U.S. Patent 2,983,717, 1961.
32. **Billmeyer, Jr., F. W.,** *Textbook of Polymer Science,* 2nd ed., Interscience, New York, 1971, 291.
33. **Sibert, F. J.,** Process for separation of red mud from dissolved alumina, U.S. Patent 3,445,187, 1969.
34. **Phalangas, C. J.,** Polymerization of water-soluble monomers with radiation and chemical initiator, U.S. Patent 3,948,740, 1976.
35. **Suen, T. J. and Schiller, A. M.,** Process of preparing polyacrylamide, U.S. Patent 2,820,777, 1958.
36. **Scanley, C. S.,** Simultaneous hydrolysis and polymerization of acrylamides, U.S. Patent 3,414,552, 1968.
37. **Chamot, W. M. and Burke, J. T.,** Method of flocculating suspended solids using copolymers as flocculating agents, U.S. Patent 3,479,282, 1969.
38. **Connelly, L. J. and Ballweber, E. G.,** Method of hydrolyzing polyacrylamide, U.S. Patent 3,998,777, 1976.
39. **Young, L. J.,** Copolymerization reactivity ratios, in *Polymer Handbook,* 2nd ed., Brandrup, J. and Immergut, E. H., Eds., John Wiley & Sons, New York, 1975, II-105.
40. **Mallett, A. S. and Craig, R. L.,** The effect of the molecular weight of sodium polyacrylate on pigment dispersions, *Tappi,* 60(11), 101, 1977.
41. **Greenfield, G. L.,** Acrylic Compositions, U.S. Patent 3,787,488, 1974; **Greenfield, G. L.,** Process of inhibiting scale using an acrylic composition, U.S. Patent 3,904,522, 1975; **Rice, H. L., Cizek, A., and Thaemer, M. O.,** Acrylic compositions for water treatment and process for making same, U.S. Patent 3,665,035, 1972.
42. **O'Brien, J. T. and White, W. W.,** Bisulfite terminated oligomers to prevent scale, U.S. Patent 3,965,028, 1976; **O'Brien, J. T. and White, W. W.,** Bisulfite terminated oligomers as dispersing agents, U.S. Patent 4,004,939, 1977.

43. **Booth, R. B. and Mead, L. C.,** Prevention of and removal of scale formation in water systems, U.S. Patent 3,463,730, 1969.

44. **Dannals, L. E.,** Cyano containing oligomers, U.S. Patent 3,646,099, 1972.

45. **Godlewski, I. T., Schuck, J. J., and Libutti, B. L.,** Polymers for use in water treatment, U.S. Patent 4,029,577, 1977.

46. **Lang, J. L., Pavelich, W. A., and Clarey, H. D.,** The homopolymerization of maleic anhydride, *J. Polym. Sci.,* 55, S31, 1961.

47. **Lancelot, C. J., Blumbergs, J. H., and Mackellar, D. G.,** Poly (maleic anhydride), British Patent 1,349,769, 1974.

48. **Cochrane, C. C.,** Electrolyte polymerization of maleic anhydride, U.S. Patent 3,427,233, 1969.

49. Food and Drug Administration, Components of paper and paperboard in contact with dry food, 21CFR176.180, 1977.

50. **Anon.,** Gulf PA-18, *Chem. Eng. News,* May 16, 1977, 38.

51. **Shirai, M., Nagatsuka, T., and Tanaka, M.,** Interaction between dyes and polyelectrolytes. VI. Metachromatic behavior of methylene blue induced by potassium poly (vinyl sulfate) and its homologues, *J. Polym. Sci.,* 15, 2083, 1977.

52. **Hoke, D. I.,** Water-soluble sulfonate polymers as flocculants, U.S. Patent 3,692,673, 1972.

53. **Persinski, L. J., Ralston, P. H., and Gordon, Jr., R. C.,** Inhibition of scale deposition, U.S. Patent 3,928,196, 1975; **Lange, K. R., Schiesser, R. H., Tonkyn, R. G., and Dean, R. T.,** Acrylamidosulfonic acid polymers and their use as rust and tubercle removing agents, U.S. Patent 3,806,367, 1974; **Lange, K. R., Schiesser, R. H., Tonkyn, R. G., and Dean, R. T.,** Acrylamido-sulfonic acid polymers and their use, U.S. Patent 3,898,037, 1975.

54. **Miller, L. E. and Murfin, D. L.,** Preparation of acrylamido-alkane sulfonic acids, U.S. Patent 3,506,707, 1970.

55. **Smith, B. R. and Alexander, A. E.,** The effect of additives on the process of crystallization. II. Further studies on calcium sulphate, *J. Colloid Interface Sci.,* 34, 81, 1970.

56. **Kutner, A. and Breslow, D. S.,** Ethylene sulfonic acid polymers, in *Encyclopedia of Polymer Science and Technology,* Vol. 6, Bikales, N. M., Ed., Interscience, New York, 1967, 455.

57. **Sarig, S. and Tartakovsky, F.,** Inhibition of strontium sulfate precipitation by soluble polymers, *Isr. J. Chem.,* 12, 905, 1974.

58. **Bleyle, M.,** Scale prevention agents of methacrylic acid-vinyl sulfonate copolyers for saline water evaporation, U.S. Patent 3,682,224, 1972.

59. **Siegele, F. H.,** Copolymers of fumaric acid and allyl sulfonic acid, U.S. Patent 3,706,717, 1972.

60. **Teot, A. S.,** Crystal habit modification of inorganic salts using polymeric sulfonates or sulfates, U.S. Patent 3,770,390, 1973; **Pye, D. J.,** Very dilute aqueous solutions of linear polymers as waterflooding secondary petroleum recovery agents, U.S. Patent 3,399,725, 1968.

61. **Monagle, D. J.,** Use of water-soluble copolymers in the flocculation and setting of inorganic particles in salt solutions, U.S. Patent 3,617,573, 1971.

62. **Vanderkooi, W. N. and Mock, R. A.,** Polymerization of styrenesulfonic acid and salts thereof, U.S. Patent 3,123,589, 1964.

63. **Goethals, E. J.,** Sulfur-containing polymers, in *Encyclopedia of Polymer Science and Technology,* Vol. 13, Bikales, N.M., Ed., Interscience, New York, 1970, 448.

64. **Perricone, A. C. and Young, H. F.,** Composition and process for drilling subterranean wells, U.S. Patent 3,730,900, 1973.

65. **Turbak, A. F.,** Sulfonation with organic phosphorous compound-sulfur trioxide adducts, U.S. Patent 3,072,618, 1963.

66. **Wurzburg, O. B., Szymanski, C. D., and Kruger, L.,** Pepsin inhibition, German Patent 2,450,148, 1975.

67. National Starch and Chemical Corporation, Plainfield, N.J., Versa-TL, Technical Service Bulletin.

68. **Cuisia, D. G. and Hwa, C. H.,** Method of inhibiting scale, U.S. Patent 4,048,066, 1977; **Persinski, L. J., Martin, F. D., and Adams, S. L.,** Well cementing method using a composition having improved flow properties, containing sulfonated copolymers of styrene-maleic anhydride, U.S. Patent 3,952,805, 1976.

69. **Brownell, G. L.,** Acids, maleic and fumaric, in *Encyclopedia of Polymer Science and Technology,* Vol. 1, Bikales, N. M., Ed., Interscience, New York, 1964, 67.

70. **Quinlan, P. M.,** Methylene phosphonates of amino-terminated oxyalkylates and uses therefor, U.S. Patent 4,080,375, 1978.

71. **Quinlan, P. M.,** Methylene phosphonates of poly-diepoxidized polyalkylene polyamines, U.S. Patent 4,035,412, 1977.

72. **Quinlan, P. M.,** Methylene phosphonates of polymerized polyalkylenepolyamines, U.S. Patent 4,051,110, 1977.

73. **Quinlan, P. M.,** Polyquaternary ammonium methylene phosphonates and uses thereof, U.S. Patent 3,792,084, 1974.

74. **Bersworth, F. C.,** N-Alkyl substituted alkylene polyamine methylene phosphonic acids, U.S. Patent 2,841,611, 1958.

75. **Peck, D. R. and Hudson, D.,** Ethylenediaminetetrakis (methylphosphonic acid), British Patent 1,230,121, 1971.

76. **Weiss, R. A., Lenz, R. W., and MacKnight, W. J.,** Properties of polyethylene modified with phosphonate side groups. I. Thermal and mechanical properties, *J. Polym. Sci.,* A-2, 15, 1409, 1977.

77. **Prentice, J. B.,** Oligomeric ester chain condensates of ethane-1-hydroxy-1,1-diphosphonic acid, U.S. Patent 3,621,081, 1971.

78. **Prentice, J. B.,** Detergent compositions containing oligomer ester chain condensates of ethane-1-hydroxy-1,1-diphosphonic acid as builders, U.S. Patent 3,562,169, 1971.

79. **Hoke, D. I.,** Quaternized N-aminoalkyl acrylamide polymers in flocculation of suspended solids from water, U.S. Patent 3,761,407, 1973.

80. **Boothe, J. E.,** Synthesis of dimethyl diallyl ammonium chloride, U.S. Patent 3,461,163, 1969.

81. **Butler, G. B. and Angelo, R. J.,** Preparation and polymerization of unsaturated quaternary ammonium compounds. VIII. A proposed alternating intramolecular-intermolecular chain propagation, *J. Am. Chem. Soc.,* 79, 3128, 1957.

82. **Lancaster, J. E., Baccei, L., and Panzer, H. P.,** The structure of poly (diallyldimethylammonium) chloride by ^{13}C-NMR spectroscopy, *J. Polym. Sci.,* B, 14, 549, 1976.

83. **Boothe, J. E. and Hoover, M. F.,** Electroconductive paper, U.S. Patent 3,544,318, 1970.

84. **Schuller, W. H. and Thomas, W. M.,** Composition comprising a linear copolymer of a quaternary ammonium compound and an ethylenically unsaturated copolymerizable compound, U.S. Patent 2,923,701, 1960.

85. **Hurwitz, M. J. and Aschkenasy, H.,** Vinylimidazoline and vinyltetrahydropyrimidine polymers, U.S. Patent 3,406,139, 1968.

86. **Schiller, A. M. and Suen, T. J.,** Ionic derivatives of polyacrylamides, *Ind. Eng. Chem.,* 48, 2132, 1956.

87. **Field, J. R. and Smalley, G.,** Flocculating agents, U.S. Patent 3,897,333, 1975.

88. **Barron, B. G.,** Cationic methacrylamide monomers, U.S. Patent 3,652,671, 1972; **Moss, P. H. and Gipson, R. M.,** N-(Tertiaryaminoalkyl) acrylamides, U.S. Patent 3,878,247, 1975.

89. **Reynolds, D. D. and Kenyon W. O.,** The preparation of polyvinylamine, polyvinylamine salts, and related nitrogenous resins, *J. Am. Chem. Soc.,* 69, 911, 1947.

90. **Gless, Jr., R. D., Dawson, D. J., and Wingard, R. E.,** Polyvinylamine and salts thereof, U.S. Patent 4,018,826, 1977.

91. **Harrison, J. R. and Monagle, D. J.,** Sewage treatment, U.S. Patent 3,479,283, 1969.

92. **Tanaka, H.,** Cationic modification of ultrahigh-molecular-weight polyacrylamide by the Hofmann degradation, *J. Polym. Sci.,* B, 16, 87, 1978.

93. **Dawson, D. J., Gless, R. D., and Wingard, Jr., R. E.,** The synthesis of polyvinylamine and its use in the preparation of polymeric dyes, *Polym. Prepr. Am. Chem. Soc. Div. Polym. Chem.,* 17, 779, 1976.

94. **Sobolev, I.,** Cationic hydroxy-containing polymers, preparation and use, U.S. Patent 3,428,617, 1969.

95. **Lloyd, W. G.,** Water-soluble vinylbenzyl quaternary nitrogen polymers, U.S. Patent 3,178,396, 1965.

96. **Vitkuske, J. F. and Rutledge, F. C.,** Method of preparing aqueous latex of high molecular weight vinylbenzyl halide polymer and resulting product, U.S. Patent 3,072,588, 1963.

97. **Wilson, L. H., Thomas, W. M., and Padbury, J. J.,** Paper of improved dry strength and method of making same, U.S. Patent 2,884,057, 1959.

98. **Wheaton, R. M.,** Halogenated polymers of nuclear methylated aromatic hydrocarbons, their quaternary ammonium salts and method of making the same, U.S. Patent 2,823,201, 1958.

99. Dow Chemical Company, Polymer compositions and method, British Patent 832,949, 1960.

100. **Sprague, R. H. and Brooker, L. G. S.,** Quaternary salts of vinylpyridine and vinylquinoline polymers, U.S. Patent 2,484,430, 1949.

101. **Shyluk, W. P.,** Poly (1,2-dimethyl-5-vinylpyridinium methyl sulfate). I. Polymerization studies, *J. Polym. Sci.,* A, 2, 2191, 1964.

102. **Suen, T. J. and Schiller, A. M.,** Flocculation of sewage, U.S. Patent 3,171,805, 1965.

103. **Maclay, W. N. and Fuoss, R. M.,** Polyelectrolytes. VII. Viscosities of derivatives of poly-2-vinyl pyridine, *J. Polym. Sci.,* 6, 511, 1951.

104. **Jones, G. D.,** Treatment of acrylamide polymers, U.S. Patent 2,831,841, 1958.

105. **Grimm, O. and Rauch, H.,** Nitrogenous condensation product, U.S. Patent 2,328,901, 1943.

106. **McDonald, M. T.,** Quaternary adducts of polyepihalohydrin and use thereof, U.S. Patent 3,591,520, 1971.

107. **Walker, G. B. and Cambre, C. M.**, Polyquaternary ammonium salts of polymerized epichlorohydrin, U.S. Patent 3,428,680, 1969.

108. **Rogers, W. A. and Woehst, J. E.**, Quaternary ammonium adducts of polyepichlorohydrin, U.S. Patent 3,320,317, 1967.

109. **Billmeyer, F. W., Jr.**, *Textbook of Polymer Science,* 2nd ed., Interscience, New York, 1971, 255.

110. **Paschall, E. F.**, Starch ethers containing nitrogen and process for making the same, U.S. Patent 2,876,217, 1959; **McClure, J. D. and Williams, P. H.**, Production of epoxy ammonium salts, U.S. Patent 3,475,458, 1969.

111. **Rembaum, A., Baumgartner, W., and Eisenberg, A.**, Aliphatic Ionenes, *J. Polym. Sci. B,* 6, 159, 1968.

112. **Panzer, H. P. and Dixon, K. W.**, Polyquaternary flocculants, U.S. Patent 28,807 (reissued), 1976; **Phillips, K. G. and Zarnecki, W. E.**, Linear dimethylamine-epichlorohydrin copolymer, U.S. Patent 3,975,347, 1976; **Noren, G. K.**, 2-Hydroxy-3-ionene chloride polymers, German Patent 2,426,365, 1975.

113. **Heywood, D. L. and Phillips, B.**, The reaction of epichlorohydrin with secondary amines, *J. Am. Chem. Soc.,* 80, 1257, 1958.

114. **Noren, G. K.**, 2-Hydroxy-3-ionene chloride, *J. Polym. Sci.,* 13, 693, 1975.

115. **Panzer, H. P. and Dixon, K. W.**, Polyquaternary flocculants, U.S. Patent 28,808 (reissued), 1976.

116. Components of Paper and Paperboard in Contact with Aqueous and Fatty Food, 21CFR176.170 Food and Drug Administration, 1977.

117. **Tonkyn, R. G. and Vorchheimer, N.**, Cationic chlorine-resistant polymeric flocculants and their use, U.S. Patent 4,088,613, 1978.

118. **Nagan, L. E.**, Filtration Process, U.S. Patent 3,131,144, 1964.

119. **Coscia, A. T.**, Process for flocculating aqueous suspensions, U.S. Patent 3,493,502, 1970.

120. **Nagy, D. E.**, Water-soluble cationic polymers from epichlorohydrin and methylamine, U.S. Patent 3,567,659, 1971.

121. **Panzer, H. P. and Rabinowitz, R.**, Polyquaternary flocculants and processes of preparing them by quaternizing alkylene polyamine resin polymers from epihalohydrin and monoalkylamines, U.S. Patent 3,741,891, 1973.

122. **Tonkyn, R. G., Vorchheimer, N., Fowler, W. F., Jr., and Heberle, R. A.**, Water-soluble cationic polymeric materials and their use, U.S. Patent 3,915,904, 1975.

123. **Coscia, A. T.**, Alkylene polyamine resin, U.S. Patent 3,248,353, 1966.

124. **Buckman, J. D., Buckman, S. J., Mercer, G. D., and Pera, J. D.**, Amine-epichlorohydrin polymeric compositions, U.S. Patent 4,054,542, 1977.

125. **Green, J.**, Method of making paper, U.S. Patent 2,969,302, 1961.

126. **Bolger, J. C., McCollum, H. E., and Hausslein, R. W.**, Condensation products of amines with epihalohydrins, U.S. Patent 3,577,313, 1971; **Kirkpatrick, W. H. and Seale, V. L.**, Epichlorohydrin polyalkylene polyamine polycondensates, U.S. Patent 3,251,882, 1966.

127. **Keim, G. I.**, Wet-strength paper and method of making same, U.S. Patent 2,926,116, 1960.

128. **Keim, G. I. and Schmalz, A. C.**, Cationic thermosetting quaternized polyamide-epichlorohydrin resins and method of preparing same, U.S. Patent 3,240,761, 1968.

129. **Rembaum, A., Rile, H., and Somoano, R. V.**, Kinetics of formation of high charge density ionene polymers, *J. Polym. Sci. B,* 8, 457, 1970.

130. **Casson, D. and Rembaum, A.**, VII. The effect of ionene polymers on a dilute clay suspension, *J. Polym. Sci. B,* 8, 773, 1970.

131. **Rembaum, A. and Yen, S. S.**, Novel polyelectrolytes, U.S. Patent 3,898,188, 1975.

132. **Gibbs, C. F., Littmann, E. R., and Marvel, C. S.**, Quaternary ammonium salts from halogenated alkyl dimethylamines. II. The polymerization of gamma-halogenopropyldimethylamines, *J. Am. Chem. Soc.,* 55, 753, 1933.

133. **Ballweber, E. G., Tai, W. T., and Selvarajan, R.**, Dichlorobutene/dimethylamine ionene polymer, U.S. Patent 3,928,448, 1975.

134. **Markhart, A. H. and Santer, J. O.**, Electroconducting resin and its use as a coating, German Patent 2,323,886, 1973.

135. **Taube, C. and Böckmann, K.**, Quaternary ammonium compounds, U.S. Patent 3,009,761, 1961.

136. **Witt, E.**, Flocculation of particles dispersed in aqueous media and flocculants used therein, U.S. Patent 3,632,507, 1972.

137. **Witt, E.**, Chain extended polyelectrolyte salts and their use in flocculation processes, U.S. Patent 3,663,461, 1972.

138. **Buckman, S. J., Pera, J. D., Raths, F. W., and Mercer, G. D.**, High molecular weight ionene polymeric compositions, U.S. Patent 3,784,649, 1974.

139. **Schaper, R. J.**, Functional ionene compositions and their use, U.S. Patent 4,075,136, 1978.

140. **Tai, W. T.**, Ionenes, U.S. Patent 4,038,318, 1977.

141. **Hart, A. W.,** Diamines and higher amines, aliphatic, in *Encyclopedia of Chemical Technology,* Vol. 7, Standen, A., Ed., Interscience, New York, 1965, 22.

142. **Phillips, K. G.,** Method of preparing polyamines, U.S. Patent 3,372,129, 1968; **Phillips, K. G. and Geerts, M. J.,** Method of preparing nontoxic polyamines, U.S. Patent 3,751,474, 1973; **Phillips, K. G., Ballweber, E. G., and Selvarajan, R.,** Production of water-soluble polyamine condensate polymers having greater linear characteristics, U.S. Patent 4,057,580, 1977.

143. **Jordan, A. D., Jr.,** Cellulosic fibrous products and methods of producing them, U.S. Patent 2,765,228, 1956.

144. **Jen, Y. and Moore, S. T.,** Polyalkylene-polyamine resinous composition, method of making paper using same and paper containing same, U.S. Patent 2,834,675, 1958; **Garms, D. C. and Norton, F. E.,** Polymeric flocculants, U.S. Patent 3,210,308, 1965; **Schiegg, D. L.,** Method of flocculation with water-soluble condensation polymers, U.S. Patent 3,523,892, 1970.

145. **Jones, G. D., Langsjoen, A., Newmann, M. M. C., Sr., and Zomlefer, J.,** The polymerization of ethylenimine, *J. Org. Chem.,* 9, 125, 1944.

146. **Weidner, C. L. and Dunlap, I.R.,** Clarification process, U.S. Patent 2,995,512, 1961.

147. PEI Polymers . . . Infinite Modifications, Practical Versatility, Dow Chemical Company, Midland, Mich., 1974.

148. **Goldstein, A.,** Alkylenimine polymers, in *Encyclopedia of Polymer Science and Technology,* Vol. 1, Bikales, N. M., Ed., Interscience, New York, 1964, 743.

149. **Garms, D. C.,** Ethyleimine-polyalkylenepolyamine-polyepihalohydrin terpolymer flocculants, U.S. Patent 3,275,588, 1966; Oxide-Aziridine Copolymers, #113-1092-70, Dow Chemical, Midland, Mich.

150. **Feeman, J. F.,** Polymeric quaternary ammonium compound, U.S. Patent 3,334,138, 1967.

151. **Capozza, R. C.,** Regular alternating copolymers of styrene and imidazolines and their derivatives, U.S. Patent 3,907,677, 1975.

152. **Coscia, A. T., Panzer, H. P., and Sedlak, J. A.,** Cationic acrylamide-styrene copolymers and flocculation of sewage therewith, U.S. Patent 3,956,122, 1976.

153. **Barabas, E. S. and Grosser, F.,** Novel flocculant terpolymers, U.S. Patent 3,929,739, 1975.

154. **Suen, T. J. and Schiller, A. M.,** Flocculation of Sewage, U.S. Patent 3,171,805, 1965.

155. **Bufton, R. G., Semancik, J. R., and King, G. G.,** Reaction products of polynitriles, water and amines, U.S. Patent 3,647,769, 1972.

156. **McClendon, J. C.,** Cationic polyacrylamide terpolymers, U.S. Patent 3,478,003, 1969.

157. **Pratt, R. J. and Diefenbach, R. K.,** Method of flocculating an aqueous suspension of solid inorganic particles, U.S. Patent 3,507,787, 1970.

158. **Buckman, J. D., Raths, F. W., Boggs, P. P., and Donerson, R. L.,** Method of flocculating with tertiary aminohydroxyalkyl esters of carboxylic acid polymers, U.S. Patent 3,730,888, 1973.

159. **Reinwald, E. and Galinke, J.,** Clarification of aqueous suspensions with oxyaminated polyacrylamide flocculating agents, U.S. Patent 3,707,465, 1972.

160. Tsuk, A. G., Flocculation and demulsification using cationic polyvinyl alcohols, U.S. Patent 3,761,406, 1973.

161. **Webb, F. J. and Tate, D. P.,** Process and composition for water-soluble polymers, U.S. Patent 3,647,774, 1972.

162. **Jones, G. D., Geyer, G. R., and Hatch, M. J.,** Cationic 2-methylene-1,3-butadiene polymers, U.S. Patent 3,544,532, 1970.

163. **Dick, C. R. and Ward, E. L.,** Amine-modified polyalkylene oxides, U.S. Patent 3,746,678, 1973.

164. **LaCombe, E. M. and Bailey, F. E., Jr.,** Interpolymers of ethylenically unsaturated sulfines, U.S. Patent 3,280,081, 1966.

165. **Price, J. A. and Schuller, W. H.,** Copolymers of a bis ethylenically unsaturated sulfonium compound, U.S. Patent 2,923,700, 1960.

166. **Hatch, M. J. and McMaster, E. L.,** Vinylbenzylsulfonium monomers and polymers, U.S. Patent 3,078,259, 1963.

167. **Rassweiler, J. H. and Sexsmith, D. R.,** Polymeric sulfonium salts, U.S. Patent 3,060,156, 1962.

168. **Rushton, B. M.,** Process of water clarification, U.S. Patent 3,509,047, 1970.

169. **Annand, R. R., Redmore, D., and Rushton, B. M.,** Cyclic amidine polymers as water clarifiers, U.S. Patent 3,576,740, 1971.

170. **Gershberg, D. B.,** High molecular weight acrylamide polymer production by high solids solution polymerization, U.S. Patent 3,929,751, 1975.

171. **Ohshima, I., Chiba, S., Ariyama, K., Ogawa, Y., and Kawamura, Z.,** Polymerization of acrylamide in the presence of water-soluble nitrogen compounds, U.S. Patent 3,951,934, 1976.

172. Carbowax® Polyethylene Glycols; Polyox® Water-Soluble Resins, Union Carbide Corporation, New York.

173a. **Stone, F. W. and Stratta, J. J.,** 1,2-Epoxide polymers, in *Encyclopedia of Polymer Science and Technology,* Vol. 6, Bikales, N. M., Ed., Interscience, New York, 1967, 103.

173b. **Smith, K. L.,** Coagulation of dispersed solids, U.S. Patent 3,020,230, 1962.
 174. **Ting, R. Y. and Hunston, D. L.,** Polymeric additives as flow regulators, *Ind. Eng. Chem., Prod. Res. Dev.,* 16(2), 129, 1977.
 175. **Varveri, F. S., Jula, R. J., and Hoover, M. F.,** Polyamphoteric polymeric retention aids, U.S. Patent 3,639,208, 1972.
 176. **Summers, R. M.,** Method for preparing a polyampholyte, U.S. Patent 2,967,175, 1961.
 177. **Leavitt, R. I.,** Method for coagulating a colloidal suspension, U.S. Patent 4,072,606, 1978.
 178. **Colwell, C. E., and Miller, R. C.,** Polymeric flocculating agents and process for the production thereof, U.S. Patent 3,014,896, 1961.

Chapter 2

NATURAL POLYMER SOURCES

Dr. Nathan M. Levine

TABLE OF CONTENTS

I. TYPES AND STRUCTURES OF NATURAL POLYMERS

The natural polymers used in water and waste treatment systems include starches, galactomannans, cellulose derivatives, microbial polysacharrides, gelatins, and glues. These are water soluble polymers, mainly nonionic. They vary in structure, molecular weight, biodegradability, ease of dissolution, and temperature of preparation. A brief discussion of each is given below.

A. Starches[1,2]

Starches come from such sources as potato, corn, tapioca, arrowroot, and wheat. They differ in size and shape of granules, temperature of gelatinization, rates of swelling in solvents, and ratios of amylose and amylopectin. Figure 1 shows the composition and structures of amylose and amylopectin.

B. Galactomannans[3,4]

Galactomannans are branched polysaccharides composed of D-galactose and D-mannose commonly found in endosperms of Leguminosae where they serve as reserve foods. Ratios of galactose to mannose vary in different species, e.g., guar (1 to 2), locust bean gum (1 to 4), tara (1 to 3). The most important species for water and waste treatment is guar. Derivatives of guar include hydroxy propyl, hydroxy ethyl, carboxy methyl, carboxy methyl hydroxy propyl, cationic guar. Figure 2 shows the structure of guar.[5]

C. Cellulose Derivatives[6]

Cellulose is a carbohydrate polymer composed of anhydroglucose units having the empirical formula $C_6H_{10}O_5$. It is the chief structural element of the cell walls of trees and other higher plants. Cellulose derivatives include cellulose nitrate, cellulose acetate, methyl cellulose, carboxy methyl cellulose (CMC), hydroxy ethyl cellulose, and hydroxy propyl cellulose. The derivative most used in water treatment is carboxy methyl cellulose.

Cellulose is a polydisperse polymer of high molecular weight comprised of long chains of D-glucose units joined together by Beta 1-4 glycosidic bonds. The anhydroglucose unit contains three hydroxyl groups — one primary and two secondary. Sodium carboxy methyl cellulose results from an ether type reaction with mono-chloro acetic acid. Solubility in water increases with degree of substitution (DS). Figure 3 shows structure of cellulose.

D. Microbial Polysaccharides

Microbial polysaccharides are high-molecular-weight polymers synthesized from simple sugars such as glucose by fermentation in an aqueous medium containing strains of nonpathogenic bacteria, enzymes, nutrients, oxygen, and traces of metallic catalysts. A generalized flow diagram for production and recovery is shown in Figure 4. The structure of phosphomannan is shown in Figure 5 and a table of anionic, extracellular microbial polysaccharides is shown in Table 1.

E. Glue and Gelatins[7]

The glue used in water and waste treatment is derived from collagen, the primary constituent of animal connective tissue found in skins, hides, sinews, and bones. Animal glue is closely related to gelatin and is a hydrolysis product of collagen comprising many fragments which vary in size from simple polypeptides to large colloidal molecules varying in molecular weight from 3000 to 80,000. Functional groups involved

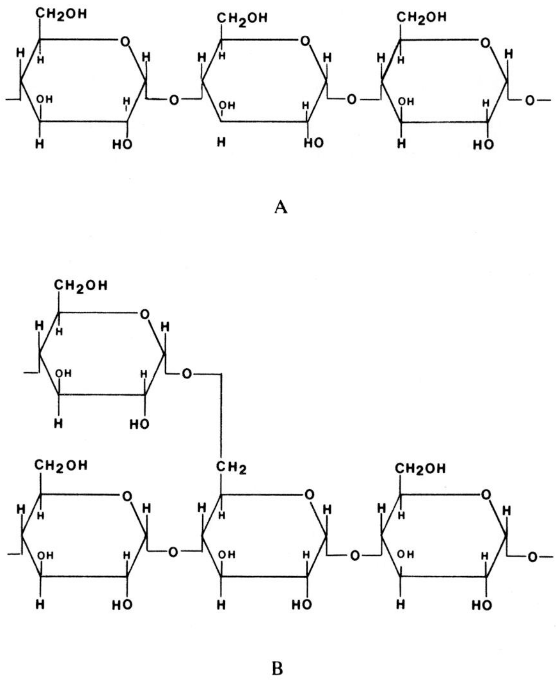

FIGURE 1.   Structure of (A) amylose and (B) amylopectin.

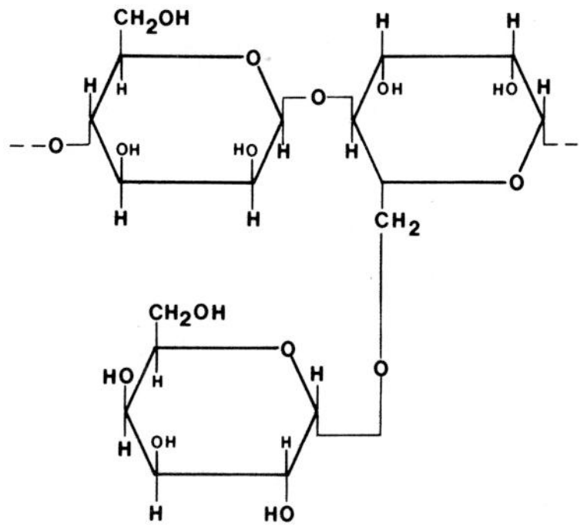

FIGURE 2.   Structure of guar.

with flocculation are amino and carboxyl. Glue is insoluble in cold water but imbibes six to eight times its weight of water at temperatures below its congealing point. The swollen particles pass into solution at about 60°C. Heavy metal ions react with animal glue forming an irreversible gel. Glues are characterized by gel strength in grams (Bloom test) and viscosity. Gelatin is a mixture of water soluble proteins of high molecular weight derived from collagen by hydrolysis.

Gelatin molecules are large and complex proteins ranging in molecular weight from 15,000 to 250,000. They may be composed of 18 different amino radicals linked together in an ordered fashion. Table 2 lists the amino acids obtained by complete hydrolysis of gelatin. Linkages are polypeptide types.

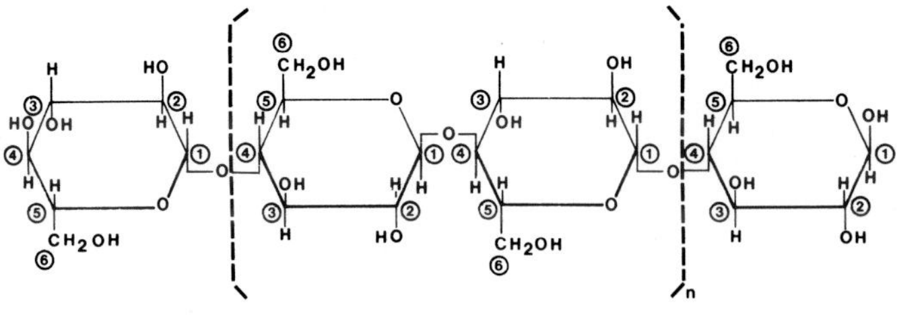

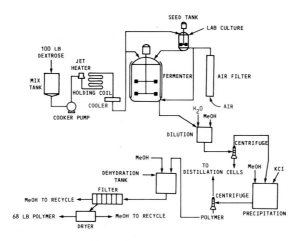

A

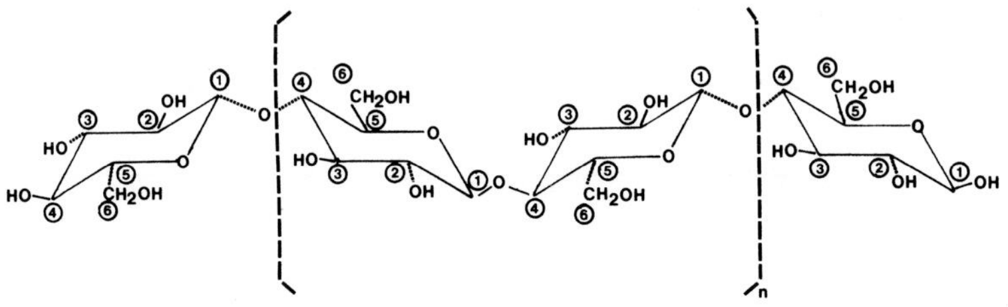

B

FIGURE 3.   Structures of cellulose. (From Kirk, R. E. and Othmer, D. F., Eds., Encyclopedia of Chemical Technology, Vol. 4, 2nd ed., John Wiley & Sons, New York, 1979, 594. Reprinted by permission of John Wiley & Sons, Inc.)

FIGURE 4.   Polysaccharide NRRL-B-1459; flow diagram for pilot plant production and recovery. (From Northern Regional Research Laboratory, U.S. Department of Agriculture, Peoria, Ill.)

II. RECENT U.S. PATENTS OF NATURAL POLYMERS, BLENDS, AND NOVEL COMPOSITIONS

U.S. Patent 3,082,173 (3/19/63) describes a stable liquid product comprising a hydrophilic colloid (guar, starch, dextrine) together with sodium aluminate, to assist floc clarifying action with coagulants such as alum or ferric sulfate.

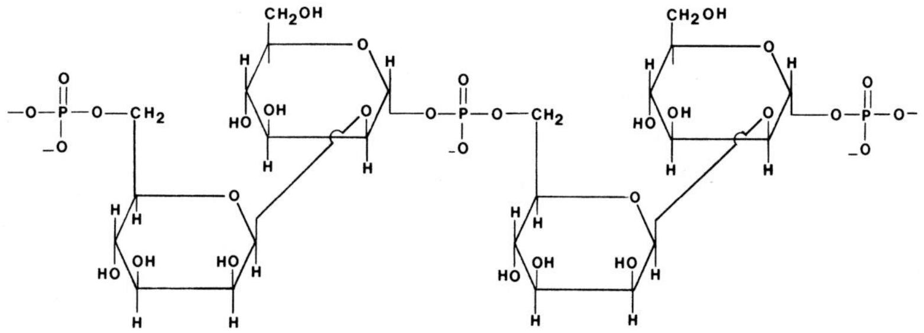

FIGURE 5. Partial structure of phosphomannan produced by Hansenula capsulata Y-1842. (From Northern Regional Research Laboratory, U.S. Department of Agriculture, Peoria, Ill.)

U.S. Patent 3,157,594 (11/17/64) describes use of starch derivatives such as those resulting from reaction of starch and N-alkyl methylol triazone as flocculants for turbid water and sewage treatment.

U.S. Patent 3,406,114 (10/10/67) deals with microbial polysaccharides as flocculating agents for finely divided solids suspended in aqueous media.

U.S. Patent 3,285,849 (11/15/66) describes a flocculant for fine coal dust by reacting polysaccharides or derivatives with N-containing resins, e.g., urea resins, and adding inorganic salts to the suspension.

U.S. Patent 3,423,312 (1/21/69) describes a process for settling raw municipal sewage by conjoint use of coagulants such as ferric chloride or iron and aluminum salts plus cationic starch or other cationic polymers. The starch is derived from corn, has a degree of substitution of 0.05 and a molecular weight of about one million. The combined action appears to be synergistic.

U.S. Patent 3,561,933 (2/9/71) describes use of graft polymers of starch and vinyl monomers, e.g., acrylamide, to flocculate clays, coal, titanium dioxide, carbon black.

U.S. Patent 3,979,286 (9/7/76) describes use of insoluble starch xanthates prepared from cross-linked starch to remove heavy metal ions from industrial effluents.

U.S. Patent 4,051,316 (9/27/77) describes use of alkali metal-magnesium starch xanthate compositions which are capable of removing most heavy metal ions from industrial aqueous effluents.

U.S. Patent 4,083,783 (4/11/78) describes the use of alkali metal-magnesium cross-linked starch xanthates to remove heavy metal ions from aqueous industrial effluents.

III. END USES OF NATURAL POLYMERS IN WATER AND WASTE TREATMENT

Polymers are used in water and waste treatment primarily as flocculants to aid in solid-liquid separations. These include such unit operations as settling, clarification, filtration, centrifugation, and flotation. The majority of such applications are treated with synthetic polymers mainly due to their higher charge densities and higher molecular weights when required. Natural polymers and their derivatives are limited more or less to requirements of low charge density, low molecular weights, and good biodegradability. Applications for synthetics are covered elsewhere in this book. The major application areas for natural polymers will be enumerated and described below.

A. Red Muds

The major source of aluminum is bauxite which is treated by the Baeyer process (high caustic, high temperature) to produce alumina. The alumina is then mixed with

Table 1
ANIONIC, EXTRACELLULAR MICROBIAL POLYSACCHARIDES

Polysaccharide from strain XRRL	Microorganism	Composition of polysaccharide			Refs
		Components	Molar ratio	Linkages[a]	
B-1459	Xanthomonas campestris	D-Mannose	3.0	GA(β, 1→2)M	17-19
		D-Glucose	3.0	M(1→4)G	
		D-Glucuronic acid (K salt)[b]	2.0	G(1→4)GA	
		O-Acetyl	1.7		
		Pyruvic acid	0.63	G(β,1→4)G	
B-1073	Arthrobacter viscosus	D-Glucose	1	MA(β, 1→4)G	20-22
		D-Galactose	1	G(β, 1→4)Gal	
		D-Mannuronic acid (K salt)	1	Gal(β, 1→4)MA	
		O-Acetyl	4		
Y-1401	Cryptococcus laurentii var. flarescens	D-Mannose	4	M(1→3)M	23
		D-Xylose	1	X(1→ᶜ)M	
		D-Glucuronic acid (K salt)	1	GA(β, 1→2)M	
		O-Acetyl	1.7		
Y-244S	Hansenula holstii	D-Mannose	5	M(α, 1→)PO K(→6)M	24-26
		Potassium	1	M(α, 1→3)M	
		Phosphorous[d]	1	M(α, 1→2)M	

[a] Identification of symbols: Gal, D-galactopyranose; G, D-glucopyranose; GA, D-glucopyranosyl uronic acid; M, D-mannopyranose; MA, D-mannopyranosyl uronic acid; X, D-xylose.

[b] The potassium salt is made for experimental convenience. Other salts could be made as desired.

[c] Position of linkage is not established.

[d] Orthophosphate.

Reprinted with permission from *Encyclopedia of Polymer Science and Technology*, Vol. 8, Baakeles, N., Ed., John Wiley & Sons, New York, 1968, 696. Copyright by John Wiley & Sons.

Table 2
AMINO ACIDS OBTAINED BY COMPLETE
HYDROLYSIS OF GELATIN

Amino Acid	% by wt	Amino acid	% by wt
Alanine	11.0	Lysine	4.5
Arginine	8.8	Methionine	0.9
Aspartic acid	6.7	Phenylalanine	2.2
Glutamic acid	11.4	Proline	16.4
Glycine	27.5	Serine	4.2
Histidine	0.78	Threonine	2.2
Hydroxyproline	14.1	Tyrosine	0.3
Leucine and isoleucine	5.1	Valine	2.6
		Cystine	Trace

Reprinted with permission from *Encyclopedia of Chemical Technology*, Vol. 10, 2nd ed., Kirk, R. L. and Othmer, D. F., Eds., John Wiley & Sons, New York, 1967, 500. Copyright by John Wiley & Sons.

cryolite and electrolized to produce aluminum cathodes. In the process of producing alumina a liquid-solid separation is required to remove the leach solution consisting of sodium aluminate, caustic soda, silicates, from a fine solids residue of iron oxides, some dolomite, and various silicas and silicates. The residue is called red mud because of the color given to it by the iron oxides.

The predominant polymer in use to separate the red muds is starch. By varying time, temperature, and caustic content of preparation, almost any variety can be used including corn, potato, sorghum, and tapioca. Because of price and availability, corn starch is preferred in the U.S.

The functions performed by starch include rapid settling of solids, clarification of aluminate liquors, aiding recovery of caustic by counter-current washing in six to nine stages, thereby also minimizing the amount of caustic discharged to the environment.

Dosages of starch required range from 1 to 2 lb/ton of red muds on treating Surinam bauxites to as much as 40 lb/ton of red muds on Jamaican ores. Despite high usage levels starch has been the only acceptable flocculant until recently, when synthetics such as poly (sodium acrylates) began challenging its position at low usage levels and the ability to produce higher red mud densities. However, poorer clarities and high prices (ca $1.50/lb) appear to keep starch in use.

The starch is generally prepared in recycle barrens after removal of the aluminate. Energy is required to cook the starch. Concentrations are generally about 1% and time to gelatinize and become fully mature for use 1 to 2 hr. The stock solutions may be further diluted and fed to the heads of settlers and wash thickeners.

B. Tissue Mill Effluents

Tissue paper mills manufacture such items as paper towelling, toilet tissues, cigarette papers, facial tissue, etc. The products generally consist of fibers from wood pulp without clay fillers except for cigarette paper which may contain calcium carbonate. In the processing of paper, white water is recycled to the furnish to recover a large portion of the fibrous materials through savealls. The excess white water is plant effluent which in the past has been discharged directly to rivers.

However, due to high suspended solids and biological oxidation demand (BOD), the U.S. Environmental Protection Agency (EPA) required a substantial lowering of such pollution to protect fish life, drinking water sources, etc. As a result such plants have incorporated effluent treatment prior to discharge. This includes use of dissolved air flotation (DAF), sometimes referred to as Sven-Pedersen, to remove fine fiber from the effluent.

The best reagent for the above type of effluent treatment to date has been reported to be a cationic guar of fairly low charge density which causes the flocculation of the fibers by bridging following adsorption, by hydrogen bonding, and by electrostatic attraction. This is facilitated by the anionic and hydrated character of the fibers. The resultant aqueous flocs are pressurized with air-saturated water and then floated on release of air bubbles when exposed to normal atmospheric pressure. The suspended solids remaining in the effluent after this treatment are generally less than 20 ppm and the BOD is acceptable.

The cationic guar used is in form of a powder. Examples of two commercial products are Jaguar C-13, manufactured by Celanese Plastics and Specialties Company, and Gendriv, manufactured by Henkel. They are required to satisfy U.S. Food and Drug Administration (FDA) specifications. Dosages range from 1 to 5 mg/ℓ based on weight of liquor to be treated. Preparation of the powders for use is generally at 0.25 to 0.50% concentration in demineralized or plant makeup water. They are hydrated for at least 1 hr prior to use and then diluted with recycle or process water to 0.05% or less, prior to contracting the effluent to be treated. Competitive products including natural and synthetic polymers are not as effective as the cationic guars in producing the needed effluent clarity.

C. Potable Water

Drinking water must meet very stringent standards with regard to pH, hardness, bacteria count, suspended solids, etc. Polyelectrolytes may be used as coagulant aids in conjunction with coagulants such as acid, lime, potassium aluminum sulfate (alum), and sodium aluminate to remove suspended solids and/or precipitates associated with removal of hardness. Sources of drinking water include artesian wells, rivers, lakes, glaciers, and in some cases seawater. Polyelectrolytes are used mainly with river water sources.

Natural polymers most commonly used with river waters include nonionic and cationic guars and pregelatinized potato starches. These and other polymers have the approval of the Department of Health, Education and Welfare in amounts of the order of 1 mg/ℓ or less.

The major applications are as settling aids in basins of conventional systems and in cylindrical units where they aid in forming coherent and permeable sludge blankets, thereby functioning somewhat as filters. In other applications the polymer may be added to the feed of a gravity sand filter, a multi media filter, or a pressure filter. In these cases the turbidity of the filtrate is reduced below 1 mg/ℓ and the time between backwashing of the filter may be extended.

The polyelectrolyte is usually added to the influent subsequent to the addition of the coagulant in amounts ranging from 0.1 to 0.5 mg/ℓ. Preparation of polymers are usually at 0.25 to 1.0% as stock solutions with dilution to 0.05% with raw or process water prior to contacting the water to be treated.

Cationic guar may be especially effective where significant quantities of clays are present in the influent, especially during spring runoff. Major competition for natural polymers are high-molecular-weight, nonionic polyacrylamides having a free monomer content below 0.5%.

D. Iron Ore Beneficiation

Most iron ores are processed by methods employing copious amounts of water, e.g., 1000 to 2000 gal/ton. These include gravity, magnetic separation, and froth flotation processes. The fine tailings produced have to be separated from the water to provide for 85 to 95% recycle and also to minimize pollution through ground seepage. Most operations in the U.S. and Canada use polymers which are fed to fine tailings thickeners to recycle water and discharge fine solids with coarse tailings in tailings disposal areas.

Until about 1977, a cationic guar was used by a large Canadian iron ore plant (commencing about 1961). This product had a low charge density and a fairly low molecular weight (less than 200,000). This product was supplanted by a high charge density polyamine. Usage levels were about 0.5 to 1.0 mg/ℓ. Its use not only aided the liquid-solid separation of slimes, but it was sufficiently effective to clarify the water sufficiently to permit good filtration of the iron ore concentrates without blinding of the filter cloths. It was displaced by the liquid polyamine because of ease of handling and price. Most of the remaining operations in the U.S. and Canada are using synthetics, e.g., polyamines and polyacrylamides.

Several plants on the Mesabi Range use a combination of lime and causticized starch. Dosages of starch range from 0.2 to 0.5 lb/ton of solids. The lime provides coagulation of colloidal iron oxides and silicates while the starch yields floc size. Competition is being felt from synthetics in both performance and ease of handling (less labor in preparation). The starch requires a caustic treatment and sometimes heat in preparation (cook to 180°F). Generally, starch is prepared as a 1 to 5% solution and then diluted with process water prior to use.

E. Soda Ash Processing From Trona Ores

Soda ash (sodium carbonate) is derived by synthetic means by the Solvay process and from natural trona ores by leaching, purification, and recrystallization. Such natural polyelectrolytes as nonionic guar and in some instances cationic guar, are used to flocculate the clay-like impurities and clarify the saturated carbonate leach liquors prior to filtration and recrystallization. This is accomplished at elevated temperatures following digestion of calcined ore by addition of the flocculant to the feed going to a clarifier, where the clays become flocclated and settle out. After thickening of the unleached residue it goes to waste on a disposal site. The liquors following digestion are recrystallized to form soda ash. It is most important that the flocculant used for purification not interfere with filtration or recrystallization. This has been a factor in favor of guar, whereas synthetics have shown bad effects on certain ores.

One of the major drawbacks of natural polymers is the inability to prepare them in saturated brine. At present, they are prepared at 0.1 to 0.5% in demineralized water or partially saturated brine and then used at 4 to 10 mg/ℓ based on digest liquor. This results in lower soda ash solution concentration necessitating more evaporation and energy in recrystallization. If saturated brine could be used in preparation there would be a gain in energy conservation. The proportion of soda ash made from trona is on the increase and should eventually replace the Solvay process almost entirely. This should be a plus for the environment.

F. Coal Washing

During the last 15 years there has been increasing emphasis on producing finer-sized coals by dispersed air flotation. Increasingly, recycling of water is being practiced in washeries to meet requirements and to minimize the volume of effluents which have to meet increasing standards of pH, BOD, ferrous, ferric, and turbidity concentrations. In addition, there is a need for dewatering tailings to the point where storage space and seepage can be minimized. Polymers have been used increasingly since the early 1960s especially as settling aids for fine tailings, filtration, and centrifugation.

In the early sixties flocculation in West Virginia and Pennsylvania was dominated by crude corn and potato starch which required cooking to 180°F to prepare at concentrations of 2.0 to 5.0%. Dosages ranged from 0.25 to 1.0 lb/ton of dry solids. Performance left much to be desired. The crude starches gave way to pregelatinized potato starches in the early and mid sixties and then to blends with polyacrylamides.

Today most washeries use anionic and cationic polyacrylamides. Natural polymers are on the wane because synthetics require one fifth or less dosage and yield more rapid settling rates with denser thickener underflows.

In the western U.S. growth of large strip mines containing large amounts of clay has given rise to reports of possible use of a nonionic guar serially with anionic polyacrylamides to keep costs down and enable better treatment of fine clays. Such systems require two preparation and distribution setups.

G. Uranium Ore Processing

Uranium is derived from sandstones and shales in the U.S., e.g., Colorado, Wyoming, New Mexico, Texas, and Utah. In South Africa uranium is associated with gold, and in Canada it is obtained from brannerite ores. Processing is by acid or carbonate leaching followed by liquid-solids separations such as counter-current decantation (CCD) in thickeners or filters. This is accomplished to separate the pregnant, uranium-bearing solutions from the solid residues for further processing to recover the uranium values. The solid residues are disposed on a tailing site from which radioactive wastes may be lost by seepage.

Guar or guar derivatives are used alone or in conjunction with polyacrylamides to achieve the liquid-solids separations. Other natural polymers have been ineffective in competiton with polyacrylamides except glue which has been used in Canada, the U.S., and South Africa to produce good clarity. Its propensity to fungus growth is a deterrent to use of glue.

Guar is especially effective as a filter aid because of the regularity of floc shape. The synthetics produce a large, fast-settling floc which tends to hold too much water to be a good filter aid for uranium ores.

Dosages employed in thickening may range from 0.1 to 1.0, and in filtration from 0.1 to 1.5 lb/ton dry solids. These materials are generally prepared in mine or demineralized waters at 0.25 to 0.50% concentration, hydrated at least 1 hr prior to use and diluted to 0.1% for filtration and 0.05% or less for thickening prior to use. Process water is generally used for dilution.

In a number of plants, blends of guar and polyacrylamides are used to develop synergisms and/or reduce costs. In others guar and polyacrylamides are used in sequential order. This procedure is very effective but does require two preparation systems for the flocculants.

H. Textile Mill Wastes

In textiles, printing thickeners are used to aid in suspending dyes and producing clear print definition and tone value. Upon completion of its function, the thickener is dispersed and washed into plant effluent with other chemicals. Due to standards set by the U.S. EPA regarding BOD, color, suspended solids, and high organic (plus 1000 mg/ℓ, secondary treatment is required.

In one plant a combination of alum and an anionic polyacrylamide was being used to flocculate the activated sludge to aid clarification prior to filtration of the water-activated sludge in a mixed media gravity filter. Improved performance of the effluent with respect to BOD and suspended solids (which were in excess of requirements) was desired. It was found that replacement of the alum/anionic polyacrylamide with a cationic guar improved performance and pointed to alum as possibly fouling the aerobic bacteria. Microbial polysaccharides had been tried but found to be more resistant to biological oxidation and less effective regarding color removal. In general, although the market is small, synthetics are preferred for textile waste treatment.

I. Recovery of Magnesium From Seawater

Indirectly, seawater becomes a natural resource for producing magnesium metal for refractories. This is accomplished by various means to precipitate the magnesium in form of a hydroxide at high pH. In this form alone the settling rate is too low to produce the density required to make satisfactory refractories via sintering. A natural polymer such as Jaguar MD7A, a nonionic guar, has been used for this purpose at dosages of 0.1 to 0.3 lb/ton of solids.

The overflow from the thickener should be clear since it is returned to the sea. In recent years anionic polyacrylamides of high molecular weight have supplanted use of guar.

In the case of magnesium metal production the thickened slurry of $Mg(OH)_2$ is redissolved and electrolyzed to produce a magnesium cathode. In this application one might look at the sea as a waste product and recovery of Mg from the sea as waste treatment.

J. Oxidized Copper Ores

In processing oxidized or roasted copper ores by leaching, purification, and electrowinning, a waste product is produced which is stockpiled in a tailings disposal area. This waste product results from acid leaching of the oxidized or roasted copper ore followed by a liquid-solid separation in thickeners and filters. Such natural polymers as guar alone or in combination with polyacrylamides are used to accomplish these objectives.

The use of guar in thickeners produces a clear overflow which is electrolyzed for cooper after removal of iron by lime precipitation. (More modern methods use solvent extraction to purify the solution). The high density underflow is filtered using guar and polyacrylamides to remove as much of the values as possible, and the residue is deposited in the tailing area. One advantage in using guar for filtration is that it permits thorough washing of the filter cake, thereby reducing the metal content of the seepage liquid of the tailings. Many operations recover copper by precipitation with iron scrap.

Dosages of guar in thickeners range from 0.05 to 0.30 lb/ton of dry solids and in filters from 0.1 to 0.2 lb/ton. Concentration of stock solutions normally prepared in recycled barrens is 0.50%. Dilution prior to use is to less than 0.05% in the thickeners and about 0.1% in filters.

K. Removal of Heavy Metals from Water

Due to recent requirements by the EPA, it is no longer possible to freely discharge heavy metals into the waterways of the nation. Work by the U.S. Department of Agriculture at Peoria, Ill., has shown that starch xanthates can be used to precipitate heavy metals such as Cd, Cr, Cu, Fe, Pb, Mn, Hg, Ni, Ag, and Zn. When used in conjunction with a polyamine most of these metals can be easily removed. Performance is superior to that obtained by lime or caustic alone. The volume of sludge is smaller and less gelatinous than that obtained with the base precipitation method alone. The effective pH range is 3 to 11.

Recent patents mentioned earlier in this chapter indicate further improvement in the removal of metal ions to acceptable standards using insoluble cross-linked starch xanthates without a need for the cationic polymer. A requirement for the cross-linked polymer is that its degree of swelling range from 65% to 450%. While there are no known commercial applications of thesé products known to the author at this writing, they indicate an interesting approach to an increasingly important problem.

L. Removal of Talcose and Carbonaceous Shale Wastes in Base Metals Ore Processing

In concentrating base metal ores containing copper, lead, zinc, nickel, and iron (pyrites), presence of talcose and carbonaceous shale gangue minerals make it very difficult to obtain good metallic grades by present methods of treatment, such as froth flotation with sulfhydryl collectors. Use of natural polymers and derivatives such as guar, locust bean gum, carboxy methyl cellulose, dextrines, and combinations of these have been found to be effective at most operations as depressants for the talcose and carbonaceous shale constituents of such ores. They are not all equally effective at a given operation but one of them or a combination of them will work at all operations. By depressing such gangue materials and adding them to the wastes, they are disposed in tailing sites.

There is no way of predicting apriori which products will do the best job per unit cost. This is usually estimated by laboratory flotation testing and then tried in the plant. In most applications products are prepared by mechanical agitation in water. Dextrines, guar, and CMC are usually prepared at ambient temperatures as 0.5 to 1.0% solutions and allowed to hydrate several hours prior to use. Dosages range from 0.1 to 1.5 lb/ton dry feed for roughers and 0.1 to 0.5 lb/ton for cleaner feeds. Dilution of stock solutions is usually not required. Addition point is generally prior to the collector. Conditioning times can range from 0 to 5 min. Synthetics do not seem to be as effective as the natural products.

M. Radioactive Wastes

A very special problem exists in transporting radioactive wastes to burial sites for disposal. Since these are usually in the form of acid liquors, there is danger of leakage developing enroute which is a potential hazard. Efforts are being made toward stable liquid gels to partially solidify the liquor. Natural polymers such as guar have been considered because of their ability to form stable gels. The gels do not have to remain stable indefinitely, only as long as necessary for completing transportation.

REFERENCES

1. Whistler, R. L. and Smart, C. L., *Polysaccharide Chemistry,* Academic Press, New York, 1953, chap. 10.
2. Kerr, R. W. L., *Chemistry and Industry of Starch,* Academic Press, New York, 1944, 1.
3. Goldstein, A. M. and Alter, E. N., Guar gum, in *Industrial Gums,* Whistler, R. L., Ed. Academic Press, New York, 1959, chap. 14.
4. Whistler, R. L. and Smart, C. L., *Polysaccharide Chemistry,* Academic Press, New York, 1953, chap. 12.
5. Goldstein, A. M. and Alter, E. N., Guar gum, in *Industrial Gums,* Whistler, R. L., Ed., Academic Press, New York, 1959, 331.
6. Hamilton, J. K. and Mitchell, R. L., Cellulose, in *Encyclopedia of Chemical Technology,* Vol. 4, 2nd ed., Kirk, R. L. and Othmer, D. F., Eds., John Wiley & Sons, New York, 1967, 593; Klug, E. D., Cellulose Derivatives, in *Encyclopedia of Chemical Technology,* Vol. 4, 2nd ed., John Wiley & Sons, New York, 1967, 616.

7. **Marks, E. M.,** Gelatin, in *Encyclopedia of Chemical Technology,* Vol. 10, 2nd ed., Kirk, R. L. and Othmer, D. F., Eds., John Wiley & Sons, New York, 1967, 500.
8. **Wing, R. E., Swanson, C. L., Doane, W. M., and Russell, C. R.,** Heavy metal removal with starch xanthate-cationic polymer complex, *J. Water Pollu. Control Fed.,* 46, 2043, 1974.
9. U.S. Patents 3,979,286, 1976, and 4,051,316, 1977, and 4,083,783, 1978.

Chapter 3

ELECTROKINETICS

John G. Penniman

TABLE OF CONTENTS

I. INTRODUCTION

Electrokinetic considerations can provide a heretofore ignored basis of optimizing operation of essentially all unit operations used for purification of wastewaters, including granular media filtration, dissolved air flotation (DAF), clarification, and the activated sludge process.

The suspended constituents of sewage and industrial wastewater, including microorganisms, dispersed oily colloids, and inert suspended matter, (such as inorganic sulfides, silt, coke fines, and the like) have a negative electrical surface charge. The stability of colloidal and very slightly flocculated suspensions relates to the fact that the individual particles carry like electrical charges causing their mutual repulsion.

When optimizing unit operations used in wastewater purification, it is of utmost importance to recognize the impact of colloid scientists' work. Application of basic principles established by colloid scientists has led to a better understanding of fundamentals operative in wastewater processes, and yielded improved operating results.

A. The Electric Double Layer

The electric double layer may be regarded as consisting of two regions: (1) an inner region, which may include adsorbed ions and (2) a diffuse region in which ions are distributed according to electrical forces and thermal motion.

Stern proposed a model (Figure 1) which the boundary of the inner region (Stern layer) was located by a plane (the Stern plane) at approximately a hydrated ion radius from the surface. Adsorbed ions attached to the surface by electrostatic or Van der Waal's forces may be dehydrated in the direction of the surface. A certain amount of solvent will also be bound to the charged surface in addition to the adsorbed ions. The shear plane, therefore, is probably located farther from the surface than the Stern plane. Ions with centers beyond the Stern plane are considered to be in the diffuse part of the double layer.

Electrokinetic potentials relate to the mobile part of the particle; therefore, the electrokinetic unit consists of the volume enclosed by the shear plane, which is rather inexactly known. The potential difference between the surface of shear and the solution is called the zeta potential (ZP).

B. DLVO Theory*

Unit operations used for wastewater purification essentially all involve solids in water. The DLVO theory quantifies particle stability in terms of energy changes when particles approach one another. The total energy is determined by summation of the attraction (London-Van der Waal's forces) and repulsion (overlapping of electric double layers) energies in terms of interparticle distance. The general character of the resulting interaction energy-distance curve illustrates the very significant conclusions: (1) attraction will predominate at small and large distances, and (2) repulsion may predominate at intermediate distances, depending on the actual values of the two forces.

An important purpose of the chemicals used for destabilization is to reduce or eliminate the repulsion force at intermediate distances, so that attractive forces will predominate, and the particles will aggregate. This is a key consideration in optimizing physical separation operations. The reason is that the negative ZP of waterborne solids causes the particle to be repulsed by negative surface charges of granular filter media (such as sand and coal), and those of the gas-liquid interface of dissolved air and induced air flotation systems.

---

* Named after investigators Derjaguin, B., Landau, L.D., Verwey, E.J.W., and Overbeek, J.T.H.G.

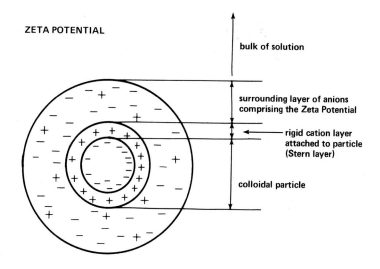

FIGURE 1. Representation of a colloid suspended in a solution, showing the electric double layer.

C. Colloid Destabilization Mechanisms

Destabilization of the waterborne suspended solids may involve four mechanisms:

1. Colloid entrapment or removal via the sweep floc mechanism
2. Reduction in surface charge by double-layer repression
3. Charge neutralization by adsorption
4. Bridging by polymers

Colloid entrapment involves chemical treatment with comparatively massive amounts of primary coagulant. The amount of coagulant used is typically so great in relation to the amount of colloidal matter, that the nature of the colloidal material is not relevant. The amount of primary coagulant used may be 5 to 40 times as much as is used for charge neutralization by adsorption. The rate at which the primary coagulants form hydrous metal oxide polymers is relatively slow, and depends chiefly upon water temperature and pH.

Coupled with the high concentration used, all negatively charged colloidal material is initially exposed to charge neutralization by the transient cationic species. The polymer matrix is three dimensional and voluminous, providing for entrapment of solids. As the polymer contracts (freeing solvent water molecules) and settles, the suspended solids remain enmeshed in the settling floc and appear to be swept from the water; hence, the description of the process as a "sweep floc" mechanism.

However, this destabilization mechanism can result in the generation of large amounts of wet alum (or iron) sludges, which are difficult and costly to dewater. Even though it is by far the most widely used mechanism for water clarification, it is not recommended because of the sludge problem and because the use of other mechanisms result in significantly lower operating and capital costs.

D. Charge Neutralization and Bridging

Charge neutralization by adsorption of the destabilizing chemical to the colloid is a key mechanism for optimizing removal of waterborne solids from waters. The colloidal charge may not only be reduced to zero, but actually beyond zero; that is, reversed. Charge neutralization by adsorption infers that the colloid-water interface is changed, and thus, so are its physiochemical properties.

It does not require much extension of one's imagination to see how this destabilization mechanism can explain those cases in which excessive chemical dosages were found to result in overdosing that resulted in a deterioration in, or failure of, direct filtration. This phenomenon is more typically experienced with very low-molecular-weight polyelectrolytes, or surfactant-type molecules with little bridging properties.

Bridging by organic and inorganic polymers describes the destabilization mechanism by which the molecules of the added chemical attach onto two or more particles, causing aggregation. There are two kinds of bridging. These comprise polyelectrolyte bridging between dissimiliarly, and similarly charged materials. Bridging of negatively charged colloids by high-molecular-weight cationic and anionic polyelectrolytes are examples of the first and second kinds, respectively.

E. Separation

Approximately 10 to 25% of the suspended solids in the waste stream from an integrated pulp and paper mill consist of very finely divided organic chemicals which can be termed "colloidal".[1] The remainder is larger particles which are readily settlable (plus dissolved organic compounds). For efficient operation of the activated sludge process, it is necessary to remove the colloidal suspended solids as the microbes can much more readily digest the dissolved organic waste.

Nonionic chemicals such as starch, which is widely used in the paper industry, form a protective colloid around particulates, thereby creating a steric hindrance which interferes with ionic destabilization. However, the starch can be decomposed microbiologically, and wastes containing pregelatinized starch which have been allowed to age for three or four days undergo a sufficient electrokinetic change that they can be efficiently flocculated.[2]

Colloidal particles are so small that Brownian Motion and thermal convection currents tend to keep them dispersed and prevent them from settling or rising (as a function of their density) so that they can be removed. Stokes' Law states that the speed of settling or rising is proportional to the square of the particle radius. To increase the speed of vertical movement, particles must be coalesced to increase their effective radius. Whether the particles coalesce depends on the electrokinetics of the colloidal system, assuming no steric hindrance.

Colloids in nature are negatively charged. These charges cause the particles to repel one another. The more negative, or the higher the ZP, the greater the repulsion and, therefore the greater the stability. As the ZP approaches zero, the negative charge becomes less effective and conditions approach optimum for flocculation and consequent precipitation.

The effect of ZP on separation is shown in Table 1.

II. ELECTROKINETIC INSTRUMENTATION

Traditionally, electrophoresis measurements have been reported in terms of "mobility," expressed in $\mu m/sec/V/cm$. However, ZP can be calculated directly in volts, using the Helmholtz-Smoluchowski formula. Because the dielectric constant and viscosity are temperature dependent, ZP must be measured at a standard temperature such as 25°C; otherwise a compensating correction must be applied.

Table 1
THE RELATIONSHIP BETWEEN ZETA
POTENTIAL AND EFFLUENT QUALITY

| | Zeta potential (mV) | |
Effluent quality	Primary clarifier	Secondary clarifier
Good	−8 to 0	±7 to 0
Fair	−15 to −8	±15 to ±7
Poor	> −15	> ±15

The expression for zeta potential is

$$\zeta = K\eta \, v/\epsilon \, E \tag{1}$$

where ζ = zeta potential (mV), K = constant, η = viscosity (P), V = transfer velocity (cm/sec), ϵ = dielectric constant, and E = field gradient (V/cm).

An electric field is applied across a cell containing the colloid, called an electrophoresis cell (Greek — electro + phoresis, a carrying). Negatively charged particles migrate to the anode at a transfer velocity directly dependent on their charge and the applied electric field. The higher the charge on the particle, the greater its velocity at a given applied voltage. Several instruments have been developed and commercialized which employ the microelectrophoresis method. They are described below.

A. "Stop Watch" Technique
The following description is based on material furnished by Zeta-Meter, Inc.[3]
The equipment consists of:

1. A steroscopic microscope with ocular micrometer; 15 × WF eyepieces; 2,4,6, and 8 × (adjusted magnification) objectives; and a special mechanical stage (a standard compound laboratory microscope cannot be employed with the cell)
2. A special illuminator, producing a thin beam of intense blue-white light, with heat-absorbing filter
3. A DC power supply, continuously variable from 0 to 300 V
4. A clear plastic electrophoresis cell, equipped with platinum-iridium electrodes
5. A cell holder consisting of a thick and highly reflective mirror for reflecting the light (45°) upward through the cell tube
6. An interrupted-type, cumulative-reading electrical timer reading in seconds and tenths

The cell is a clear plastic block with a small polished tube extending through the center. The tube is connected by tapered ports to two solution chambers, one of which is sealed with a direct-coupled, platinum-iridium electrode of the closed type, while the other contains an open-type electrode.

The closed electrode is positioned in one chamber, preventing movement within the cell tube other than that induced by the impressed voltage. The other chamber is left open (or covered with a loose-fitting plastic cap) for convenience in use. This arrangement is normally employed for measuring the electrophoretic velocity of all colloids. If floc particles are to be evaluated, then two open-type platinum electrodes are preferred to permit tilting the cell back and forth until a floc particle is properly positioned for timing. The cell is supported on the holder, so that the thin light beam reflects upward through the cell tube at an angle. This removes direct light from the optics of the microscope and permits particles too small for normal observation to be readily seen if they reflect light. The cell is positioned somewhat above the top of the mirror to provide an air-gap heat barrier, and passage of the light through the thick mirror further reduces heat transmission.

The 4,6, and 8 × objectives are preferable for precise work, but the 2 × objective permits viewing the entire width of the tube for evaluation of precipitates. A special ocular micrometer is employed in one eyepiece, and the cell is positioned with the zero micrometer line set at the apparent (not actual) front wall of the tube. The front wall is clearly visible as a long sharp line when the microscope is focused to exact middepth of the cell. The difference between actual and apparent diameter is due to refraction.

If a suspended (solid) particle moves by electrophoresis in relation to the liquid, then it follows that the liquid must also move with relation to a fixed solid, in this case, the walls of the cell tube. To compensate for this movement of the liquid (electroendosmosis, manifested as a flow along the walls of the tube with return at its center), it is necessary to measure the rate of travel of discrete particles at a distance of 15% of the tube diameter from the tube wall.

B. Rotating Prism Technique; The Laser Zee® Model 500[4]

An electric field is applied across the cell, causing the negatively charged particles to migrate to the anode. The migrating particles are observed through a Nikon® binocular microscope, equipped with an eyepiece reticle. Illumination of the viewing plane is provided by a collimated helium-neon laser beam. The plane is defined by a fine-focusing adjustment to an accuracy of ±0.001 mm (1 μm).

Built into the microscope is a prism mounted on a galvanometer. The prism rotates at a rate determined by an operator-controlled ZP knob. The particles are observed through the microscope using the reticle lines for reference, and the ZP knob is adjusted until the prism rotation exactly cancels the transfer velocity of the particles and they therefore appear stationary. A digital readout of ZP is provided, expressed in millivolts.

To maximize reproducibility, an averaging computer can be actuated. It operates on a 30-sec cycle, and provides an average of the setting of the ZP knob over that period, with readout announced by an audible "beep".

The Laser Zee® automatically solves the equation for aqueous systems at 20°C when the operator has matched the rate of the prism to the speed of movement of the particles.

The rapidity of measurement minimizes effects which, in other instruments, might be troublesome and produce errors. Since the measurement is typically made in less than 10 sec, particles do not have a chance to settle out and thereby degrade measurement accuracy by altering the charge on the chamber surface. For the same reason, it is easier to keep the chamber clean. When the chamber does require cleaning, it is simply filled with detergent and immersed in an ultrasonic bath.

The Laser Zee®, dark field illumination system, employing a 2 mW helium neon laser, allows comfortable viewing of particles as small as 100 A (0.01 μm). Systems not employing laser illumination typically require that particle size be an order of magnitude larger.

The Model 500 provides a direct digital readout of ZP without the use of tables or graphs, and automatically compensates for different applied field strengths. Older instruments require the operator to record time intervals for the transit of each particle between grid lines and then refer to tables based on time interval and applied voltage to determine zeta potential. With earlier instruments, the applied voltage is determined with an analog meter which is seldom accurate to better than 2 to 3%. The Model 500 electronics are calibrated with a digital voltmeter to provide a precision of better than 1%.

The Laser Zee® is thermodynamically stable, and the ease of use ensures that the measurement is completed before any appreciable joule heating. As a result, the Model 500 does not experience "thermal overturn", which either reduces the number of particles that can be measured at one filling or limits the voltage that may be applied.

This unit is highly sensitive as the ZP approaches zero, where particle movement is minimal and accurate measurements are most difficult to make. The operator is free to double or triple the applied voltage, thus doubling or tripling the apparent particle movement and ease of measurement of colloids close to zero charge. This increased sensitivity is possible because of the thermodynamic stability of the cell-electrode system.

C. Semi-Automatic Technique

The Komline-Sanderson Zeta-Reader makes it possible to continuously monitor the zeta potential of a fluid stream.[5] In the Zeta-Reader a microscope image of the particles in the waste stream is fed via a television camera to a small digital computer which calculates the ZP and displays it digitally on the front of the instrument. The Zeta-Reader samples, analyzes, flushes out the sample, and samples again semiautomatically at any frequency desired. The ZP value is also converted to a 4 to 20 mA DC signal for recording and/or control purposes. In addition, the Zeta-Reader will diagnose and report malfunctions in its own electronic or mechanical systems.

The Zeta-Reader combines modern optical analysis techniques with the most advanced image analysis methods to rapidly and accurately measure the electrophoretic mobility of colloidal suspensions in water.

The Zeta-Reader is designed to be used either as a bench top instrument in the laboratory or as an on-line sensor and proportional control signal generator for closed-loop automatic control. The heart of the Zeta-Reader is an electrophoresis cell where particles suspended in the sample liquid are viewed by means of a scanner. The electrophoresis cell is automatically flushed clean every 30 sec (or at longer intervals if desired) and filled with fresh sample. The scanner views the motion of the suspended particles in the cell and relays the information to the data processing section where the ZP is computed. The ZP value appears digitally on the front of the instrument in large red figures.

The Zeta-Reader is a semiportable unit built in two sections. For on-line monitoring, the section housing the computer and data module is located in the control room while the pump and sample module section is located close to the source of the sample. For use in a laboratory, the two sections can be placed side by side on a bench.

The Zeta-Reader includes three major modules. The Pump Module houses the sampling pump, the flush water, pressure regulator, and a container for reference standard colloid which is used for calibration of the instrument. The Sample Module contains the scanner, sample cell, and solenoid valves to control the sequence of liquids through the sample cell. The Data Module contains the computer and electronics to process the video signal, to detect and locate the particles, and to communicate with the computer. Sequence timing circuits and power supplies are also located here. Selector switches are located on the outside of the Data Module for displaying sample temperature for manual, external, or automatic measurement initiation, and for sample cell voltage. A 4 to 20 mA analog signal proportional to ZP is available at the back of the Data Module.

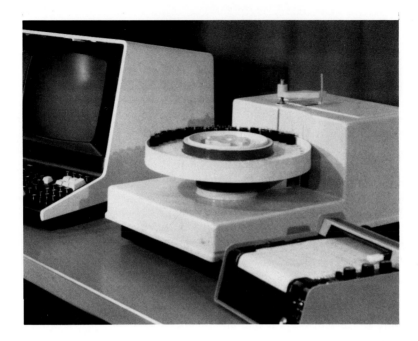

FIGURE 2.   The System 3000 Electrokinetics Analyzer consisting of a laboratory sampler tray, a sensor unit, and a computer terminal. (Courtesy Penkem, Inc., Croton-on-Hudson, N.Y.)

The Zeta-Reader diagnoses and reports malfunctions in its own electronic or electromechanical systems by lighting the green malfunction light and displaying a number from 1 through 17 indicating the type of malfunction which has occurred. It uses the Data General Nova 1210 minicomputer. Usually a malfunction can be corrected immediately thanks to the unique self-monitoring system of the Zeta-Reader.

D. Automatic Technique[6]

The System 3000 Automated Electrokinetics Analyzer is a sophisticated, state-of-the-art instrument for making fully automated measurements of the key electrokinetic parameters of colloidal systems, namely: specific conductance, pH, and electrophoretic mobility.[4] In addition, the System can optionally determine other important parameters such as: the mobility distribution function, turbidity, the mean settling or creaming velocity, the settling velocity distribution function, and particle size. All of these measurements are obtainable using a sample volume as small as 1 mℓ utilizing a wide variety of suspending media, from hydrocarbon solvents to physiological strength saline solutions. Data on particles as small as 100 Å and as large as 100 μm are easily obtainable.

A typical System is illustrated in Figure 2 and consists of a Laboratory Sampler Tray, a Sensor Unit, and a Computer Terminal.

The Laboratory Sampler provides means for storing up to 40 colloidal samples for subsequent analysis. (For process monitoring applications, the Laboratory Sampler is, of course, replaced by a suitable on-line sampler.)

The Sensor Unit loads the appropriate samples, makes the electrophoretic and other measurements, and outputs the data to the peripheral devices such as the Computer Terminal.

The Computer Terminal is a light-weight portable unit which can be located in an office or laboratory up to 30 m from the Sensor Unit. It is used for input of commands

E,25

SAMPLE	CONDUCTANCE	MOBILITY	ERROR
0001	1.503 E+00	−1.872 E−08	1.059 E−10
0002	1.503 E+00	−1.903 E−08	1.941 E−10
0003	1.504 E+00	−1.890 E−08	1.464 E−10
0004	1.505 E+00	−1.906 E−08	2.322 E−10
0005	1.504 E+00	−1.894 E−08	2.422 E−10
0006	1.507 E+00	−1.927 E−08	1.859 E−10
0007	1.507 E+00	−1.898 E−08	1.225 E−10
0008	1.509 E+00	−1.889 E−08	1.637 E−10
0009	1.512 E+00	−1.901 E−08	1.717 E−10
0010	1.513 E+00	−1.911 E−08	8.363 E−11
0011	1.514 E+00	−1.869 E−08	1.253 E−10
0012	1.515 E+00	−1.896 E−08	9.160 E−11
0013	1.515 E+00	−1.889 E−08	1.550 E−10
0014	1.521 E+00	−1.889 E−08	8.487 E−11
0015	1.520 E+00	−1.901 E−08	9.782 E−11
0016	1.523 E+00	−1.897 E−08	1.433 E−10
0017	1.522 E+00	−1.874 E−08	5.378 E−11
0018	1.523 E+00	−1.925 E−08	8.849 E−11
0019	1.525 E+00	−1.877 E−08	1.569 E−10
0020	1.525 E+00	−1.872 E−08	2.672 E−10
0021	1.527 E+00	−1.886 E−08	1.408 E−10
0022	1.533 E+00	−1.865 E−08	1.735 E−10
0023	1.533 E+00	−1.819 E−08	2.499 E−10
0024	1.535 E+00	−1.834 E−08	1.408 E−10
0025	1.538 E+00	−1.835 E−08	1.162 E−10

FIGURE 3. A typical System 3000 Electrokinetics Analyzer computer print out shows sample number or location, specific conductance, mean mobility, and an estimate of the standard error of the mean mobility. (Courtesy Penkem, Inc., Croton-on-Hudson, N.Y.)

to the System and for output of test data. To obtain laboratory data using the Laboratory Sampler, the operator initiates the test by simply typing a "Start" command. (For on-line process monitoring, the data-taking proceeds continuously without human intervention.)

Operation of the Electrokinetics Analyzer is completely controlled by a self-contained microcomputer which loads the samples, sets up the conditions for measurement, takes the data, tests it for accuracy, and formats the output for display on the Terminal. If the operator wishes to alter a particular parameter, it is done from the Terminal. For example, measurements are typically performed at a thermostatically controlled temperature of 25°C, but this can be modified by a simple "Change" command. In this way, the user has the best of two worlds. For routine measurements, the System specifies all of the operating conditions. On the other hand, the sophisticated user with unique requirements can alter the test conditions to suit his needs.

Figure 3 shows a typical data printout of 25 measurements of a given colloid, for purposes of checking reproducibility. The first column in Figure 3 contains the reference number assigned to each sample in the Laboratory Sampler. (For on-line monitoring from several points in the process, the number in the "sample" column designates the sampling location.)

The second column contains the specific conductance of the sample in mho/m. (If you are more accustomed to working with μmho/cm, multiply the value by 10^4.)

The third column contains the measured value of the mean mobility of the colloid

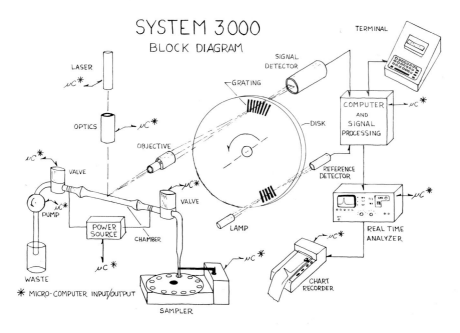

FIGURE 4.   System 3000 Block Diagram. (Courtesy Penkem, Inc., Croton-on-Hudson, N.Y.)

in units of m² sec⁻¹ volt⁻¹. (If you prefer to use cm² sec⁻¹ volt⁻¹, multiply by 10⁴; if you prefer μm/sec/v/cm, multiply by 10⁸; and if you want ZP in mV, multiply by 12.9 × 10⁸.)

The fourth column contains an estimate of the standard error of the mean mobility, also in units of m² sec⁻¹ volt⁻¹.

Those Systems equipped with a pH sensor, turbidity readout, or other options, would have additional columns for printout of these parameters.

Note the high degree of precision in the mobility readout of Figure 3. The computed standard error for the set of 25 measurements is approximately 1%. Note that the error predicted by the System based on just a single measurement is also approximately 1% of the mean mobility. Not only does the System give a mobility value, it predicts the measurement precision. The above tests were made on 0.5-μm diameter latex particles dispersed in physiological strength saline (0.15 *M* NaCl), typically a difficult sample to measure on conventional instruments because of the errors normally encountered at such high ionic strength.

Although the *mean* electrophoretic mobility is an extremely important parameter, in many applications it is also important to understand how the charge is *distributed about the mean value.* A mobility histogram capability is optionally available on the System 3000.

A simplified block diagram of the System 3000 is shown in Figure 4. The System is comprised of:

1.   A sampling device which delivers colloidal samples to the Sensor
2.   A 2 mW helium neon laser and associated optics for illuminating the colloidal particles
3.   An electrophoresis chamber which contains the colloid and provides means for applying an electric field across the fluid

4. A microscope assembly for projecting a magnified image of the laser-illuminated particles onto the surface of a grating
5. A rotating glass disk on which the grating has been engraved
6. A photomultiplier tube for converting the modulated light passing through the grating to an electrical signal
7. A frequency tracker for determining mean mobility
8. A computer and auxiliary electronics for controlling the overall operation of the System
9. A computer terminal for input and output
10. (Optionally) A spectrum analyzer and chart recorder for measuring and plotting mobility histograms

The method for determining electrophoretic mobility can be described as follows: The electrophoresis chamber is first rinsed and a small amount of colloid is automatically transferred from the sampler to the electrophoresis chamber. Valves on the chamber then close to seal off the contents from outside disturbances. The chamber consists of 1 mm inner diameter cylindrical quartz tube having two bubbles spaced approximately 20 mm apart which serve as electrode compartments. The chamber is mounted horizontally in a thermostatically controlled water bath. A vertically mounted 2 mW helium neon laser and associated optics provide a curtain of illumination which ensures that only particles at the stationary layer are illuminated. For test purposes the particles may be observed by the operator, using a built-in eyepiece. A beam splitter provides a second image which is projected onto the surface of the grating on the rotating glass disk.

In the absence of current at the electrodes, the electric field is zero and the particles are stationary. Since the image of the particles is focused on the grating, and since the size of each particle image is normally smaller than the width of each grating segment, the light from each particle will be alternately blocked and transmitted by the grating as the disk turns. Since the photomultiplier tube (PMT) responds to the light transmitted through the grating, the PMT output voltage will have a signal component at a frequency proportional to the speed of the disk. If a current is now applied between the two electrodes, an electric field will be established across the sample and the particles will move electrophoretically. The electrophoretic velocity of each particle produces a change in the tangential velocity of the image of that particle with respect to the grating. If the particle image is moving in the same direction as the grating, the image will cross fewer line pairs per unit time and hence the signal component for that particle will be at a slightly lower frequency. If the particle image moves in the opposite direction, the image will cross more line pairs per unit time and the signal will be at a higher frequency. Normally there will be many particles in the field of view at one time, each particle producing a signal component at a frequency shift related to its electrophoretic velocity. The mobility is determined by measuring the amount of frequency shift, with respect to a reference signal derived from a separate light source and photodetector which senses the motion of the grating alone.

The magnitude of the frequency shift due to electrophoresis is given by the following equation:

$$\Delta f = \left(\frac{M \times E}{L}\right) \times \mu \qquad (2)$$

where δf = the frequency shift due to electrophoresis, M = the optical magnification of the microscope assembly, E = the electric field across the fluid in the region of measurement, L = the periodicity of the grating, and μ = the electrophoretic mobility of the colloidal particles.

All of the parameters are accurately measured to provide an absolute calibration of the instrument. The mean mobility printed on the computer terminal is measured using a frequency tracker which determines the mean frequency shift due to the electrophoretic movement of the particles.

For optional measurement of mobility histograms, the System employs a Fast Fourier Transform spectrum analyzer which measures the frequency spectra of the PMT signal which in turn is a direct analog of the mobility distribution function.

Settling velocities are measured by rotating the image in the microscope by 90° using a dove prism. The vertical velocity caused by gravity then appears as a horizontal velocity tangential to the movement of the grating. Mean settling velocity and settling velocity histograms are then measured in the same way as previously described for the electrophoretic velocity.

To quantify the Brownian motion and thereby calculate mean particle size, the System 3000 measures the frequency spectra of the PMT output signal with the electric field across the sample set to zero. Figure 5 shows the relatively narrow spectra obtained with latex particles having a diameter of 0.5 μm, whereas Figure 6 shows the much larger bandwidth for AgCl dispersion having a mean diameter of only 0.1 μm.

The bandwidth of the spectra can be related to particle size in the following manner:

$$\Delta B = 4\pi D \left(\frac{M}{L}\right)^2 \tag{3}$$

where δB = bandwith of spectra at one half power points, D = diffusion coefficient of the particles, M = optical magnification of the microscope assembly, L = periodicity of the grating.

For aqueous systems:

$$D = \frac{2.15 \times 10^{-13}}{a} \text{ cm}^2/\text{sec} \tag{4}$$

where a = radius of the particle.

Thus the particule radius can be related to the spectrum bandwidth by the simple equation:

$$a = \frac{K}{\Delta B} \tag{5}$$

where K is an instrument constant equal to

$$4\pi \times 2.15 \times 10^{-13} \left(\frac{M}{L}\right)^2$$

All aspects of the system operation are controlled by the System 3000 microcomputer. The 14,000 bytes of program memory contain all instructions necessary to load the samples, set up the measurement conditions, accumulate the data and test it for errors, and then output the results. The following description of a portion of the software package is given to convey an appreciation of its level of sophistication. For example, after the sample is loaded, the next step is to measure specific conductance. In doing this, the software corrects the data for polarization effects of the electrodes,

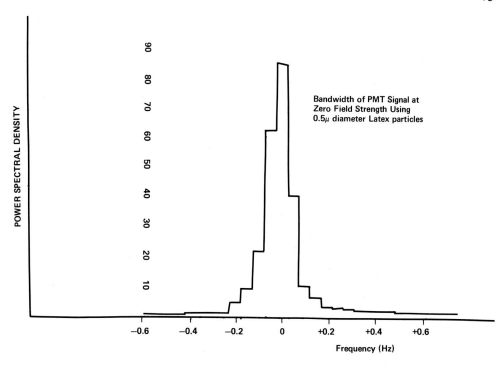

FIGURE 5.   Band width of photo multiplier tube signal at zero field strength using 0.5μ diameter latex particles.

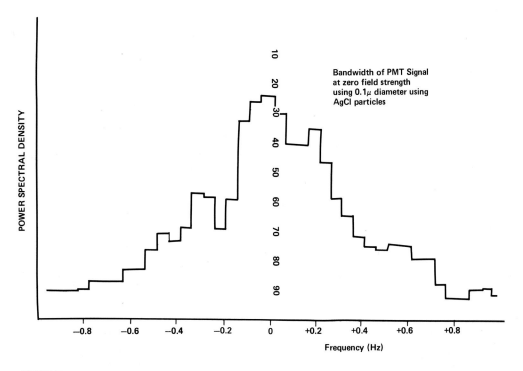

FIGURE 6.   Band width of photo multiplier tube signal at zero field strength using 0.1μ diameter AgCl particles.

joule heating of the chamber contents, and small temperature deviations in the thermostatically controlled temperature bath. After the specific conductance has been accurately measured, the software calculates an electrode current which will provide the correct field strength at the measurement portion of the chamber. During the mobility determination, the software reverses the polarity of the applied field at an optimum rate so as to avoid artifacts due to electrode reactions, and constantly tests the data for evidence of drift due to leaks in the valves or gassing due to electrode degradation. To a first order approximation, these errors are not only detected but also calibrated out. If the error sources exceed preset tolerances, an error message is printed alerting the operator to the exact nature of the system degradation. As the mobility data is collected, a standard error is also calculated and printed out so that the user has a measure of the validity of the data.

III. THE PARAMETERS WHICH CONTROL ELECTROKINETIC DESTABILIZATION

A. Introduction

Particles in nature are usually negatively charged. The reason lies partly in evolutionary anthropology and is probably still quite conjectural. In any event, all naturally occurring organic cells are negative as are most inorganic compounds. A principal exception is asbestos, which is positively charged. Both the polarity (i.e., whether positive or negative) and the extent of charge on a given particle depend on three factors:

1.   The nature and number of functional groups on the surface of the particle
2.   The type and amount of dissolved ions
3.   Any reaction between (1) and (2) above

For example, $-COO^-$ (carboxyl) groups on the surface of a particle would tend to make it negative; whereas $-NH_3^+$ (amino) groups would make it positive. The degree of ionization of these functional groups and, therefore, the net particle charge, depends on the solution pH, as will be shown later.

Dissolved ions which confer a positive charge are typically trivalent cations such as Al^{+++} or Fe^{+++}. They are roughly ten times more effective than the bivalent cations Ca^{++} or Mg^{++} and are preferentially used for cost efficiency. On the other hand, PO_4^{---} confers a high negative charge which accounts for its successful soil dispersing effectiveness as a detergent ingredient.

If NaOH were added to particulates containing $-COO^-$ groups, a sodium carboxylate soap would be formed which has a very high negative charge, perhaps exceeding $-100 \, mV$.

Figure 1 schematically represents a negatively charged particle. It is surrounded by a layer of positive ions called the Stern layer.[7] The second of the electric layers is the one measured via electrical transport in a microelectrophoresis instrument (as described earlier) in making a ZP measurement. It was first described in 1879 by Helmholtz[8] and subsequently Guoy and Chapman[9] described a diffuse electric double layer model that permitted more quantitative treatment of electrokinetic data.

B. pH Dependency

The ZP of a colloid varies as a function of the pH, as shown in Figure 7 for cellulose. The point of zero ZP is called the isoelectric, at about pH 2.5 for cellulose. The isoelectric points of bacteria are at very low pH values, (pH 2 to 4) and are significantly lower than those of proteins (pH 5 to 8). In adding a cationic chemical to a colloid in

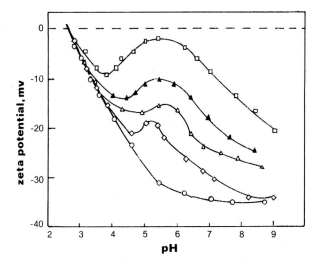

FIGURE 7. The effect of alum concentration on the zeta potential vs. pH curves of cellulose fibers. O, no alum, ◇, 1.006 × 10⁻⁶ mol alum, △, 5.674 × 10⁻⁶ mol alum, ▲, 9.964 × 10⁻⁶ mol alum, and □, 1.011 × 10⁻⁵ alum. (Courtesy Jaycock, M. J., Nazir, B. A., and Pearson, J. J., 2nd Int. Seminar on Paper Mill Chemistry, New York, September 10 to 13, 1978. With Permission.)

order to coagulate it, one always first adjusts the pH to the optimum value for the system to be coagulated. Figures 7, 8, and 9 illustrate the importance of pH control in the use of Al^{+++} and Fe^{+++} salts. Jaycock et al.[10] shows in Figure 7 that a pH for Al^{+++} just under 6.0 is best in working with cellulose fibers. Moffett[11] demonstrates in Figure 8 that a pH for Al^{+++} of 6.4 is optimum in the chemical pretreatment of Darby River Water. Tenney and Stumm[12] show in Figure 9 that a pH for Al^{+++} in the low end of the 5.0 to 6.0 range is optimum for flocculating microorganisms in sludge conditioning, but that a pH of 5.0 is needed for Fe^{+++}. The efficiency of organic cationic chemicals is also highly pH dependent. As is shown in Table 2[13], the effectiveness of an assortment of polyelectrolytes in treating refinery waste effluent is shown to differ substantially depending on the pH at which they are evaluated, in this case at pH 8.0 and 10.0.

For these reasons, the first step in maximizing the coagulation chemistry of a given system is to determine the pH: ZP relationship over a wide range, e.g., pH 2 to 12, and to plot the data, thereby defining the pH isoelectric. Subsequent evaluation of cationic chemicals should be made at a pH as close to the isoelectric as is compatible with other constraints, because this is the cheapest and the surest way of operating close to zero ZP and thereby maximizing the coagulation chemistry of the system.

C. Zeta Potential

ZP represents the electrokinetic charge which exists at the solid-liquid interface of particles in suspension. The ZP can be either positive or negative, and its magnitude is expressed in millivolts. The higher the charge on these particles, the greater their mutual repulsion. Generally, at a ZP of ± 10 mV or higher, the mutual repulsion becomes great enough to inhibit coalescence of the particle (even though Brownian movement causes many interparticle collisions). If two particles are prevented from "sticking" after a collision because the ZP is too great, the effective particle radius does not increase. Recalling Stokes' Law, previously described, it is obvious that a

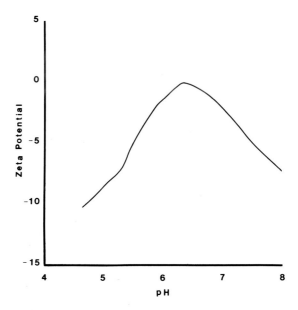

FIGURE 8.   ZP as a function of pH for cellulose. (Courtesy Jaycock, M. J., Nazir, B. A., and Pearson, J. L., 2nd Int. Seminar on Paper Mill Chemistry, New York, September 10 to 13, 1978. With permission.)

high ZP will cause very poor clarifier or dissolved air flotation (DAF) efficiency. Conversely, reducing the ZP permits the everpresent Van der Waal's forces to cause the particles to agglomerate, maximizing the settling/rising rate and clarifier/DAF efficiency (assuming steric repulsions are absent). Thus minimum ZP is generally assumed to give maximum clarifier/DAF efficiency.

Clarifier and DAF efficiency are used as examples to illustrate the importance of reducing the ZP to a minimum. Under ± 10 mV all coagulation chemistry processes are electrokinetic dependent. Such processes include clarification, sedimentation, DAF, sludge dewatering, wet end paper mill chemistry, and the chemical pretreatment of potable water.

Table 3[13] shows how cationic polyelectrolytes are evaluated for effectiveness in reducing the ZP of a waste colloid. Typically, a ZP in the range of −5 mV to 0 mV is desirable in order to destabilize the colloid. It is sometimes useful to go as far as + 3 mV, especially when the "dual polymer" process is used, i.e., the addition of a low-molecular-weight, high charge density cationic polyelectrolyte is followed by a high-molecular-weight, low charge density anionic polyelectrolyte.

D. Specific Conductance

The third and last of the primary electrokinetic parameters is that of specific conductance, which is a measure of the soluble salts in the system, both organic and inorganic, expressed in μmho/cm. Soluble salts which are balanced in valency such as NaCl (a 1:1 salt) and $CaSO_4$ (a 2:2 salt) function principally to alter the ZP by compressing the double electrical layer around the particle, thereby ultimately reducing the charge towards zero. They impair the effectiveness of the more electrokinetically active 2:1 and 3:1 salts, as well as polyelectrolytes, because they are also adsorbed on the available colloidal surface in direct proportion to their concentration.

This effect is illustrated in Figure 10.[13] The colloid is silicon dioxide of 5 μm size

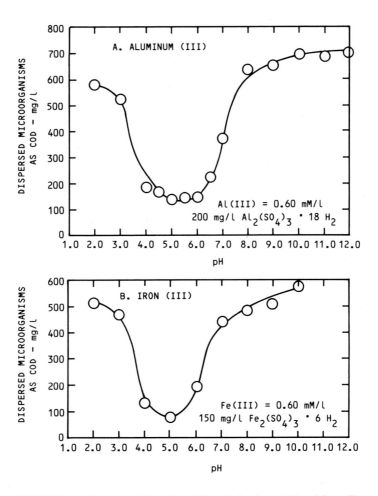

FIGURE 9. pH greatly affects the ability of aluminum (A) and iron (B) salts to coagulate dispersed microorganisms found in river water. (Courtesy Jaycock, M. J., Nazir, B. A., and Pearson, J. L., 2nd Int. Seminar on Paper Mill Chemistry, New York, September 10 to 13, 1978. With permission.)

dispersed in deionized water at a concentration of 100 mg/ℓ. Curve A depicts the electrokinetic charge as the result of increasing addition of NaCl and shows an asymptotic approach to zero ZP. Curve B represents the same silicon dioxide dispersion to which 2.5 ppm of alum (Al$_2$[SO$_4$]$_3 \cdot$ 18H$_2$O) has been added, prior to the addition of the same increasing quantities of NaCl. The addition of alum results in a change in polarity, from -30 mV to $+30$ mV. In both cases, a maximum charge occurs in the range of 2000 to 3000 μmho specific conductance. This probably coincides with the development of a mono-molecular layer of adsorbed ions on the surface of the colloidal particles, hence maximizing the charge.

The significance of this sort of relationship is that the competing neutral ion concentration should be minimized, if other constraints permit, to that amount which maximizes the ZP. High neutral salt concentrations depress the ZP, and reduce the scale factor against which the significance of the absolute zeta potential value should be estimated. After a threshold value is reached (a monomolecular layer is adsorbed) the neutral salts compete with and tend to decrease the efficiency of ionically active chemicals such as 2:1 and 3:1 salts and polyelectrolytes.

Table 2
THE COMPARATIVE EFFECTIVENESS OF CATIONIC POLYELECTROLYTES FOR
CHARGE NEUTRALIZATION OF SUSPENDED MATTER IN API SEPARATOR EFFLUENTS
AS DETERMINED BY ZETA POTENTIAL TITRATION CURVES

Conditions	Polyelectrolyte[a,b]	mg/l to achieve indicated ZP endpoint				Rank at indicated ZP endpoint			
		-5	-3	0	+3	-5	-3	0	+3
pH = 8; specific conductance = 560, and zero sulfides	C-31	.25	.5	.5	1	1	1	1	1
	581	.5	1	1.25	3.75	2	2	2	5
	431	1.5	1.75	2.5	3	6	4	3	2
	7132	1.25	1.75	2.5	3.25	5	4	3	3
	1180	1.5	2	2.5	3	6	5	3	2
	1190	1	1.5	3	6	4	3	4	8
	2860	2.5	2.75	3	3.25	8	6	4	4
	2870	2.25	2.75	3.25	3.75	7	6	5	5
	863	.75	1	3.25	6.75	3	2	5	9
	7134	2.5	3	3.75	4.5	8	7	6	6
	2640	2.25	3	4	4.5	7	7	6	6
	751	2.25	3	4	5	7	7	7	7
	FA	2.5	3.25	4	5	8	8	7	7
	864	5.25	7	10	c	9	9	8	10
	860	6.75	7.25	c	c	10	10	9	10
pH = 9.8; specific conductance = 680; and zero sulfides	1180	1	1.25	1.5	2	1	1	1	1
	1190	1	1.25	1.5	2	1	1	1	1
	581	1	1.5	2	2.75	1	3	2	2
	7132	1	1.25	2	3.5	1	2	2	3
	2870	1.75	2.25	3	3.5	2	3	3	3
	2860	2.25	2.5	3	3.5	3	4	3	3
	863	1.75	2.25	3.5	4.75	2	3	4	4
	431	3	3.5	4+	4.75	5	5	5	4
	751	2.5	3.5	4.5	5.25	4	5	6	5
	C-31	4.25	4.5	4.75	5.25	7	6	7	5

pH = 10; specific conductance = 4,100; and zero sulfides

[a]								
2640	5	5.5	6	6.5	8	8	8	6
7134	4	4.75	6	7.25	6	7	8	7
FA	3	4.5	6.5	c	5	6	9	8
2870	2	2.75	4	5.25	1	1	1	2
431	3	3.75	4.5	5	3	3	3	1
751	2.5	3.5	5	6.75	2	2	2	3
581	4	4.75	5.75	6.75	4	4	4	3
2860	5	6	6.5	7.5	5	5	5	4
7132	6.5	7.25	7.25	9	5	6	6	5
2640	7	8.5	10	c	6	6	7	6
863			10	c	7	7	7	6
C-31	4	c	c	c	4	8	8	6

[a]   C-31 (Dow); 581 (Cyanamid); 431 (Dearborn); 7132 (Nalco); 1180, 1190 (Betz); 2860, 2870, 2640 (Calgon); 860, 863, 864 (Hercules); 751 (Mazer), and FA (BASF Wyandotte).

[b]   Arranged in order of performance using zero ZP as endpoint.

[c]   Indicated ZP not achieved.

From Grutsch, J. F. and Mallatt, R. C., *Hydrocarbon Process.*, 55(6), 115, 1976. With permission.

Table 3
THE INFLUENCE OF pH ON THE CATIONIC DEMAND OF POLYELECTROLYTES

	Polyelectrolyte	mg/l of polyelectrolyte for zero zeta potential at		mg/l increase or decrease (−)
		pH = 8	pH = 10	pH = 10
Sensitivity to pH. Specific conductance = 570 to 770	C-31	1	5	4
	581	1.75	3	1.25
	7132	2.75	2.25	−.5
	431	3	5	2
	2860	3	3	0
	2870	3	3	0
	1180	3	1.25	−1.75
	1190	3.25	1.25	−2
	7134	4.5	5.5	1
	751	4.5	5	.5
	2640	4.5	6.5	2
	FA	4.5	10	5.5
	863	5	3.75	−1.25

From Grutsch, J. F. and Mallatt, R. C., *Hydrocarbon Process.*, 55(6), 115, 1976. With permission.

Table 4[13] shows that the effectiveness of polyelectrolytes in treating refinery waste effluent is dependent on the NaCl concentration. All evaluations of the efficiency of ionically active chemicals should be made at a specific conductance close to that of the actual application. When other constraints permit, the neutral salt concentration level should be selected so as to maximize the efficiency of the ionically active chemicals.

E. Other Factors
1. Temperature
ZP varies by about 2% per degree C at ambient temperatures. pH and specific conductance are also temperature dependent. Surface active agents including polyelectrolytes can vary in solubility, and therefore in activity, as a function of temperature. It is important, therefore, that electrokinetic data be taken at a constant temperature or corrected to a standard temperature, such as 25°C.

2. Agitation
Flocculated or coagulated colloids tend to redisperse under agitation. The greater the amount of agitation or shear, the larger will be the tendency to redisperse. Indirect coagulation chemistry parameters such as subsidence rate, sludge volume index, and turbidity should always be measured under experimental conditions in which the degree of agitation is known and controlled. Sensitivity to redispersion under shear can be decreased by use of the "dual polymer" polyelectrolyte addition technique, which will be discussed later.

IV. OPTIMIZING THE ELECTROKINETICS OF A WASTEWATER TREATMENT PROCESS

All waste effluents have one thing in common — they are heterogeneous messes, impossible to fully characterize because they are continuously changing in solids content, fluid flow, and nature of the colloid to be treated. Therefore the definitive dis-

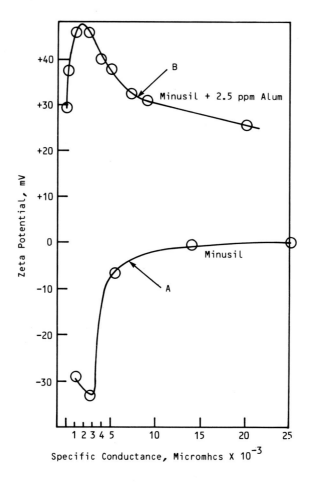

FIGURE 10. The ZP on silicon dioxide (Curve A) and silicon dioxide and alum (Curve B) varies considerably with specific conductance. (From Grutsch, J. F. and Mallatt, R. C., *Hydrocarbon Process.*, 55(6), 115, 1976. With permission.)

cussion of maximizing the coagulation chemistry of a waste must be approached with timorous trepidation. The U.S. EPA has chosen the petroleum industry as the pathfinding domestic industry to pioneer, develop, and optimize the best practical and best available technology economically achievable (BATEA) for "end-of-the-pipe" waste effluent treatment. Since the coagulation chemistry principles are identical regardless of the source of the waste and because they have been put into most effective practice in the petroleum industry, it naturally follows that this experience can be instructive to other industries.

Effluent from the activated sludge process (ASP) in a fresh water aerated lagoon contains bio-colloid which has an isoelectric point in the pH range of 1 to 2, yielding a rather characteristic curve, as shown in Figure 11. Following is a brief discussion of the steps which must be taken in order to maximize coagulation chemistry.

A. Necessary Determinations

Determine the isoelectric — This is accomplished by taking a large sample, adjusting the temperature to 25°C and measuring the pH, ZP, and specific conductance. An appropriate acid or alkali is then added to aliquots of the sample, and the measurement

Table 4
THE INFLUENCE OF SPECIFIC CONDUCTANCE ON THE CATIONIC DEMAND OF POLYELECTROLYTES

		mg/ℓ of polyelectrolyte required to reach zero zeta potential at indicated specific conductance		
Conditions	Polyelectrolyte	680	4,100	11,000
Sensitivity to salinity (salt) at pH = 10	1180	1.25	^{a}	6
	7132	2.25	7	3.75
	581	3	5.75	4
	2860	3	7	3.25
	2870	3	5	2.75
	863	3.75	^{a}	^{a}
	431	5	4.25	3
	751	5	6.25	4
	C-31	5	^{a}	^{a}
	2640	6.5	^{a}	3.75

^{a} Data missing.

Grutsch, J. F. and Mallatt, R. C., *Hydrocarbon Process.*, 55(6), 115, 1976. With permission.

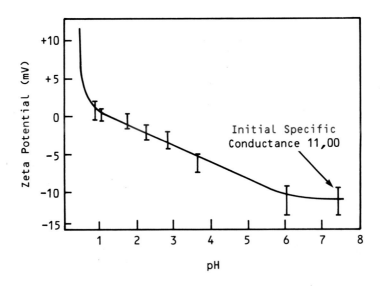

FIGURE 11.   Charge on suspended matter in aerated lagoon effluent as a function of pH.

repeated until the isoelectric is reached and several data points are obtained on the other side. It is sometimes necessary to add increasing amounts of acid to one series of aliquots and of base to another series in order to fully define the pH/ZP relationship.

Determine the cationic demand of the process and its variability — This is accomplished partly by periodic determination of the ZP distribution of the colloid, as suggested by Figure 12 over a period of hours and days and weeks. By titration with

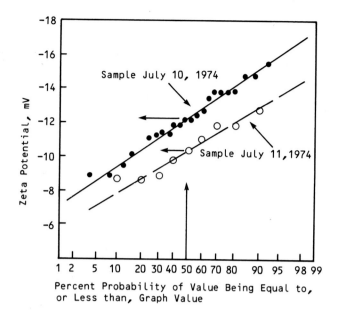

FIGURE 12. Optimize the effluent system — ZP probability distribution for aerated lagoon effluent samples.

cationic chemicals, as shown in Figure 13 one simultaneously determines the cationic demand of the system and its variability with time.

Determine the most effective combination of chemicals to treat the effluent — This is accomplished by the addition of ladder series of chemicals and combinations of chemicals and the assessment of their effect.

The next section represents an illustration of the use of these principles, and provides a format for actually maximizing the coagulation chemistry in a clarifier.

B. How To Plan A Coagulation Chemistry Experiment

It is the purpose of this section to present an experimental format which can be used in optimizing coagulation chemistry. While the example selected is that of sedimentation, similar principles would apply to save-all DAF operation, clarification and sludge dewatering. Three separate procedures are followed, as described below:[14]

Procedure 1 — Vary the polymer dosage to find the most cost efficient amount
  A. Preparation of 1% activated polymers
    1. To make up a 1.85% activator solution:
      a. Add 1.0 mℓ of activator to 53 mℓ of tap water which has been heated to 50 to 70°F.
      b. Mix well with a glass stirring rod.
    2. Add 10 mℓ of 1.85% activator solution to 1000 mℓ of tap water.
    3. Add 10 mℓ of polymer.
    4. Mix for 10 minutes at 800 to 1500 rpm. (You now have a 1% activated polymer solution.)
    5. To obtain a 0.1% activated polymer solution, add 20 mℓ of 1% solution to 180 mℓ of tap water. Mix well.
  B. Test procedure
    1. Add a selected dosage of polymer to 750 mℓ of clarifier influent.
    2. Rapid mix for 1 min, then slow mix for 3 min with a propellor mixer.
    3. Pour into an Imhof cone.
    4. Record:
      a. The time in minutes for 10 mℓ of sludge to settle.
      b. The number of mℓ of settled sludge after 10 min.

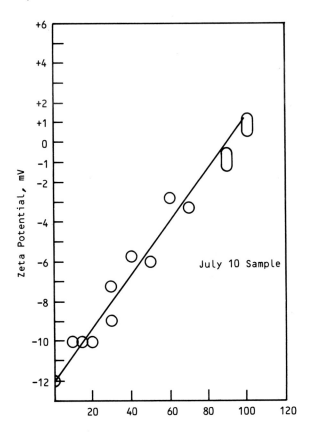

FIGURE 13.   Titration of a colloid surface charge by a polyelectrolyte (cationic).

5.   After 10 min remove a 100 mℓ aliquot, measure, and record:
    a.   ZP
    b.   Specific conductance
    c.   pH
C.   Conclusion
    Refer to Figure 14. Approximately 8 mg/ℓ of polymer is needed to settle the most sediment in the least time.
Procedure 2 — How to Determine the Optimum pH: using the 8 mg/ℓ of polymer determined to be most appropriate, vary the pH
    A.   Prepare activated polymer solution as previously.
    B.   Test procedure
        1.   Titrate 1000 mℓ of clarifier influent with dilute (0.01 to 0.1 N) sulfuric acid (H_2SO_4) or caustic soda (NaOH) to the desired pH.
        2.   Measure out 1000 mℓ of each pH-adjusted sample of clarifier influent.
        3.   Add 8 mg/ℓ of polymer.
        4.   Rapid mix for 1 min, then slow mix for 3 min with a propellor mixer.
        5.   Pour into an Imhof cone.
        6.   Record:
            a.   The time in minutes, for 10 mℓ of sludge to settle
            b.   The number of mℓ of settled sludge after 10 min
        7.   After 10 min, remove a 100 mℓ aliquot, measure and record:
            a.   ZP
            b.   Specific conductance
            c.   pH
C.   Conclusion
    See Figure 15. A pH of 6.0 is needed to settle the most sediment in the least time.

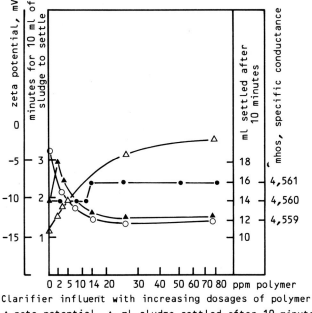

FIGURE 14.   Titration of clarifier influent with a polyelectrolyte.

Procedure 3 — How to determine optimum polymer dosage

Earlier work (beyond the scope of this chapter) showed that 300 mg/ℓ of sodium aluminated (NaAlO$_2$) and 350 mg/ℓ of alum (Al$_2$(SO$_4$)$_3$ · 18 H$_2$O) resulted in a system close to zero ZP at a pH of 6.0. The sodium aluminate:alum system was used in a series of experiments to screen polymers for efficiency by determining the optimum amount to use.

    A.   Prepare the activated polymer solution as previously.

    B.   Test procedure

          1.   Titrate 1000 mℓ of clarifier influent to a pH of 6.0 by first adding a sufficient amount of 1 % sodium aluminate to bring the pH to 8.0; then adding a 1% alum solution to obtain a pH of 6.0.

          2.   Measure out 1000 mℓ of the pH-adjusted sample.

          3.   Add the selected dosage of polymer.

          4.   Rapid mix for 1 min, then slow mix for 3 min with a propeller mixer.

          5.   Pour into an Imhoff cone.

          6.   Record:

              a.   The time, in minutes, for 10 mℓ of sludge to settle.

              b.   The number of mℓ of settled sludge after 10 min.

          7.   After 10 min remove a 100 mℓ of aliquot, measure and record:

              a.   ZP

              b.   Specific conductance

              c.   pH

    C.   Conclusion

        See Figure 16. The polymer tested was inappropriate to the sodium aluminate/alum system. The fastest settling rate and the largest number of mℓ settled in 10 min were both obtained at zero polymer dosage. The use of this particular polymer would be counterproductive.

The use of sodium aluminate and alum represents an application of the time-honored principle that charge reduction is accomplished much more cost-effectively by inorganic chemicals, as they are inherently much cheaper than organics (alum can be about 6¢/lb and sodium aluminate 11¢/lb, whereas polymers can cost $/lb or more).

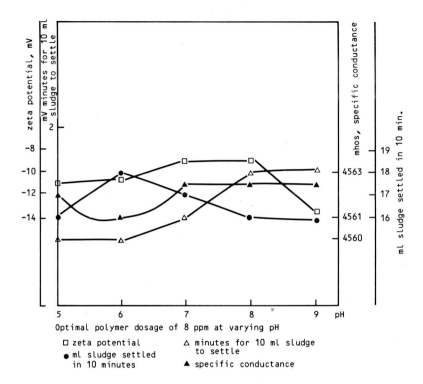

FIGURE 15.   Plotting solids settling rates and settled volumes (at optimum polymer dosage of 8 mg/ℓ) as a function of pH, determines the optimum pH for settling.

In measuring sedimentation, it is important to keep track of the time as the settling *rate* is critical. (One quick and easy way to make a settling rate measurement is to stick masking tape on the graduated cylinders and mark the sedimentation level every 10 min.) It is often useful to observe and record the properties of the supernate (e.g., turbidity, dissolved solids, BOD_5) and of the sediment (per cent suspended solids).

Finally, the two-step process should not be overlooked, in which the charge is first brought close to zero with cationic chemicals and then a high-molecular-weight anionic polymer is added. It has the effect of "floccing the flocs" and can create massive flocculation, thereby greatly enhancing the separation process.

It is often desirable to adjust the pH as close to the isoelectric as other constraints will allow as this can be a cheap and sure way of reducing the zeta potential towards zero. The effluent is next titrated with an assortment of commercially available cationic polyelectrolytes in order to determine which one is most cost effective in reducing the ZP towards zero, as shown in Figure 17. The efficiency of cationic polyelectrolytes often benefits synergistically from the prior addition of 3 to 20 mg/ℓ of alum. Each biocolloid system seems to have a threshhold alum pretreatment requirement which can easily be determined. Figure 18 illustrates that pretreatment with 20 ppm of alum reduces the amount of cationic polyelectrolyte required to reach zero charge from 10 mg/ℓ to 4 mg/ℓ.

Up to this point, we have adjusted the pH if appropriate, pretreated the biocolloid with alum, and added sufficient cationic polyelectrolyte to reach the range of −5 mV to 0 mV ZP. The next step is to add a high-molecular-weight anionic polyelectrolyte which functions to agglomerate the destabilized aggregates. This constitutes the dual polymer process and has the additional effect of substantially increasing the resistance

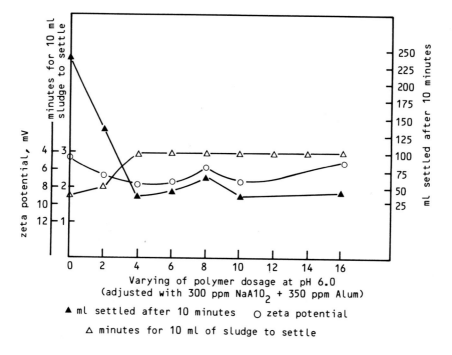

FIGURE 16.   Plotting the same settling data as in Figure 15, but against polymer requirement, shows that a cationic polymer was harmful rather than helpful when added to the same
type waste. In this test, alum and sodium aluminate were first added to the waste, and the pH
adjusted to 6.0

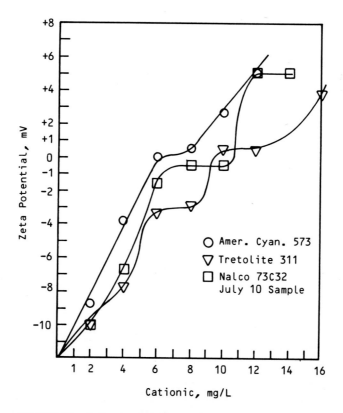

FIGURE 17.   A titration is used to evaluate various commercial cationic polymers for ZP neutralization efficiency.

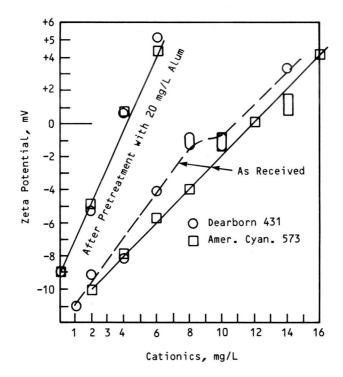

FIGURE 18.   Titration curves showing the polymer savings obtained when the suspended solids were pretreated with 20 mg/ℓ of alum.

Table 5
WASTEWATER

Test[a]	Polymer dosage	mg/ℓ (mℓ 0.1% per 500 mℓ)	ZP(mV)	SC (μmho)	BOD (mg/ℓ)
1	Blank	—	−3	2656	2450
2	6PD-855	100	−3	2660	2290
3	6PD-855	300	−2	2660	1900
4	6PD-855	500	0	2661	1800
5	6PD-855/625	100/33	−3	2660	2030
6	6PD-855/625	100/67	−5	2660	2170
7	6PD-855/625	300/100	−3	2500	1710
8	6PD-855/625	300/200	−3	2400	1670
9	6PD-855/625	500/167	−3	2400	1450
10	6PD-855/625	500/333	−2	2200	1250
11	609	200	−3	2700	2550
12	609	400	−4	2600	2310
13	609	800	−4	2500	2170

Note: The polyelectrolytes 6PD-855, 625, and 609 are manufactured by Nalco.

[a]   Test conditions: cooled to 24°C, pH 5.8.

of the aggregates to deflocculation under shear. 0.02 to 0.05 mg/ℓ of anionic polyelectrolyte is required if the next step is filtration, and 0.2 to 0.8 mg/ℓ is required for DAF because of the greater shear/deflocculation forces.

It has been reported in literature dealing with the chemical pretreatment of potable water[15,16] that anionic organic solubles can be effectively removed by maximizing the coagulation chemistry under careful ZP and pH control. The tabulation of specific conductance measurements in Table 5 indicates that this thesis is equally applicable in waste effluent treatment. There appears to be a direct correlation between specific conductance and increasing quantity of the "dual polymer" addition level. Test 10, for example, at 2200 μmho is 17% lower than the control, or blank, at 2656 μmho despite the addition of ionizing chemicals. Quite clearly a substantial precipitation of soluble anions has resulted. This is clearly confirmed by the BOD$_5$ data, which shows in Test 10 a reduction of almost 50% in biodegradable dissolved organics.

The presence of inorganic salts in the waste quite likely accounts for the specific conductance being reduced by a lesser percentage than the BOD$_5$ by the "dual polymer" process. An effective control strategy concludes with a fine tuning of the process by adjusting chemical feed rates to minimize specific conductance.

REFERENCES

1. **Penniman, J. G.**, *Pulp Pap.*, 1977.
2. **Rebhan, M., Sperber, H., and Saliternik, C.**, *Tappi*, 50(12), 62A, 1967.
3. **Zeta-Meter, Inc.**, New York, N.Y.
4. **PenKem, Inc.**, Croton-on-Hudson, N.Y.
5. **Komline-Sanderson Engineering Corp.**, Peapack, N.J.
6. **Goetz, P. J.**, An Automated Zeta Potential Instrument for Optimizing the Electrokinetics of Papermaking, 2nd Int. Seminar Paper Mill Chemistry, New York, September 10 to 13, 1978.
7. **Stern, O. Z.**, *Elektrochem. Z.*, 30, 508, 1924.
8. **Helmholtz, H.**, *Ann. Phys. (Wiedemann)*, 7:337, 1879.
9. **Guoy, G. and Chapman, D. L.**, *J. Phys. Theor. Appl.*, 9, 457, 1910; *Phil. Mag.*, 25, 475, 1913.
10. **Jaycock, M. J., Nazir, B. A., and Pearson, J. L.**, The Papermaking Furnish as A Colloid System, 2nd Int. Seminar on Paper Mill Chemistry, New York, September 10 to 13, 1978.
11. **Moffett, J. W.**, *J. Am. Water Works Assoc.*, 60, 11, 1968.
12. **Tenney, M. W. and Stumm, W.**, Chemical flocculation of microorganisms in biological waste treatment, *J. Water Pollut. Control Fed.*, 37, 1965.
13. **Grutsch, J. F. and Mallatt, R. C.**, Optimize the effluent system, *Hydrocarbon Process.*, 55(6), 115, 1976.
14. **Penniman, J. G.**, Three experimental procedures for optimizing coagulation chemistry, *Pap. Trade J.*, 163(1), 30, 1979.
15. **Black, A. P. and Christman, R. F.**, *J. Am. Water Works Assoc.*, 55, 753, 1963.
16. **Haff, James D.**, Removal of Humic Acid Using Alum and Synthetic Polyelectrolytes, abstract prepared for Universities Forum Am. Water Works Assoc. Annu. Meeting, June, 1976.
17. **Penniman, J. G.**, proposal to Boxboard Research and Development Assoc., March 22, 1978.

Chapter 4

COAGULATION FOR GRAVITY TYPE CLARIFICATION AND THICKENING

Richard M. Schlauch

TABLE OF CONTENTS

I. INTRODUCTION

When a liquid contains a suspension of particles that are too small to settle out, even when subjected to the relentless pull of gravity over long periods of time, it is said to be a stabilized colloidal suspension. The liquid may be natural surface water and the stabilized suspension minute clay particles no more than a few micrometers or less in diameter. In this case the colloid is a solid suspended in a liquid. We often refer to the solid clay suspension as turbidity because it tends to scatter light that passes through the liquid obscuring vision and making it appear turbid.

Wastewaters too, often contain large numbers of suspended colloidal particles making them appear turbid. These colloids can be a suspension of either solids, liquids, or gases, or a combination of these physical states of matter dispersed in water. An emulsion of oil in water is an example of a liquid colloid stabilized in liquid. More common are the many forms of solid suspensions that exist in waste waters. Examples include metal precipitates, bioorganisms, silicates, organic and inorganic fibers, proteins, polymers, pigments, ash, and many others. Existence of gaseous colloidal suspensions in waste water are far less common than liquid or solid suspensions. Such suspensions are usually not very stable. The gaseous particles (bubbles) either dissolve in the water or they slowly rise out of the water because their specific gravity is too low and buoyancy overcomes the forces that cause a colloid to become stabilized.

Some colloids are hydrophilic, meaning that their stability results from chemical interactions between hydroxyl groups adsorbed onto the particles and water. These colloids are reported to be so thermodynamically stable that from 10^{15} to 10^{18} particles per centimeter^3 can exist in a true suspension.[1] Examples of hydrophilic colloids are organic macromolecules such as starches, gums, resins, proteins, and polyelectrolytes.

Other colloids do not interact chemically with water and are not thermodynamically stable. However, they can remain in suspension indefinitely due to electrostatic interactions betwen the particle surface and water. Such colloids are termed hydrophobic (also called suspensoids). They are usually inorganic elements or compounds such as finely divided metals, metal hydroxides, and metal sulfides. The stabilizing electrostatic surface charges can be weakened or strengthened by changes in the ionic strength of the surrounding liquid. The stability of these colloids is also influenced by the concentration of colloidal particles themselves and an upper limit of 10^8 particles per centimeter^3 has been reported for colloids of this type.[1]

In natural water and in waste water systems it is uncommon to find pure colloidal suspensions of either type. It is common to have colloids that possess both hydrophobic and hydrophilic characteristics. Clays are finely divided metal silicates that adsorb hydroxyl ions and become somewhat hydrophilic. Metal hydroxides or oxides can adsorb organic molecules and become partially hydrophilic. It is not surprising, therefore, that many of the commonly found colloidal suspensions have their stability dependent to a great degree on the pH of the surrounding liquid. This is, of course, considering that pH influences ionic strength and is a measure of hydroxyl ion concentration of the suspending medium.

The more stable suspensions normally found in water and wastewater occur at particle diameters that are less than $1/100$ μm up to about 1 μm. On the other hand, unstable suspensions can be found with particle diamters up to 100 μm if the specific gravity of the particle is low and hydraulic turbulence exists.

It is the objective (function) of the application of coagulants to reduce or interfere with the existing chemical and/or electrostatic surface interactions between the suspended particles. The destabilization of particle surface charges by chemical addition is often required before effective coagulation of suspended particles can take place.

Once destabilized, there is a significantly greater tendency for the individual particles to collide through the work of hydraulic gradients. The collisions result in the gathering of particles into increasingly larger groups because strong, although short range, chemical attractions between particles become dominant. The larger the masses become, the less they are likely to be jostled about by the rapid and random motion of the surrounding liquid molecules and the more they tend to respond to the pull of gravity. Coagulation is, as La Mer put it,[11] "a kinetic process involving the transition of a quasi-stable phase to two more stable phases."[1]

Certain polyelectrolytes exist that neutralize the electrostatic surface charge of colloidal particles and by doing so, make them coagulate more easily. Also, there are other types of polyelectrolytes that adsorb colloidal particles regardless of surface charge and bind them into larger groups that settle more rapidly. Both types are useful in settling and gravity thickening operations. The selection of the type to use depends on the characteristics of the water or wastewater suspensions, the hydraulics of the treatment systems, and the final use or disposal of the separated liquid and solid phases.

II. POLYELECTROLYTE APPLICATION AND TESTING TECHNIQUES

Before applying a particular polyelectrolyte to a specific coagulation application, it is almost always necessary to first test it. There are hundreds of different polyelectrolytes available under various trade names that are sold for water and wastewater coagulation applications. Depending on the water or wastewater characteristics and the results desired from its treatment, it may be possible to find either many, few, one, or even none that will do an acceptable job. Therefore, in order to find polyelectrolytes that are useful for a specific application and, also, to determine the approximate concentration requirements, it is necessary to do preliminary bench tests.

The many different polyelectrolytes that are available can be classified under relatively few general groups or headings. However, within each group, the differences that exist can show appreciably different effects on the coagulation and removal of a particular suspension. A knowledge of the characteristics of each group of polyelectrolytes may be useful in narrowing down a list of candidates to be tested for a particular application. However, unknown characteristics of the water or wastewater often mean the best selection of a polyelectrolyte can be accomplished more readily by trial and error testing (with the use of some general rules of thumb) rather than relying solely on accepted coagulation theory and principles without performing bench tests.

Preliminary bench testing of polyelectrolytes is not only important where their use as primary coagulating agents is concerned. When polyelectrolytes are to be used as flocculating aids and settling aids in conjunction with inorganic coagulants, such as aluminum and iron salts, it is important to determine how they will behave beforehand. If an untested polyelectrolyte is added to a coagulation-sedimentation process, without knowledge of its dosage requirements and proper time of injection, it can create severe problems which mandate shutdown, draining and cleaning of all tanks, settling basins, piping, filters and other surfaces that the solids and/or clarified liquid have contacted. On the other hand, a properly applied dosage of an effective polyelectrolyte can significantly increase effluent clarity and liquid throughput, while reducing the sludge volume without creating equipment maintenance problems.

A. Function of Polyelectrolytes in Settling Operations

The mode of application of polyelectrolytes to settling operations may be quite variable depending on how they function as particle settling agents. The specific function of each of the various types of polyelectrolytes varies depending on its molecular weight and structure, and the predominant charges and chemical characteristics of the electrolyte molecule when dispersed in solution. In addition to understanding the function of a polyelectrolyte in terms of its characteristics, it is important to know what problems or side effects these characteristics can cause. Also, the function of a given polyelectrolyte in particle sedimentation may be different when used by itself than if used in combination with inorganic coagulants.

1. Classification of Polyelectrolytes

In discussing the various functions of polyelectrolytes it is helpful to first classify them into different groups according to an array of characteristics on which their individual functions depend. Polyelectrolytes are typically classified according to a number of different characteristics they possess, but in many cases, not all of these properties are included in manufacturers' published descriptions. Typically, they are classified as to the sign of the charge of their ionic groups (i.e., cationic, anionic, or nonionic). Other characterizing features used include molecular weight, molecular structure, and the source of the material from which they are formed. In selecting polyelectrolytes for specific applications, it is extremely helpful, if not absolutely necessary, to know the charge, a relative assessment of molecular weight (i.e., high, medium, or low) and something of the molecular structure. Knowing the source of the original polymeric material may also be important in some critical applications. Another type of polyelectrolyte classification used is a designation of its toxicity to living organisms and the resulting implication of acceptability for human consumption. In other words, is the material acceptable for use in potable water treatment or in products that come into contact with food such as paper wrappings or containers?

Polyelectrolyte manufacturers provide users with general information on their products. Normally, they give a description of charge and relative charge strength (i.e., high, medium, or low), relative molecular weight, source, and designate their acceptability for use in potable water. Many manufacturers also indicate the basic molecular structure and this can be very helpful to users who are trying to evaluate the endless list of choices available. In addition, many manufacturers take the trouble to provide lists of general applications for which their specific polymers are recommended, along with the range of typical dosages that are required.

Table 1
CLASSIFICATION OF ORGANIC POLYELECTROLYTES

By source	Natural			Synthetic		
By potability	Potable	NonPotable		Potable	NonPotable	
By molecular wt.	High	Medium	Low	High	Medium	Low
By charge density	High	Medium	Low	High	Medium	Low
By charge	Cationic	Nonionic	Anionic	Cationic	Nonionic	Anionic

Common Examples By
Molecular Structure
(Chemical Type)

Polysaccharide gums
Starch derivatives
Algin derivatives
Carboxymethyl cellulose
Chitin derivatives
Nucleoproteins

Common Examples By Molecular Structure
(Chemical Type)

(Cationic)	(Nonionic)	(Anionic)
Polyamines	Polyglycidyl	Polyacrylates
Polyethyleneimines	polymers	Carboxylic
Polyimideamines		polymers
Polybutadienes		
Polyamideamines		
Polyquarternaries		
Pyridine based		

(Cationic, nonionic, and anionic)
Polyacrylamides
Guanadine derivatives

It is not the attempt of this discussion to give a breakdown of each type of classification into each of its possible components. This would be too lengthy for coverage of all the specific structures available. A good text to consult for more information on the structure and formation of various polyelectrolytes has been written by Gutcho.[2]

Table 1 provides a summary of polyelectrolyte classifications with typical examples of chemical types. Observation of this table shows that, in terms of charge and charge density, as well as relative molecular weight, source, and acceptability for potable use, we can be rather definitive of its characterization. However, a multiplicity of specific structures of organic polyelectrolytes is possible under any given chemical type. Each specific structure, gives the polymer molecule a specific set of chemical characteristics which can determine how it will interact with a given substrate (i.e., suspended solid particle). It would take many years to evaluate by testing even the more common synthetic polyelectrolyte structures on all the many specific types of suspended solids that can exist in water and wastewaters.

Some day an extensive polyelectrolyte applicability list containing such information may be developed. For the present, the attempt is to explore the applicability of various polyelectrolytes for the coagulation of different types of suspended solids removal applications. The examples to be discussed are the result of either the writer's own experience or the reported work of others.

Before going into specific types of polyelectrolytes found useful for certain applications, this discussion will return to the functions of polyelectrolytes. Then, attention will be paid to the necessary practical considerations of testing procedures used to select polyelectrolytes and of techniques for feeding them to settling process systems.

The two primary functions of polyelectrolytes are (1) to reduce or neutralize colloid electrostatic surface potential which is the same as that of primary coagulants and (2) to help bind together small particles to form larger masses. This is accomplished by chemical adsorption of various polymer sites onto particle surfaces. This is the function of coagulation and flocculation aids.

In some water clarification operations, the function of a polyelectrolyte may be primarily destabilization in order to promote coagulation of natural colloidal material. However, in many waste water clarification operations the primary function is usually particle binding (or bridging). An examination of some of the more generally accepted colloid stabilization and coagulation theories serves to explain these observations.

According to the writings of many of the prominent authorities on modern coagulation theory, such as La Mer,[1] O'Melia,[3] and Stumm,[4] the diameters of the particles making up those suspensions that typically cause problems in water and waste water fall in a range from about 0.05 μm to about 100 μm. The particles smaller than 5 μm are usually considered colloidal size and their stability or resistance to coagulation is quite strong. Particles ranging from 5 or 10 μm, on up to larger diameters are usually suspended by a combination of Brownian movement, hydraulic gradients, low particle density, and formation of weak or fragile flocs above the sub-millimeter range.

In order for coagulation to proceed, small suspended particles must be brought into close enough contact so that attractive forces of a chemical nature (i.e., adsorption) can bring about attachment of one particle to the other. However, mutually repulsive forces between particles keep many initial particle approaches toward one another from actually becoming effective contacts. These forces may result from either of two different phenomena. The stabilizing forces may arise from (1) electrostatic surface charges or (2) chemical interactions between the solid surface and the surrounding liquid.

There is indication[1,3,4] that electric surface charges can set up a diffuse ionic atmosphere of spherical shape around the particles when it becomes stabilized. This diffuse outer layer is thought to be produced by the migration of ions, of charge opposite to that of the particle surface, toward the particle surface forming a concentrated layer of counter ions around the particles. The particle surface charges are usually negative and the counter ions are, therefore, normally cations. Since the counter ions in the vicinity of the particle have migrated to the solid surface, the initial surface charge is partially neutralized. However, the counter ions, now clustered around the particle surface, prevent the remaining counter ions in the bulk solution from diffusing into the zone that they have vacated by mutual electrostatic repulsion. The ions of the same sign as the particle surface, however, remain in the diffuse zone (i.e., diffuse layer) and are held in a relatively stable position by attraction to the counter ion layer, giving it a net negative charge. The spherical diffuse layer of liquid is held in association with or attached to the particle by attraction of the anions in the diffuse layer to the cluster of cations around the particle. The effect of this attached diffuse liquid layer is to prevent collisions by mutual repulsion between the particles and to further prevent contact of solid surfaces because the elastic-like diffuse layers deflect one particle away from another. Electric double layer theories such as this follow the Debye-Hückel theory accepted for simple electrolytes.[1]

It is, on the other hand, maintained that colloids of hydrophilic nature are stabilized by interactions between the water and hydroxyl groups adsorbed on the particle.[1] Rather than being sensitive to ions of specific charge, these colloids are sensitive to hydrating properties of the liquid, such as pH and temperature.

Solid dispersions found in both nature and in wastewaters likely possess combined properties of electrostatic surface potential, as exhibited by hydrophobic colloids, and the chemical interactive forces characteristic of hydrophilic colloids. This fact has probably been responsible for the different and conflicting results obtained, and opinions held, by many investigators of coagulation theory. In many wastewaters, as in seawater, stable colloidal dispersions of the hydrophobic type are not expected.[4] This is due to the destabilizing effect of the high ionic concentrations that can occur in wastewater.

In cases where hydrophobic colloids exist, with their diffuse double layer holding them in suspension, polyelectrolytes possessing strong ionic character, opposite in charge to the electrostatic surface charge of the colloid, are found to be effective. Polymers of this type, when dispersed in solution, are themselves charged colloidal molecules. Attractive chemical interactions between sites on the polymer molecule and the particle surface may be responsible for the ability of these polyelectrolytes to enter into the diffuse layer. With their charge of counter ion groups, they then neutralize the stabilizing effect of the particle surface charge. In such instances these charged polyelectrolytes are acting as primary coagulants. The repulsive surface charges are neutralized to a low enough degree, so particles can approach each other closely enough for contacts to occur and then, chemically attractive forces can bind the particles together. The stabilizing electrostatic surface charges are thought to be rather weak, yet long range forces.[1] On the other hand, the attractive chemical binding mechanisms are stronger but short range forces. Therefore, once interparticle distances become very short, coagulation is favored over stabilization. Traditionally, high concentrations of cations such as aluminum and ferric salts have been used to collapse the electric double layer by diffusion, thus neutralizing the colloid surface charge. Having no specific attraction to the solid particle, the effective diffusion of the coagulant ions depends largely on concentration gradients and thorough mixing of the system. Since a polyelectrolyte can have a specific attraction for a colloid surface, the concentration of polymer required is usually less than that of inorganic coagulant alone. Normally, about 10 to 100 times more inorganic coagulant is required than is found effective for the organic polyelectrolytes.

More frequently, polyelectrolytes are utilized in settling operations as particle binding or bridging agents. The best polyelectrolytes for these applications are usually found to be the very high-molecular-weight, low charged types. These polyelectrolytes may be rather ineffective in low concentrations for initiating coagulation of the highly charged hydrophobic colloids discussed in the above paragraph. However, when suspended particles are hydrophilic colloids or are larger partially coagulated masses, such polymers may be excellent flocculating agents. These polymers adsorb (or attach) to a particle at certain portions of their chain-like molecule, while other portions of the same molecule adsorb to one or more other particles. More polymer molecules continue to adsorb onto these and other particles, until thousands or even millions are tied together in large masses called flocs. Depending on the rate that polymer molecules are introduced into the suspension, a more or less tightly knit or dense network of uniformly sized flocs can be found that settle readily when liquid turbulence is minimized.

There are many different polyelectrolytes available that incorporate both high molecular weight (very large or long molecular chains of between 1 million and about 15 million molecular weight) together with rather high charge densities of one sign or the other. These find wide use in wastewater applications since many of these suspensions (such as metal oxides and hydroxides) behave as though they have both hydrophilic and hydrophobic characteristics.

By performing either as primary coagulants or as particle binding agents, polyelectrolytes are used successfully to fulfill one or more purposes in settling operations. These functions associated with primary polyelectrolyte functions include: coagulating agents, flocculating aids, floc strengthening agents, floc and particle settling aids, and filtration aids. By performing these functions, they are used in settling operations to

Table 2
CHEMICALS USED IN WATER AND WASTEWATER TREATMENT

Names		Solubilities in g/l					
Chemical formula Formula weight	Grades used in water treatment	32°F	50°F	68°F	86°F	Shipping container	Weight (lbs/ft³)
Coagulants							
Aluminum sulfate or filter alum $Al_2(SO_4)_3 \cdot 18H_2O$ FW : 666	Slab, lump, ground, or powdered 14.5 to 17.5% Al_2O_3	608	653	710	788	Bags, kegs, barrels, in bulk	Slab or lump = 57—67 Powdered = 38—45
Ammonia alum or ammonium alum $Al_2(SO_4)_3 \cdot (NH_4)_2SO_4$ $\cdot 24 H_2O$ FW : 906	Lump or crystal required for pot type feeders	39	95	151	200	Bags, kegs, barrels, boxes	64 — 68
Potash alum or potassium alum $Al_2(SO_4)_3 \cdot K_2SO_4 \cdot 24$ H_2O FW: 949	Lump or crystal required for pot type feeders	57	76	114	166	Bags, kegs, barrels, boxes	64 — 68
Ferric sulfate, ferrisul, or ferrifloc $Fe_2(SO_4)_3$ FW: 400	Granules Varies with source from 70 to 80% $Fe_2(SO_4)_3$	Very soluble: in cold water use 2 parts water to 1 part ferric sulfate				Bags, kegs, barrels, drums, in bulk	68 — 70
Ferrous sulfate or copperas $FeSO_4 \cdot 7 H_2O$ FW: 278	Crystals, granules	287	375	485	602	Bags, boxes, kegs, barrels, in bulk	63 — 66
Sodium aluminate $NaAlO_2$ FW: 82.0	Variable composit. usually 89% $NaAlO_2$ and 10% NaOH and Na_2CO_3	295	340	370	400	Bags, drums, barrels	53 — 60

From The Permutit Water & Waste Treatment Data Book, The Permutit Company, Inc., Paramus, N.J.
With permission.

increase effluent clarity, increase hydraulic loading, increase solids loading, reduce final sludge volume, and reduce dosage requirements of inorganic coagulants such as alum and ferric salts.

2. Function of Primary Coagulants

Inorganic coagulants are normally added to water and wastewaters as soluble salts. A list of common chemicals used as coagulants is presented in Table 2, 3, and 4 as they can be purchased.[5] These salts dissociate in water to form di- or trivalent metal ions of iron or aluminum. The trivalent iron and aluminum ions react with hydroxyl and other alkaline ionic species. Many of these reaction products are only slightly soluble in water under neutral pH conditions. The hydroxides of these metals are almost insoluble at neutral pH and precipitate rapidly upon addition of the soluble metal salt (i.e., coagulant) to the water. Some of the reactions that take place upon the addition of these coagulants to natural waters and wastewaters are shown in Table 5.[5]

Table 3
CHEMICALS USED IN WATER AND WASTEWATER TREATMENT

Names Chemical Formula Formula Weight	Grades used in water treatment	Solubilities in g/l				Shipping container	Weight (lbs/ft³)
		32°F	50°F	68°F	86°F		
		Alkalis					
Sodium bicarbonate NaHCO₃ FW: 84.0	Powder 99% NaHCO₃	65	75	88	100	Bags, kegs, barrels, in bulk	59—62
Caustic soda or sodium hydroxide NaOH FW: 40.0	Flake: 98.06% NaOH Solid: 98.06% NaOH Also in ground form and as 50 to 73% liquid	420	515	1090	1190	Flake: drums Solid: drums	400 lb drums 700 to 730 lb drums
Sodium silicate Na₂O 3.25 SiO₂ FW: 234.3	40° Baume solution or solid glass	40° Baume solution is miscible with water in all proportions				Drums, tank trucks, tank cars	86.1
Washing soda or crystal sodium carbonate Na₂CO₃ · 10 H₂O FW: 286	Crystal or lump required for pot type feeders	189	338	580	1050	Bags, kegs, barrels	68 — 71
Soda ash or sodium carbonate Na₂CO₃ FW: 106	Light soda ash powder 99.16% Na₂CO₃	70	125	215	388	Bags, barrels, in bulk	34 — 52

From The Permutit Water & Waste Treatment Data Book, The Permutit Company, Inc., Paramus, N.J. With permission.

Although Table 5 shows solid precipitates of aluminum hydroxide (i.e., Al[OH]₃) and ferric hydroxide (i.e., Fe[OH]₃) as the final reaction products, many intermediate species occur during the formation of solid metal hydroxide precipitates from the dissolved trivalent metal ions. These intermediate species include various proportions of partially hydrated metal ions such as MOH^{2+} and $M(OH)_2^+$ as well as $M(OH)_4^-$. In addition to these species, the intermediate hydrates can polymerize into chains of metal hydrate units that carry partial charges along the chains.

The trivalent metal ions with their high charge density act as primary coagulants for neutralizing electrostatic surface charges on colloids. However, since their existence is very short lived under neutral pH conditions, high concentrations of coagulant chemical would be required, along with very slow addition of chemical, to supply enough coagulant metal ion to do a complete job of colloid destabilization.

On the other hand, the existence of intermediate metal hydrate species is both of longer duration and for a limited time, of higher concentration. These species are thought to be very instrumental in destabilization of electrostatically charged colloids. Also, the polymeric metal hydrate species may bind or bridge the destabilized colloids into microflocs (i.e.,, initial agglomerates of solid particles still below visible size). As the polymeric species continue to grow into visible solid precpitates, they enmesh or encapsulate the colloidal solids that they contact. This process is the conventional clar-

Table 4

CHEMICALS USED IN WATER AND WASTEWATER TREATMENT

Names Chemical formula Formula weight	Grades used in water treatment	Solubilities				Shipping container	Weight lbs/Ft³	
		32°F	50°F	68°F	86°F			
		Limes and Magnesia						
Chemical lime, quick lime, calcium oxide CaO FW: 56.1	Lump, pebble, or ground High calcium lime usually contains 98% CaO	Slakes with water forming hydrated lime, Ca(OH)₂, having following solubility:				Bags, barrels, in bulk	Lump Pebble Ground Pulverized	50—65 60—65 50—70 39—71
		1.8	1.7	1.6	1.5			
		1.8	1.7	1.6	1.5			
Hydrated lime, slaked lime, calcium hydroxide Ca(OH)₂ FW: 74.1	Powder Usually contains 93% Ca(OH)₂					Bags, barrels, in bulk		25—50
Dolomitic lime CaO + MgO MgO content varies	Powder: for silica removal typical analysis = 58% CaO and 40% MgO	Slakes with water forming Ca(OH)₂ and MgO which forms Mg(OH)₂ very slowly				Bags, barrels, in bulk	Lump Pebble Ground Pulverized	50—65 60—65 50—75 37—63
Hydrated dolomitic lime Ca(OH)₂ + MgO MgO content varies	Powder: for silica removal typical analysis = 62% Ca(OH)₂ and 32% MgO	—	—	—	—	Bags, barrels, in bulk		28—52
Magnesia or magnesium oxide	Powder: various grades differ greatly in density	Hydrates very slowly in water. Solubility of hydrate, Mg(OH)₂ is very low as shown:				Bags, barrels, in bulk		8—40
		—	—	0.02	0.02			

From The Permutit Water & Waste Treatment Data Book, The Permutit Company, Inc., Paramus, N.J. With permission.

Table 5
COAGULATION REACTIONS

$$Al_2(SO_4)_3 + 3\,Ca(HCO_3)_2 = 2\,Al(OH)_3 + 3\,CaSO_4 + 6\,CO_2$$
$$Al_2(SO_4)_3 + Na_2CO_3 + 3\,H_2O = 2\,Al(OH)_3 + 3\,Na_2SO_4 + 3\,CO_2$$
$$Al_2(SO_4)_3 + 6\,NaOH = 2\,Al(OH)_3 + 3\,Na_2SO_4$$
$$Al_2(SO_4)_3 \cdot (NH_4)_2SO_4 + 3\,Ca(HCO_3)_2 = 2\,Al(OH)_3 + (NH_4)_2SO_4 + 3\,CaSO_4 + 6\,CO_2$$
$$Al_2(SO_4)_3 \cdot K_2SO_4 + 3\,Ca(HCO_3)_2 = 2\,Al(OH)_3 + K_2SO_4 + 3\,CaSO_4 + 6\,CO_2$$
$$Na_2Al_2O_4 + Ca(HCO_3)_2 + 2\,H_2O = 2\,Al(OH)_3 + CaCO_3 + Na_2CO_3$$
$$FeSO_4 + Ca(OH)_2 = Fe(OH)_2 + CaSO_4$$
$$4Fe(OH)_2 + O_2 + 2H_2O = 4\,Fe(OH)_3$$
$$Fe_2(SO_4)_3 + 3\,Ca(HCO_3)_2 = 2\,Fe(OH)_3 + 3\,CaSO_4 + 6\,CO_2$$

From The Permutit Water & Waste Treatment Data Book, The Permutit Company, Inc., Paramus, N.J. With permission.

ification process for the sedimentation of unwanted solid suspensions and its use may even reach back over thousands of years.

The extent as well as the rate of coagulation of colloidal suspensions, whether hydrophilic or hydrophobic, greatly depends on the relative concentrations of the metal ions and their intermediate reaction products. Since the existing concentration of hydroxyl ions in the system during coagulation reactions is the factor governing the rate of formation of the metal hydroxides and their intermediate hydrates, *optimum pH is the key to effective coagulation with these inorganic coagulants.*

The importance of pH is not so much that an abundance of hydronium or hydroxyl ions will, by themselves, destabilize colloid surface charges. Such an effect, however, is experienced in some systems where, merely by addition of acid to a suspension of solids, coagulation is initiated without addition of polyelectrolytes or metal ions. When inorganic metal salts (such as aluminum and iron salts) are added to water, the important effect of the system pH is to determine the ionic sign, charge density, and concentration of the metal species that form. This is critical to the concentration of coagulant required and the degree of colloid destabilization obtained.

Concentrations of solid metal hydroxide that precipitate, and soluble intermediate species that form upon addition of the metal salt to water, are pH dependent. The minimum solubility of each metal salt is associated with a specific pH value which can vary somewhat from one aqueous system to another. However, in most aqueous systems, the minimum solubility of aluminum hydroxide is usually at a pH of about 6.0 and for ferric hydroxide it is usually at a pH of about 8.0.

Solubility curves are given by O'Melia that show that the solubility of both aluminum hydroxide, $Al(OH)_3$, and ferric hydroxide, $Fe(OH)_3$, increase significantly either above or below the pH of minimum solubility.[3] Aluminum hydroxide solubility increases extremely rapidly below pH 6.0. For example, at pH 4.0 the soluble aluminum ion concentration is between 25,000 and 30,000 mg/ℓ as Al. A less steep increase occurs above pH 6.0 for aluminum hydroxide since at pH 10.0 the soluble aluminum is a still significant, but less dramatic, 30 mg/ℓ as Al. The minimum solubility of aluminum hydroxide, at pH 6.0, is about 3×10^{-3} mg/ℓ Al. Ferric hydroxide solubility is many orders of magnitude less soluble than aluminum hydroxide over a wider pH range. At pH values of both 4.0 and 12.0 the concentration of ferric ion in solution is only about 6×10^{-3} mg/ℓ as Fe. The minimum solubility of ferric iron, at pH 8.0, is only about 1×10^{-6} mg/ℓ as Fe.

In order to produce a rapid growth of floc particles and complete coagulation of colloids, it is very important to have most of the added metal coagulant precipitated out as the metal hydroxide floc at the end of the coagulation process. Then, during the precipitation process, the colloidal particles become destabilized by the metal ions and/or charged intermediate metal hydrate species. Meanwhile, the precipitating solid metal hydroxide particles (flocs) provide high surface area for enhancing the destabilization and adsorption of colloids. Most coagulation processes with metal salts occur in this manner combining physical charge neutralization with chemical adsorption for highly efficient removal of suspensions.

With aluminum salts it is evident that high concentrations of coagulant have to be added when the pH is below 5.5 or above 9.0. Therefore, it normally becomes impractical to use them outside of the pH 5.5 to 9.0 range. Ferric salts, producing very insoluble ferric hydroxide precipitate, are practical from an economic standpoint as well as the fact that they leave so little iron in solution (within the pH 4.0 to 12.0 range) when the coagulation process is completed.

As mentioned, one important way in which the pH affects the rate and degree of coagulation in aqueous systems treated with aluminum or iron salts is due to the formation of intermediate metal hydrate species. On the acid side of their minimum solubility pH values, the hydrates of these metals take on cationic charges due to the lack of hydroxyl ions. On the basic side of the minimum solubility pH, the soluble metal hydrates become predominantly anionic due to the excess of hydroxyl ions in the formation of complex metal ion species. Thus, by shifting the pH value toward one side or the other of the minimum solubility pH value, these metal coagulants can be made more effective in destabilizing a hydrophobic colloid at an optimum pH value. In the same way, pH variation can affect the stability of a hydrophilic colloid by influencing the hydroxyl ion concentration of the solution on which the colloid stability depends.

3. Problems and Influencing Factors

There are numerous factors associated with polyelectrolytes, that require consideration before applying them to various settling operations. Some of these considerations involve anticipating problems that can occur. Other considerations are associated with characteristics that some polyelectrolytes have which make them behave completely different under different sets of circumstances. Some of this changing behavior can be used to advantage and some cannot.

a. Variations in Suspended Solids Concentrations

A definite problem, often encountered in settling operations, is changes in the suspended solids concentration of the influent to the coagulation process. Frequent concentration variations of marginal intensity and short duration can be more of a problem than occasional variations of great intensity and long duration. That is, when a polyelectrolyte is used as the sole coagulant, the ratio of its weight to the weight (or, more directly, the surface area) of the suspended solid must be maintained within a certain range. If the solids concentration suddenly varies significantly, either overdosing or underdosing can occur (causing ineffective coagulation) before an effective change in polyelectrolyte dosage can be made. If, however, the influent stream is directed to an equalization basin, changes in solids concentration will be less severe and less frequent. After adequate equalization, the resulting variations may not require changes in polymer dosage or, at least, allow ample time to make tests to determine what dosage changes should be made.

Variation of influent suspended solids concentration is only one factor of concern affecting polyelectrolyte dosage requirement. Significant changes in the influent solution pH and ion concentration can also influence both dosage requirements and the charge of some polyelectrolytes. As previously mentioned, the pH and ion concentrations of the solution can influence the stability of a colloid to a very significant degree. So, although the suspended solids concentration may not vary, changes in electrical or chemical stabilizing forces on the colloid must also be dampened if the polyelectrolyte dosage is to be effectively controlled at the proper level. The equalization system must be adequate to minimize critical changes in suspended solids concentration, pH and ion concentration. If pH variations are the only significant changes, it may be more practical to add acid or alkali to stabilize pH than it is to provide a large equalization basin.

The problem of overdosing is not only that ineffective coagulation is obtained due to restabilization of colloidal material. Additionally, excess quantities of polyelectrolyte can remain in solution and bypass the coagulation-sedimentation process. These dissolved polyelectrolyte molecules can adsorb onto solid surfaces they come in contact with downstream of the settling step causing severe problems. Examples of such problems are irreversible clogging of sand filters, fouling of ion exchange resins, and clogging of internal pipe surfaces and flow meters. Also, in some wastewaters, excess polyelectrolyte can coagulate suspended solids in receiving bodies of water which, when they are lakes, rivers, or tributaries, may cause an esthetically undesirable buildup of sediment or sludge around the area of the outfall. In applications of process water treatment, excess polyelectrolyte can cause undesirable effects on manufactured products due to deposits of the organic material. In food or food product processing, undesirably high polyelectrolyte residuals may adversely affect taste or quality.

b. Polyelectrolyte Storage and Preparation

The preparation and storage of polyelectrolytes also require consideration to get the most effectiveness out of these products. Chapter 8 deals extensively with these matters; therefore, they will only be touched upon at this time. Proper storage techniques should be followed according to the manufacturer's instructions if the products are to be fully effective when they are used. This usually means storing them in a cool, dry place and tightly sealed. Some products, especially those sold as liquids, must be kept from freezing. Bacterial infestation, particularly with natural polymers, is a situation that one has to be careful to avoid.

In preparation of polyelectrolytes for feeding to a treatment process, the products are normally carefully blended with water to solubilize and/or disperse them. The dry powder polyelectrolytes are usually prepared in tanks as 0.1 to 0.5% solutions (i.e., 1 to 5 g of powder per liter of water). The necessity of careful dissolution of these products into water cannot be overstressed. Appropriate methods and devices for preparing these solutions are described by the manufacturer's technical data and in published references on polyelectrolytes. It is held sufficient here, to stress that the polyelectrolytes must be completely dispersed in solution before they are fed to coagulation or flocculation processes.

Once prepared in a 0.1 to 0.5% solution, the polyelectrolyte should be conveyed to the treatment process by a metering pump to maintain the desired dosage. Often, however, the uniform dispersal of a polyelectrolyte to the solids in a treatment process is also aided by metering it into a dilution water stream first. The size of the dilution stream should cut the polyelectrolyte solution concentration to about 0.01 to 0.05% to insure good dispersal when it comes into contact with the suspended solids. In preparing and handling the polyelectrolyte solutions, only slow speed stirring (i.e., less than 500 rpm) should be used. Also, other high speed shearing devices, such as centrifugal pumps, should not be used. Shearing action tends to cut down or fracture the polyelectrolyte chains into shorter, less effective molecules.

Synthetic, long chain, high-molecular-weight polyelectrolytes, such as the polyacrylamids, are more susceptible to molecule rupture by shearing. On the other hand, certain natural polymers, such as guar gum, have been found to be more shear resistant than other polymers. This is thought to be due to the fact that guar gum molecules are more like rigid rods than the flexible snake-like chains of synthetic polymers. The short, low-molecular-weight polymers have less inertia and a greater chance of remaining intact after colliding with a high speed object. Likewise, their chances of moving with the fluid stream around an object, such as a mixer blade, are much greater than with the longer, heavier, flexible molecules.

c. Combination Coagulation Systems

Polyelectrolytes are often successfully used with inorganic coagulants. In many cases, the result of using alum or an iron salt in combination with a polyelectrolyte is more beneficial than using either the inorganic salt or the organic polymer alone.

When variations in size, type, and concentration of suspended solids occur, use of a polyelectrolyte alone may not give continuously effective coagulation. However, an excess of inorganic coagulant can be used in an appropriate pH range to destabilize and/or adsorb whatever colloid is present and the unused metal coagulant precipitates to a settleable solid (unlike excess or unused polyelectrolyte that can remain dissolved). Therefore, excesses of inorganic coagulant can be used with a minimum of residual finding its way through the treatment process.

Unlike using polyelectrolytes, however, coagulation and subsequent flocculation with inorganic salts is a rather slow process (15 to 30 min) producing rather fragile slow settling flocs. On the other hand, many high-molecular-weight polyelectrolytes tend to enhance rapid flocculation of destabilized particles (2 to 5 min) and produce strong rapidly settling flocs. Such polyelectrolytes can be used to flocculate the metal hydroxide that precipitates with or without captured suspended solids. Since the dosage of inorganic coagulant can be held constant if influent solids concentration does not vary too greatly, a polyelectrolyte feed can also be maintained at constant dosage resulting in a very effective coagulation-sedimentation process. The use of excess inorganic coagulant will, of course, produce more sludge than polyelectrolyte treatment alone. But this disadvantage may be offset by the elimination of a large equalization basin and constantly changing charge and dosage requirements for a polyelectrolyte used as the sole coagulant.

Another circumstance where an inorganic salt and organic polymeric flocculant combination may be advantageously used is where the suspended solids are composed of a heterogenous group of differentiated particles that do not all respond to the destabilizing effects of a single polyelectrolyte. In such cases coagulation can be incomplete no matter how thorough the dispersal, mixing, and contacting of the polymer with the solid suspension is. Applications like this give only partial removal of colloid, especially in cases where a colloid requiring charge neutralization is present with hydrophilic colloidal material. In these cases, a combination of inorganic salt added first, followed by a polyelectrolyte (after initial destabilization occurs) may be used to produce complete coagulation and rapid sedimentation of the resulting agglomerated material.

In other cases, where a single polyelectrolyte can't be found that does a complete coagulation, it may also be possible to use a dual polyelectrolyte system. This is similar to adding a combination of inorganic salt and organic polymer in that each coagulant has specific functions which act in destabilizing different portions of the solid suspension. An example of a dual polymer system is one where a highly charged cationic polyelectrolyte is added to destabilize natural organic color colloids in water and then a high-molecular-weight nonionic or low charge anionic polyacrylamide is added to flocculate suspended silt along with the destabilized color as a rapidly settling floc.

Dual polymer systems are, however, usually more suited to sludge dewatering operations than to settling and clarification processes. There are various reasons for this. First, the control required to balance the effects of two different polymers in destabilizing changing concentrations of suspended solids is quite difficult. Then, there is also the increased probability of dissolved polyelectrolyte leakage in the effluent. Dual polymer systems are used successfully as conditioners for sludge dewatering because sludges are primarily composed of flocculated material and destabilized particles to begin with. The differently charged polymers perform in agglomerating the sludge particles (by enhanced bridging) into large discrete masses which allows the solids to be separated from the water more readily. Some residual polymer remaining in the water, after it is separated from the sludge, is not usually of great concern since this water is usually disposed to waste treatment systems or back into the influent of the sedimentation process.

In selecting the best coagulant system to use (whether it is a single polyelectrolyte, an inorganic coagulant only, a combined inorganic coagulant and polyelectrolyte, or a dual polymer system), the effluent quality, sludge volume, and overall operating cost are three very important determining factors. The chemical cost can be a major part of the operating cost and, therefore, the least expensive coagulant system that produces acceptable effluent quality and sludge volume should be used. In many cases involving relatively high and fluctuating suspended solids loads (e.g., 150 to 1500 mg/ℓ lower cost is achieved with a combination of inorganic salt (e.g., 10 to 100 mg/ℓ with optimum pH adjustment) along with a very low dosage of high-molecular-weight polyelectrolyte (e.g., 0.1 to 1.0 mg/ℓ). With low suspended solids loads (e.g., less than 100 mg/ℓ) a low dosage of a single polyelectrolyte (e.g., 1 to 10 mg/ℓ) may be most cost effective. In some settling operations the quantity and quality of water processed may be more important than chemical cost or sludge volume produced. In these cases, differences in coagulation and settling rates should be used as a basis for selecting between combined and single coagulant systems. However, the inorganic coagulants are normally priced from about 10 to 50 times less per unit weight, and are more readily put into solution than many organic polyelectrolytes.

When it appears that high concentrations of organic polymers are required for effective coagulation, inorganic coagulants should be investigated if it is desired to keep costs and operational handling at reasonable levels. It is also found that very low polyelectrolyte concentrations can substantially reduce the amount of an inorganic coagulant required. Therefore, it is normally important to investigate organic polyelectrolytes alone and in combination with inorganic coagulants to find the most cost effective system.

d. Floc Strength

Chemical costs, effluent quality, and sludge production are among the more obvious factors affecting the use of coagulants. Less obvious, but still important, are factors related to the nature or behavior of the floc agglomerates that grow to form settleable sized masses.

It is often observed, when polyelectrolytes are compared with inorganic coagulants, that the flocs produced by or with polymers are larger and faster settling. Results that may also be observed with polyelectrolytes, but not recognized for their importance, are that flocs grow to full size significantly more rapidly and become more uniform in size and are less fragile than flocs of inorganic coagulants. This allows more of the solid material to settle as a uniform mass leaving a highly clarified supernatant above it. This characteristic is useful in producing a sludge blanket. A sludge blanket is a relatively high concentration of flocculated particles (e.g., 2000 to 5000 mg/ℓ suspended solids is common) which are partially supported in a fluidized state, by hydraulic flow between individual particles, and partially by contact with the particles below them (i.e., hindered settling). This high solids concentration results from the fact that the particles are of approximately the same size and density and do not fracture readily when colliding with other particles or objects. Sludge blankets are extremely useful and effective in adsorbing residual colloidal material, excess polyelectrolyte molecules, and even some ionic material because the extensive solid-liquid surface area of the blanket possesses a great many chemically active sites.

Inorganic coagulants produce lighter, more fragile flocs that are more differentiated in size due to their slow growth and tendency to shear during mechanical and hydraulic stirring. Therefore, sludge blankets produced only by inorganic coagulants tend to be less concentrated and allow more solids to remain in the liquid rising through them. This is due to numerous floc fragments ("breakaway flocs") that are sheared from the larger flocs by the velocity of the moving liquid stream. The nature and surface of a floc produced by alum or ferric salts, however, tend to be very highly reactive in removal of many colloidal and some ionic species. It is, therefore, desirable to combine polyelectrolytes with inorganic coagulants to obtain the advantages of sludge blankets formed by both types of coagulants.

Floc particles tend to shear or tear apart into smaller particles as well as agglomerate to form larger masses during coagulation. This is the major factor influencing the degree of mixing needed during the flocculation step preceding a sedimentation step. If the particles are not thoroughly stirred so that individual particles are brought together for numerous contacts, the flocs remain small and will not settle at a desirable rate. If the degree of mixing agitation is too high, already agglomerated particles will break apart and the settling rate may again be too slow. The proper degree of mixing will bring the colloidal particles into effective contacts quickly enough so that after 10 or 15 min of agitation (at room temperature) the colloid is destabilized, and the flocs are of a uniform size and effective settling rate.

High shear forces, such as high mixing speeds, will tear already settleable size particles into unsettleable fragments. With suspensions that have been flocculated by inorganic salts or low-molecular-weight, highly charged polyelectrolytes, these fragments may recombine to a settleable size floc if the degree of agitation is appropriately reduced over a long enough period. However, when certain low charged, very high-molecular-weight polyelectrolytes are used during the flocculation process, and if the flocs are subsequently sheared, many of the fragments can remain uncombined and unsettleable, even after optimum stirring conditions are established. This is thought to be due to the fact that the molecular chains of the polymer have been broken or dislodged from particle-to-particle surfaces and have wrapped around themselves or fragments of other polymer molecules. The fragmented masses may then exhibit a degree of stabilization that resists coagulation even after addition of more polyelectrolyte of the same kind.

When particle shearing conditions like these occur, it is sometimes possible to reagglomerate the fragments by addition of a coagulant (polyelectrolyte) of opposite charge to the one that has been sheared. This, because it is a dual polyelectrolyte addition, will normally be somewhat difficult to accurately control in a dynamic system. It is best to avoid high shear conditions both during and after the flocculation and settling steps.

A floc formed from only inorganic coagulant, such as alum, will shear even more readily than those in combination with high-molecular-weight polyelectrolytes. These flocs will, however, tend to reagglomerate if the optimum degree of agitation is employed. These flocs can take considerably longer to reagglomerate after shearing than initial agglomeration, because of less affinity between the colloidal particles and pre-cipitated coagulant after the combination has been broken. Floc particles that are growing as a result of metal hydroxide precipitating from solution, are more reactive coagulants or flocculants than already precipitated flocs. It is, therefore, important to use only moderate stirring to avoid the shearing of floc particles produced with inor-ganic salts. At the same time, it is helpful to add a low concentration of a high-molec-ular-weight polyelectrolyte to increase flocculation rate and reduce floc shearing. The weight ratio of polyelectrolyte concentration effective for these purposes is commonly from about 1/500 to about 1/5000 of the inorganic coagulant concentration used.

It has been determined, over many applications, that the mixing conditions appro-priate for coagulation and flocculation for sedimentation processes, are not the same as conditions best for either sludge dewatering, in-line coagulation or direct filtration. In sedimentation, as described, mixing should be rather gentle after an initial dispersal of coagulant to allow flocs to grow to settleable size and over a long enough time to provide sufficient numbers of contacts between a great number of finely divided sus-pended particles. In sludge dewatering, effective mixing is usually of greater intensity for a relatively short period of time (i.e., less than 10 min). Normally concentrations of polyelectrolyte 50 to 100 times that of settling process requirements are also used. These conditions allow for very rapid coagulation of large compact particles with less entrained water and with better drainage characteristics. In in-line and direct filtration applications, the objective is to provide a very brief, high intensity mix to disperse the coagulant thoroughly. Then, however, it is desirable to maintain the flocculation and particle growth step inside the filter bed rather than on top or ahead of it. This allows for less blinding of the filter bed surface and greater utilization of the void space in the bed for longer filter runs.

B. Tests For Polyelectrolyte Selection

There are a number of different test methods that are used for determining the types and dosages of polyelectrolytes that are suitable for a specific clarification application. The four methods that are discussed in this chapter are jar tests, zeta potential tests, streaming current tests, and refiltration rate tests.

1. Description of Tests Used To Evaluate Coagulation

The general procedures for conducting the first two methods will be described. How-ever, all four methods have been used successfully and a brief description of each method is offered to point out the basic principles of each.

a. Jar Tests

Certainly the most familiar and widely used coagulation test employed by those ac-quainted with water and wastewater treatment is the jar test. Essentially, a jar test is a series of equal volume, identical samples that are exposed to a controlled variety of treatment conditions. Both during and following the administration of these treatment conditions, the effect of each variation on the clarification of the sample is observed. Observation of the degree of clarification obtained can be by simple visual comparison, or by instruments that aid in measuring sample clarity, or by a complete laboratory analysis of resulting sample quality.

Since the first reported jar tests were used by Langelier and Hyde in 1918, many variations of the jar test concept have come into use.[6] The general principle is, (what-ever the variation) to reproduce, as closely as possible, the existing or anticipated con-ditions of the treatment plant. In effect, the operation of the treatment plant is at-tempted in miniature in order to determine what effect a change in a single variable will have if all other variables are held at constant and representative levels. Jar tests are particularly useful and remain popular for controlling coagulation-sedimentation and precipitation-sedimentation processes. They are less suited to the control of direct filtration or in-line coagulation processes.

b. Zeta Potential

Zeta potential (ZP) is a measure of the electrostatic charge on the surface of particles suspended in liquid. As previously mentioned, the electrostatic surface charges set up an electrokinetic force surrounding the particles and tend to make them repel each other. Since the particles are electrically charged, they can be shown to undergo a phenomenon called electrophoretic mobility, which means they migrate toward an electrode of opposite charge when a DC voltage is applied to a sample of the liquid suspension. The rate and direction of this mobility, at a specific applied voltage, is used to determine the ZP.

In natural water systems, colloidal suspensions commonly are found to possess ZPs of 20 to 30 mV and are negatively charged. By contrast, some wastewaters are found to have negative ZPs on the order of 40 or 50 mV. In order to achieve coagulation in most systems, the ZP must be reduced in value to less than 10 mV (preferably to less than 5 mV)[7]. When a particle migrates toward the positive electrode (anode) in a DC field, the ZP is given with a negative sign (i.e., −mV). When the particle moves to the negative electrode (cathode) it is given a positive sign (i.e., +mV). The rate and direction of particle movements are determined with some type of tracking device (a microscope for instance) and an electrophoresis cell for applying the DC voltage and producing electrophoretic motion.

ZP is used in coagulation testing by finding the treatment conditions that reduce the ZP of the particles to a value within the optimum ± 5 mV range. At this level of surface charge, it is observed that coagulation of individual colloids or particles proceeds readily. The coagulant or treatment conditions used to produce the optimum ZP are those treatment conditions where effective coagulation and clarification of the sample starts.

Determination of the effect of a specific dosage of a coagulant can usually be made within a few minutes once experience with this method is gained. Although ZP measurement is very useful in estimating coagulant dosages and controlling all types of coagulation processes, it is particularly useful in controlling coagulant dosages to direct filtration and in-line coagulation processes.

c. Streaming Current

Streaming current (SC) is also a technique used to measure surface charges associated with suspended particles. This method, however, doesn't measure the surface potential on individual particles. Rather, it measures a flow of electrical current that is statistically proportional to the average surface charge on the particles.

In order to visualize the concept of SC, recall the theoretical model of a charged colloidal system (that is, a hydrophobic colloid). There is a system of suspended particles possessing rather long range surface charges that attract a layer (high concentration) of ions opposite in charge (counter ions) to cluster around the individual particles. The formation of this counter ion layer results in a diffuse layer (of greater thickness than the counter ion layer) around the particles that gives the colloid its stability. As the distance away from the particle surface into the diffuse layer increases, the counter ions become more mobile (because their attraction to the particle is weaker).

For the phenomenon of streaming current, measurement employs a device that can essentially immobilize some of the colloid particles while the fluid is moved past them at high velocity (by mechanical means).[8] The velocity of the moving fluid is such that the outer diffuse layer is apparently destroyed by intermixing with the bulk solution, and the mobile counter ions are swept downstream of the immobilized particle surface. The movement of the like-charges on these counter ions constitutes an electrical current called a streaming current. If the liquid is contained by nonconducting material, the electrical current will return through the streaming liquid by ombic conduction. Only those counter ions actually swept away contribute to the streaming current. It can be measured by placement of electrodes into the system. The difference in charge density between the two electrodes constitutes the electrical signal that can be detected by a sensitive ammeter. The electrode upstream of the immobilized particle will take on the same charge sign as the particle surface charge while the downstream electrode assumes the charge sign of the counter ions.[8]

In practice, it is found that the sign of the charge on the counter ions can be more readily determined by this method than the actual magnitude of the charge on the particles when performing coagulation tests. Using this principle, the optimum types and dosages of coagulants can be distinguished from ineffective types and insufficient or excessive dosages. This is done by determining the dosages at which the coagulants achieve charge neutralization and any additional coagulant results in reversal of the charge sign on the counter ion. Since determination of optimum coagulant dosage can be made rather rapidly, as with ZP measurement, SC is particularly suited to controlling direct filtration and in-line coagulation processes.

d. Refiltration Rate

Refiltration rate is yet another technique which is described as a method for determining optimum dosages for coagulants and appears to be particularly suited to polyelectrolyte applications.[1] This method involves controlled "dropwise" addition of dilute polymer solutions to the suspension under controlled gentle agitation and stirring until the floc has reached maximum size under these conditions. The sample is then filtered through a filter paper to produce a cake of solids on the filter while the filtrate is collected in a reservoir. When the entire liquid volume has been filtered, the filtrate is "refiltered" back through the solid filter cake. The rates of initial filtration and refiltration of the sample (i.e., volume per unit time) is measured at each polyelectrolyte concentration used to flocculate the sample.

La Mer reported that this method of optimum polymer dosage determination for a specific sample, is "surprisingly reproducible to ±2%". He found that when the refiltration rate is plotted vs. the polymer concentration, the rate initially increases rapidly with increasing concentration until it reaches a "sharp maximum". Further increases in polymer concentration then produce a gradual decline in refiltration rate which continues until the rate is ultimately less than the refiltration rate of the untreated suspension.

The decline in refiltration rate is attributed to the "overdosage" of polymer peptizing some of the flocculated solids back into fine particles. The fines then yield a turbid liquid that clogs the filter. It also seems possible, to the writer, that (particularly with high-molecular-weight polymers) the decrease in refiltration rate (at dosages above optimum) could be partially due to increased viscosity of the liquid caused by unadsorbed polymer molecules. In either case, this appears to be a very valuable tool for predicting polymer overdose with respect to applications involving filtration, whether applied directly or as a polishing step following sedimentation.

In assessing which of these four coagulation test methods is best for evaluating coagulation-sedimentation processes, consider what the test objectives are. Some typical objectives sought from a jar test analysis are listed below. These parameters are not necessarily arranged in any order of priority.

1.   Determine the types of coagulants that will effectively remove the suspended solids from (clarify) the water or wastewater.
2.   Determine treatment chemicals that will effectively remove dissolved solids or favorably alter the chemical composition of the water or wastewater.
3.   Establish effective concentration ranges of treatment chemicals/coagulants.
4.   Establish optimum dosages of treatment chemicals/coagulants.
5.   Establish order and time of addition for treatment chemicals/coagulants.
6.   Establish optimum reaction or flocculation time.
7.   Estimate the treated effluent quality.
8.   Estimate settling rate of flocs or precipitates.
9.   Estimate the sludge volume (settleable solids) produced as a result of each treatment parameter variation.

The methods that measure particle surface charges (i.e., ZP and streaming current) allow us to scan a relatively large number of coagulants at various dosages (especially polyelectrolytes) in a rather short period of time. The results of such a scan are the determination of the specific coagulants and their optimum concentrations that produce particle destabilization. These methods would not, however, indicate much about particle strength, settling rate, or sludge volume production.

Jar tests, on the other hand, can satisfy all of the above objectives if they are carefully planned and executed. But by comparison, jar tests require a relatively long period of time to run and larger volumes of sample than the other methods mentioned. For example, with ZP or SC tests, the optimum dosages of about 10 different coagulants may be determined in about 2 or 3 hr consuming a total of about 1 ℓ (maybe less) of sample. Of course, some experience is necessary to perform these tests at such a fast pace. To evaluate the same 10 coagulants, jar tests would probably take at least 8 to 10 hr and require 30 to 60 ℓ of sample.

Refiltration rate, which is probably the best way to determine floc strength in addition to estimating optimum dosage, can be performed on relatively small volumes of sample (for example, 5 to 10 ℓ to evaluate 10 coagulants). But the time requirement can vary considerably. An effective coagulant may be flocculated and filtered within 10 to 20 min per coagulant dose. However, poorly coagulated slurries may require an hour or more per coagulant dosage, so that a complete 10 coagulant evaluation could require several days to complete.

From these considerations, it appears desirable to narrow down the list of coagulants to be tested with ZP or SC tests. Then jar tests, and possibly refiltration, can be run on only the more effective coagulants to define particle settling rates, sludge volume, particle strength, and other criteria.

2. Jar Test Procedures

It may be a common misconception that there is, or should be, a "standard" jar test procedure that can be used universally to determine requirements for achieving solids removal through coagulation-sedimentation processes. In truth, there are many process methods for using coagulation (and still, about as many different devices to carry them out) that it is virtually impossible to write a single standard procedure that will fit all possible applications. Moreover, there are usually several alternative methods that can be used to achieve a desired result when dealing with water and wastewater treatment.

It is possible, however, to describe certain aspects of a jar test that are common to most coagulation-sedimentation processes which can be performed in a standardized manner. A standardized recommended practice for jar test of water is presented in the

1974 Annual Book of ASTM Standards. (Part 31, titled "Standard Recommended Practice For Coagulation-Flocculation Jar Test of Water" (Designation: D 2035-74).)[9] Also, since wastewater treatment usually requires a special emphasis on parameters such as sludge volume, and variations in hydraulic and solids loadings to the system, a procedure for coagulation-flocculation of wastewater is presented by the writer.

It is the opinion of the writer that neither of these tests should be performed without prior consideration of all the problems, implications, and objectives pertaining to the application at hand. If need be, certain steps or specifications can be changed or modified so the jar test will more closely represent the system or application being tested. The specifications and examples presented here conform (generally) to most applications, but certainly not all. In other words, it is of little value for us to allow a procedure to make us generate data that isn't needed while we neglect the pursuit of data that is.

In the treatment of water or wastewater by chemical coagulation and sedimentation, the same general principles of operation are used. Therefore, the jar tests used to test water and wastewater applications follow the same general practice. However, since wastewaters (even after a complete chemical analysis of sample) provide more uncertainty as to the chemical treatability (i.e., types, dosages, etc. of coagulants required) and equipment design parameters (i.e., surface rates, detention time, sludge volume, etc.) than natural water supplies, more information from jar tests is usually desirable. Some more common reasons for this unpredictability of wastewater are

1. Wide range of variations in wastewater suspended solids
2. Wide range of variations in wastewater dissolved solids
3. Critical variations in wastewater pH value
4. Fluctuation in wastewater hydraulic flow
5. Presence of unknown or unsuspected constituents
6. Variations in wastewater temperature
7. Variations in possible synergistic effects caused by independent variations in the above parameters

In addition to its unpredictability, problems associated with large sludge volume, high solids density, viscosity, thermal currents, and other interferences with settling operations, make it necessary to derive more information from jar tests for wastewater. Figure 1 is an example of a jar test data sheet for water, and Figure 2 is a data sheet for wastewater jar testing. In this discussion, the writer suggests the following wastewater jar test practice as a modification of the ASTM "Standard Jar Test Practice For Water".

a. Scope

As in the jar test of water, this general procedure covers the evaluation of treatment to reduce dissolved and other nonsettleable matter by chemical coagulation-flocculation followed by gravity settling. In addition, it is intended to aid in estimating the volume of sludge which must be removed (sludge blowoff as percent of influent volume) during the settling operation. Further, it evaluates the use of sludge recycle (solids contact) in reducing chemical costs and improving effluent quality. In wastewater treatment, estimating chemical dosages is important primarily for sizing chemical feed equipment and secondarily for chemical operating cost estimates.

b. Applicable Documents

See also, ASTM Standards D1192, D1293, and D1496 (*1974 Annual Book of ASTM Standards, Part 31*) with respect to sampling and testing of wastewater.

| Sample _____ pH _____ Turbidity _____ Date _____ |
| Location _____ Color _____ Temperature _____ Sample Size _____ ml |

Chemicals, mg/litre *(a)*	JAR NUMBER					
---	1	2	3	4	5	6
Flash Mix Speed, rpm						
Flash Mix Time, min						
Slow Mix Speed, rpm						
Slow Mix Time, min						
Temperature, °F						
Time First Floc, min						
Size Floc						
Settling rate						
Turbidity						
Color						
pH						

(a) Indicate order of addition of chemicals.

FIGURE 1.   Jar test data. (From *1974 Annual Book of ASTM Standards,* Part 31 American Society for Testing and Materials, Philadelphia, 1974, 818. With permission.)

c. Summary of Method

Various chemical additions are made to a representative wastewater sample in order to estimate the chemicals, dosage requirements, and the reaction time requirements (as well as other conditions) to achieve desired results. The parameters investigated in this suggested practice include: (1) additives, (2) pH value, (3) temperature, (4) order of addition and mixing requirements, (5) effect of sludge recirculation, and (6) sludge blowoff volume.

d. Definitions

Refer to "Terminology" at the end of this chapter. Other pertinent definitions may be found in D1129 of 1974 ASTM Standards, Part 3.[9]

e. Interferences

The same interferences affecting the jar test of water (i.e., D2035-74)[9] should be avoided with wastewater jar tests. These include: temperature changes, gas release, and changes in wastewater characteristics due to sample handling, transporting, and holding time.

f. Apparatus

Multiple stirrer — The same type of multiposition stirrer as used in the Standard ASTM Jar Test (D2035-74) is appropriate for wastewater.[9]

Jars (or beakers) — These should all be identical in size and shape. Normally recommended size is 1000 ml to 1500 ml Griffin beakers.[9] When the volume of wastewater sample is limited, 600 ml beakers containing 500 ml of sample will suffice.

Measuring and dispensing reagents — Reagent racks, such as described in ASTM Standard D2035-74,[9] are useful when simultaneous addition of chemical solutions is essential. However, in wastewater treatment, the writer has usually found that as long as the different solutions are applied within 10 to 15 sec of each other (assuming a fast mix at 100 rpm* for 30 sec to 1 min follows the last addition to insure complete disper-

---

*   Velocity gradient jar test studies showed that minimum threshold speed for turbulence is 100 rpm for unbaffled jars and 40 to 50 rpm for baffled vessels.[10]

Sample _____   Date _____

Time _____ pH _____ Turb. or S.S. _____ Z.P. _____

Location _____ Color _____ Temp. _____ Sample Size _____ ml.

	JAR NUMBER					
	1	2	3	4	5	6
Chemicals, mg/l (a)						
Fast Mix Speed, rpm						
Fast Mix Time, min.						
Slow Mix Speed, rpm						
Slow Mix Time, min.						
Temperature						
First Floc Seen, min.						
pH During Slow Mix						
2 min. Settled Supernate, Turb.						
Settled Solids Vol. 2 min.						
Settled Solids Vol. min.						
Settled Solids Vol. min.						
Final Supernatant, min.						
Turbidity						
pH						
Color						
Zeta Potential						
etc.						

FIGURE 2.   Wastewater jar test data.

sal of all reagents) the final results will not be significantly affected. Disposable plastic syringes (without needles) have been used quite successfully for both measuring and dispensing solutions or slurries to the samples in a jar test. Useful sizes available are 1 ml, 3 ml, 5 ml, and 12 ml graduated volumes. These can be washed many times between uses and are, therefore, relatively inexpensive. By using two hands to dispense from two syringes simultaneously, it is possible to add six reagent aliquots to six beakers in less than 10 sec. For adding slurries such as lime, the syringes are superior to test tubes, and the like, in that the entire contents of the syringes are discharged, whereas some solids often remain behind on the walls of other types of containers.

When dispensing polymer solutions it isn't advantageous to dump the whole aliquot into the sample under test in one slug. This causes local supersaturation of the polymer and results in poor polymer performance. Syringes, however, enable a slow dropwise addition of polymer solution to the sample which can be readily controlled. Pipettes are also useful for measuring and dispensing (instead of test tubes, graduated cylinders, etc.). However, when several different reagents have to be added, pipettes become less desirable because of the excessive time required for reagent addition.

g. Reagents

Purity of reagents — When possible, the grade of chemicals to be used in the full scale treatment process should be used. Normally, these are commercial grades which are of lower purity and produce more sludge than reagent grade chemicals. Use of commercial grade chemicals should then give chemical dosage requirements and sludge volume production closer to actual process conditions than will reagent grade chemicals. If commercial grade chemicals are not available, reagent or technical grade chemicals can be used.

Purity of water — For reagents prepared daily, potable tap water is usually sufficient. However, for reagents kept for longer periods, distilled or deionized water should be used.

Typical chemicals and additives — Generally, the same chemicals and additives used in water treatment (listed in ASTM D2035-74)[9] are used for wastewater treatment. It is sometimes necessary to prepare the concentration of prime coagulants, precipitating agents, oxidizing agents, alkalis, weighting agents, adsorbents, etc., at concentrations greater than 10 ($\pm$0.1) g/ℓ. For wastewater jar tests, concentrations of 20 ($\pm$0.1) g/ℓ to 100 ($\pm$0.1) g/ℓ (i.e., 1 mℓ/ℓ = 20 mg/ℓ to 100 mg/ℓ respectively) are often required to keep the total volume of reagent solution at 10 mℓ or less to avoid critical overdilution of the wastewater, especially when several different agents are added to a sample.

In addition to those chemical additives listed for water treatment, agents sometimes used in wastewater treatment include:

Precipitating Agents

Zinc salts (i.e., zinc chloride, zinc sulfate, etc.)
Calcium chloride ($CaCl_2$)
Sodium Phosphates (i.e., $Na_3PO_4 \cdot 12 H_2O$, etc.)
Sulfides (i.e., H_2S, $Na_2S \cdot 9 H_2O$, NaHS, FeS)

Reducing Agents

Sodium sulfite (Na_2SO_3)   Ferrous sulfate ($FeSO_4 \cdot 7 H_2O$)
Sodium metabisulfite ($Na_2S_2O_5$)   Sulfur dioxide (SO_2 gas)
Sodium bisulfite ($NaHSO_3$)

Acids

Sulfuric acid (H_2SO_4)   Carbon dioxide (i.e., $CO_2 + H_2O = H_2CO_3$)
Hydrochloric acid (HCl)   carbonic acid

Adsorbents

Powdered activated carbon (for dissolved organics)
Diatomaceous earth (for grease and oil)

Coagulant aids — Either anionic, cationic, or nonionic polyelectrolytes are found useful in flocculating or agglomerating suspended material in wastewater. Available polyelectrolyte manufacturers' technical data should be consulted for selecting specific polymers to evaluate. Powdered polymers should be prepared as 0.1% solutions. The liquid polymers can usually be prepared and handled very easily at concentrations well above 0.1%. (Always consult manufacturers' specifications for the proper concentration of stock solutions).

h. Sampling

Proper sampling of wastewater is a critical and important step in the success of a wastewater jar test. With wastewater, make sure that the average wastewater conditions are represented by the sample under test. If not, the results of the jar test could be very misleading. A major effort is required in deciding what samples should be taken, how much, and how often.

It is usually helpful to collect grab samples at close intervals (about 15 to 60 min apart) that are proportional to the flow rate and then combine them after the end of the day into a composite sample. This sample, representing the average wastewater condition for the day, can be jar tested to determine chemical requirements, etc. If the wastewater conditions are expected to change on different days, a jar test should be performed on a sample composited for each day a change in wastewater conditions is known or expected to occur. When the treatment requirements are evaluated for the average conditions represented by the daily composites, jar tests should be performed on individual grab samples collected at frequent intervals (about every 1 to 2 hr) to determine the range in wastewater treatment requirements that occur during the day. In this way, the amount of necessary flow volume equalization can be evaluated.

If the test is to be performed for a nonexistent or hypothetical wastewater, samples of an existing wastewater source that has conditions like or similar to the hypothetical waste should be tested. Additionally, jar tests should also be run on simulated or synthetically prepared wastewater representing conditions expected in the hypothetical wastewater. Such samples should be prepared with varying composition of expected constituents so that the impact of each constituent variation on treatability can be evaluated.

In all cases, the jar tests should be conducted as soon after the samples are obtained as possible. This is necessary to avoid critical chemical or physical changes from occurring in the sample upon standing. Wastewaters containing biological activity are particularly susceptible to changes in suspended solids upon standing. Such changes can and will produce misleading results.

i. Procedure

Begin by measuring equal volumes (e.g., 1000 mℓ) of the representative wastewater sample into each of the beakers. Normally, four to six samples are tested simultaneously depending upon the number of positions on the stirrer apparatus. Off-center location of mixing blades, i.e., about 6 mm (¼ in.) from the beaker wall provides better (more thorough) mixing conditions in cylindrical beakers. Sample temperature and pH value should be noted and recorded at this time.

The next step is to load the test reagents in the syringes and place each one near the beaker that is to receive it. If several different reagents are to be added, place the syringe to be used first closest to the beaker, and arrange the rest in order of addition sequence next to this, in a like manner for each beaker. If at all possible, limit the volume of each reagent addition to a maximum 10 mℓ. When larger volumes than 10 mℓ of reagent have to be added, dilute the smaller reagent volumes with water to the same volume to minimize differences due to dilution of the wastewater sample.

Start the multiple stirrer at 100 to 120 rpm (fast mix). Then, add the test reagents to each beaker in the desired order. Be sure the first reagent in the sequence has been added to each beaker before the second is added. If three or more reagents are to be tested, continue to add in the manner that all beakers receive the same reagent in the sequence before any of the beakers receive the next reagent in the sequence. In other words, the second reagent, in sequence, should be added to all beakers before any of the beakers receive the third. It may be desirable to vary the order of reagents in the sequence to test the effect of this. However, it is normally found that best results are attained when the sequence of order of addition is as follows:

1. Add alkali or acid for desired pH adjustment.
2. Add adsorbents (such as powdered carbon) and/or weighting agents (such as clays).
3. Add inorganic coagulants or precipitants to destabilize colloids, suspended solids, or dissolved (i.e., ionic) species.
4. Add coagulant aids such as polyelectrolytes.

It is desirable to add pH adjusting agents first so they have time to completely dissolve before the coagulant or precipitant is added. Since pH influences the effectiveness of coagulation or precipitation that takes place when using inorganic reagents, the process is usually more effective when the pH is preadjusted before coagulant addition.

Adding the polyelectrolyte last insures that no colloids are added to or produced in the system after the polymer molecules have all been adsorbed on suspended solids initially present in the wastewater. In other words, the polymer addition is held back until all possible suspended solids have been introduced, and then the polymer is added to aid agglomeration of all the suspended solids. Polymers referred to here are the high-molecular-weight (nonionic and anionic) bridging types that increase floc size and strength. Polymers used as primary coagulants (highly charged cationics) are usually added in place of or just after the inorganic coagulants.

After the last chemical addition, the fast mix is continued for about 30 to 60 sec (the exact amount of fast mix time may be a variable studied). The fast mix time and speed are recorded.

Next, reduce the stirring speed to about 30 rpm. Then, when the flocs start to get visibly larger in size record the time (usually this occurs in 5 to 10 min) and readjust the speed to the minimum that will keep the flocs evenly suspended throughout the sample. Every 5 to 10 min of slow mix record the time, relative floc size, and stirring speed (rpm). After 20 to 30 min of total slow mix time stop the stirring in all beakers (the optimum amount of slow mix time may be variable of the study).

Following the slow mix period, withdraw the paddles from each sample. Estimate the time it takes the bulk of the particles to settle to a level below a point 5 cm (approximately 2 in.) below the liquid surface. The objective here is to estimate if the solids will settle at normal clarifier surface rates of 2.5 to 3.0 $m^3/hr/m^2$ (i.e., approximately 1.0 to 1.25 gal/min/ft^2). The solids should settle 5 cm in about 1.0 to 1.25 min or less to allow for operation at these surface rates.

After 15 min of settling, record the relative height of the settled solids layer in each beaker. Carefully, withdraw enough supernatant liquid from each beaker to measure turbidity, pH, and other required analysis. The samples should be withdrawn (with a syphon or pipette) from a point one half the depth of the supernatant liquid. The height (depth) of the settled solids layer can be measured and recorded again after 20 min, 30 min, and 60 min. (Note: the depth of the settled solids layer divided by the initial depth of the untreated sample gives an estimate of the relative percentage of sludge blowoff required under those conditions. By allowing increased time to settle and concentrate the solids, the blowoff volume is reduced.)

Repeat the first six steps until all the unknown variables under test have been evaluated. A typical approach to evaluating the type and dosages of chemical agents to add is as follows: evaluate several different anionic, nonionic, and cationic polyelectrolytes at various dosages to see if they will work as primary coagulants or effective flocculating agents without inorganic coagulants. If the pH of the untreated wastewater is outside of desired limits, adjust with acid or alkali to various values within the allowable pH range to determine an optimum pH for coagulation with the polymers. If polymers alone are not effective enough proceed as follows: select several coagulants (i.e., alum, ferric salts, etc.) and prepare solutions as specified previously. The optimum pH value for coagulation with these salts will be evaluated first. This is done by adding the coagulant to be tested to a sample of the untreated wastewater at an arbitrary dosage (for example, 100 mg/ℓ). Then, while rapidly stirring the sample, add acid or alkali in small increments while measuring the pH with a laboratory pH meter. A plot or table of acid or alkali dosage vs. pH can be prepared showing the corresponding dosages of pH adjusting agent required to adjust the sample (with 100 mg/ℓ of coagulant) to specific pH values. The jar test is then initiated by adding the predetermined dosages of pH adjusting agent to the beakers containing the wastewater sample. These dosages correspond to the desired specific pH values that are to be evaluated. The coagulant is then added to the mixing sample in each beaker (at 100 mg/ℓ dosage for example) and the test is carried out to determine which pH values give the best results.

After the optimum pH value is determined for the coagulants, the optimum coagulant dosages are determined. This can be facilitated by again taking a sample of the untreated wastewater and adding increasing dosages of coagulant from lower to higher than anticipated requirements while rapidly stirring it and measuring the pH. With each dosage increase (for example, six different coagulant dosages) add and record the dosage of pH adjusting agent required to maintain the pH in the optimum range previously determined for that coagulant. If precipitants or adsorbents are to be added, they should be present when the optimum coagulant dosage is being determined. Polyelectrolytes can then be evaluated as coagulant aids after the optimum pH value and coagulant dosages have been determined.

In many cases it is advantageous, from an economic and quality standpoint, to have sludge recirculation (or sludge blanket contact) to increase the solids contact during the coagulation process. Sludge recirculation can be evaluated, after determining the optimum coagulating agents (in the first seven steps) as follows: treat 1500 mℓ of raw wastewater in each of five beakers by the optimum coagulant conditions already established, without sludge recycle. After the mixing steps are complete, allow the sludge to settle about 30 to 60 min. Then, decant all supernatant material off and save the settled solids. Transfer the settled solids in each beaker to individual test tubes or graduates and allow to stand about 10 min. Pour off all free water.

Treat 500 mℓ of raw wastewater in each of six to 600-mℓ beakers at the optimum pH value but vary all other chemical dosages as follows:

Beaker number	1	2	3	4	5	6
Percent of optimum dosage	20	40	60	80	100	100

Then, during fast mix, add up to 50 mℓ of the sludge collected in the previous step to beakers 1 through 5 and none to beaker 6. Continue to fast mix for 30 sec and proceed with the jar test as described in the fourth through sixth steps. If sludge recycle is advantageous, it should be apparent by an improved effluent quality in beaker 5 from that observed in beaker 6. Also, the other beakers may show equal or better effluent quality than beaker 6, indicating chemical dosages can be reduced when using sludge recycle.

The time values given in this procedure are only general suggestions. The times may be adjusted or modified to fit special situations more suitably.

j. Reproducibility

If desired, the degree of the errors caused by the time differences in chemical addition and supernatant withdrawal can be evaluated by rerunning certain tests in duplicate or triplicate. Figure 2 is a suggested jar test data form that can be used for recording wastewater jar tests.

3. Zeta Potential Test Procedure
a. Scope

This suggested procedure describes only one of many possible test methods for ZP measurements of suspended solids in aqueous liquids. The specific objective here, is to evaluate a wide range of coagulants, particularly polyelectrolytes, for the destabilization of suspended particles. It is assumed here, and generally accepted, that reduction in ZP (as measured by electrophoretic mobility) to ±5 mV constitutes a destabilization of particles so they can be coagulated more readily by mechanical agitation.[7,11]

This procedure when used alone can enable the selection of a few effective coagulants that can be tested on stream in either a pilot plant or full scale plant operation. However, for settling operations, it is suggested that the selected coagulants be evaluated by jar tests before applying them on-stream. Whereas ZP measurements enable rapid determination of effective types and dosages of coagulants, jar tests can indicate the relative settling rates of the flocs and the volume of sludge produced.

b. Summary of Method

A representative sample of a water or wastewater source is rapidly stirred under turbulent conditions. A reagent is added in increments to the stirring sample. Between the addition of each increment, a small portion of (approximately 40 to 50 mℓ) the sample is transferred to an electrophoresis cell and the average ZP is measured. For each dosage of reagent, a ZP is measured and both are recorded as the measurements are made.

In addition to coagulants (such as alum, ferric salts, and polyelectrolytes) other chemicals can affect the value of the ZP. These are acids, alkalis, oxidizing and reducing agents, precipitating agents, adsorbents, etc. Therefore, when a sample is to be treated for more than just suspended solids removal by coagulants, all chemical additives should be added or reacted in the sample before evaluating the coagulants. The primary variables evaluated in this procedure are (1) the effect of pH on zeta potential of suspended solids and (2) the effect of coagulants on zeta potential of suspended solids.

c. Interferences

The interferences found in measurement of zeta potential of suspended solids include:

1. Large differentials between sample temperature and room temperature
2. High electrical conductivity causing thermal currents and random movement of particles in electric field, for example seawater or brine
3. High biological activity in the sample, for example protozoa and massive algal blooms
4. Very high suspended solids concentrations interfering with free movement of individual particles when in an electric field
5. Large differences in particle sizes and zeta potential on individual particles making the tracking of individual particles difficult

d. Apparatus

The apparatus required to measure ZP consists of an electrophoresis cell, a DC power supply, and a means of tracking the direction and velocity of particle movements when electric power is applied to the cell. The setting up and construction of suitable equipment (from scratch) into a functional, accurate, and easily operated instrument is usually outside the scope of the average water or wastewater testing laboratory. There are, however, complete instrument packages available for ZP measurement under various trade names. One type of instrument uses a stereoscopic microscope to view the particle while tracking the particle movements. The Zeta-Meter® (manufactured by Zeta-Meter, Inc., N.Y.) uses this method with a specially designed electrophoresis cell, power supply, and lighting system.[7] Other devices, such as one that employs a laser device instead of a microscope for tracking particles, (Lasar-Zee™ manufactured by Penn-Kem Corp., Croton-on-Hudson, N.Y.) are available. Another instrument, that uses a television unit to view electrophoretic mobility is manufactured by Komline-Sanderson, Inc., Peapack, N.J. The procedure described here is based on the writer's experience in using the Zeta-Meter® with respect to sample sizes. However, it is expected that the basic approach of the test can be used with any type of ZP measuring device.

In addition to the instrumentation needed for actual measurement of ZP, the following laboratory equipment is needed: (1) A stirring device (preferably a magnetic drive type for easy cleaning), (2) a beaker (one or more ranging from 100 to 1000 mℓ) for holding and stirring the sample while reagents are added, and (3) pipettes or syringes (e.g., 1 mℓ to 10 mℓ sizes) for slow dropwise addition of chemical reagents into the stirring sample. Also, pipettes or syringes (e.g., 50 mℓ size) for withdrawing treated sample and transferring to electrophoresis cell.

e. Reagents

The same reagents that are used in jar tests of water and wastewater also apply to ZP measurement. The coagulants and polyelectrolytes are, however, especially applicable.

f. Sampling

The same sampling procedures required in water or wastewater jar tests apply to ZP tests.

g. Procedure for Testing a Reagent (i.e., Coagulant, Polymer, Alkali)

1. Measure 100 mℓ of sample into a 200 or 250 mℓ Griffin beaker. Start magnetic stirrer with appropriately sized mixing bar to create rapid swirling of sample without causing a vortex.

2. Measure the pH, turbidity, temperature, and other relevant characteristics of the sample that may change with chemical treatment.

3. Stop stirring and withdraw enough sample to fill the electrophoresis cell (following the manufacturer's instructions). Measure the average ZP of the sample.

4. Return the sample in the cell back into the beaker with the rest of the sample. Resume stirring. Add a small dosage of reagent (about one tenth or less of the anticipated dosage required). Allow to stir about 30 to 60 sec and then stop stirring.

5. Withdraw, again, enough sample to properly fill the electrophoresis cell. Measure the average ZP and record it along with the reagent dosage used. Note: if reagent addition changes the pH value of the sample, measure and record the pH value also.

6. Start sample stirring again and return contents of cell back into beaker. Add another increment of reagent and allow 30 to 60 sec to stir. Stop stirring and transfer an appropriate volume of sample to the electrophoresis cell. Measure and record reagent dosage, average ZP, pH, temperature, etc.

7. Repeat (1) through (6) until it is apparent that the ZP is being increased, decreased, or unchanged. If the ZP is decreased, continue to add reagent in increments

until it reaches zero and, if possible, continue until a charge reversal (reversal of particle direction) occurs. If the ZP is increased or unchanged after several incremental additions of a coagulant, record the results and assume the reagent is ineffective as a coagulant for the sample under test.

8. When the ZP evaluation of each reagent is completed, wash and rinse the beaker, stirrer, electrophoresis cell, pipettes, and other equipment in contact with the sample and/or reagent.

9. Repeat steps (1) through (8) for the evaluation of other reagents.

h. Reproducibility

The reproducibility of ZP measurements with respect to reagent dosages in a given sample is normally very good. However, proper setting up, adjustment, and use of equipment is essential to achieve reproducible results. Usually some experience with ZP measurements is necessary before proficiency and reproducibility of results are obtained. However, as one becomes more proficient, the evaluations can be made more rapidly. For the novice, about ten or more individual particle measurements may be necessary to arrive at an average ZP. With some experience, repeatable results can often be obtained with as few as three individual particles representing the average ZP.

C. Techniques of Applying Polyelectrolytes

Formerly, before the past decade or two, settling rate tests (for clarification applications) that were used to derive information for the design of settling tanks had to be accurate and highly representative if optimum performance from the equipment was to be realized. One type of test for this purpose, described by Camp,[12] employs a tall settling column, fitted with sample taps at various levels, surrounded by a constant temperature water bath. Similar sedimentation tests are also to be found in other sources.[13,14]

The settling columns are used to measure the quality of the sample (usually suspended solids concentration) at various depths and different settling time periods. They are usually as tall as a settling tank with a diameter large enough so, as samples are withdrawn, the liquid depth does not change significantly. The results sought indicate the settling rate of that fraction of the suspended solids that have to be removed. Therefore, the surface rate of the settling tank is set at a value somewhat lower than the indicated particle settling rate. For flocculent suspended solids, the design overflow rate is normally set at a value somewhere between 25 and 75% lower than the settling test result, according to common wastewater engineering practice.[13,14]

The settling tank overflow rate is not the only design criterion that concerns the engineer. But it is certainly one of the most important. In the past, the reason for a good deal of concern over particle settling velocity was the variability of most flocculent suspensions due to the fragile nature of floc particles. Even those solids coagulated with clays, in addition to alum or ferric salts, are subject to easy fracture with a resulting loss of settling velocity.

With the development of high-molecular-weight polyelectrolytes for water and wastewater treatment (throughout the 50s and 60s) it was found that flocs could be produced more rapidly and result in larger, stronger, more rapidly settling masses. This changed the picture of sample testing for settling tank design in many water or wastewater clarification applications. Now, in many cases, the settling rate can be controlled by proper use of polyelectrolytes alone and/or combined with inorganic coagulants. As a result, it is now often the case that existing settling tank design rates dictate what particle settling rates must be produced, rather than the existing settling rate of particles dictating the allowable surface rate of the tank.

Table 6
TYPICAL PARTICLE DENSITIES, SIZES, AND SETTLING RATES OBSERVED IN WATER AND WASTEWATER TREATMENT WITHOUT POLYMERS

Particle description	Treatment process	Density (g/cc)	Diameter (mm)	Free Setl Vel (cm/min)
Fine sand grains	Untreated	2.6	0.2	14
Mud flocs (95% H$_2$O) ⎫ Silt	Natural	1.03	—	—
Susp. vegetable matter ⎭	River Water	1.0—1.5	—	—
Al$_2$O$_3 \cdot$ 20 H$_2$O	Rapid	1.18	0.005—0.03	—
Fe$_2$O$_3 \cdot$ 20 H$_2$O	Sand filters	1.34	0.005—0.03	—
Al(OH)$_3$ (to 99% H$_2$O by vol.)	Flocculation ⎫ Silt	1.002—1.005	1.0	3—5
Fe(OH)$_3$ (to 99% H$_2$O by vol.)	Sedimentation ⎭	1.003—1.005	1.0	4—5
CaCO$_3$ crystals (75% H$_2$O)	Lime softening	1.2	0.1	3.6
MgOH$_2$ (99% H$_2$O)	Sedimentation	1.001	0.1—1.0	<4
Grit; coal dust, coffee grounds, seeds, egg shells, nondecomposable matter	Grit chamber sewage sedimentation	1.2—2.7	≥0.2	12—14
Organic suspended solids, comprised of proteins, fats, cellulose, and entrained H$_2$O and gas	Primary sewage sedimentation tank	<1.0—1.2	<0.001 to 3	<3 to >4
Activated sludge; comprised of gelatinous flocs, zoogleal matrix, bacteria, and undecomposed organic matter (up to 98% H$_2$O by vol.)	Secondary sedimentation tank	1.005—1.1	1—2	8—16

Based on data from Camp, T. R., *Trans. Am. Soc. Civ. Eng.,* Vol. III, 900, 1946.

Settling velocity of a particle, according to Stokes' Law,[12] is primarily controlled by its diameter and density if quiescent conditions are maintained. However, as Table 6 indicates, because a particle has a large diameter it does not necessarily have a rapid settling rate. Likewise, we find that all particles having high densities do not necessarily exhibit rapid settling rates. Cases in point are Al(OH)$_3$ or Fe(OH)$_3$ flocs, having about the same particle diameter as activated sludge (i.e., approximately 1 mm) settle at one fourth to one half the rate of activated sludge. Freshly formed CaCO$_3$ crystals, having a significantly higher density than activated sludge (i.e., 1.2 vs. 1.005 respectively) settle at one fourth to one half the rate of activated sludge.

It becomes apparent, from the above that optimally particle growth must be obtained without loss of particle density. This means avoiding the entrainment of large percentages of water within the structure of the floc particle while coagulation and flocculation are progressing. With proper application of coagulation principles, this can be approached. Many polyelectrolytes show a marked advantage in this role, in many cases by at least doubling the settling rates obtained without use of polymers.

1. Principles of Control in Coagulation Processes
The factors controlling coagulation efficiency (i.e., the number of particles removed from suspension via coagulation, relative to the initial number of particles, after a given period of time) are

1.   Particle collision efficiency — the fraction of all contacts between particles that actually result in successful aggregation.
2.   Particle collision frequency — the average time at which collisions between particles occur.
3.   Velocity gradients — energy or velocity of interparticle collisions.
4.   Detention time — the time provided for coagulation to take place.

All of these factors must be established at some minimum level before coagulation can proceed. Then, increasing the degree of any one or more of these factors will improve the efficiency of the coagulation process to some extent.

Collision efficiency can be increased by addition of coagulants or changing the chemical or physical characteristics of the liquid (i.e., pH, dissolved solids, temperature, etc.). Stumm states that in fresh natural waters (without coagulants) the collision efficiency is usually low with about one out of every 10^4 to 10^6 collisions actually being successful. He also points out that very high dissolved solids, such as in seawater, suppress the stabilizing forces to the degree that one out of every ten or less collisions leads to successful aggregation.[4]

Collision frequency is controlled by at least two criteria of great importance in the treatment process. One of these factors is the mixing velocity applied during coagulation. Turbulent mixing (as opposed to laminar fluid motion) increases the rate of particle transport resulting in increased collision frequency. The other factor of importance is particle concentration (specifically, the total number of dispersed particles with respect to the total volume of the suspension). The greater the particle concentration is, the shorter the interparticle distances are which leads to increased collision frequency.

Velocity gradients are also the product of the degree of mixing applied. In addition to increasing the collision frequency, effective mixing produces turbulent hydraulic velocity gradients (a dispersion of eddys or small high velocity currents in all directions) that increases the velocity of particles with respect to each other and the bulk liquid. Higher interparticle approach velocities increase the energy of the approach so that interparticle stabilization (i.e., repulsion) is more readily overcome. It is, from the standpoint of achieving both effective velocity gradients and collision frequencies, absolutely necessary to have adequate agitation (i.e., mixing or stirring) during the coagulation process.

Sufficient detention time is also important, even if the other factors are at suitable levels when the coagulation step is initiated. Considering that the rate of coagulation of particles is the product of collision efficiency, particle concentration (i.e., number per centimeter³ of suspension), velocity gradients, and detention time,[4] in a simplified case, the rate of coagulation has to decrease as the number of particles in the system decreases. That is, initially, when the number of particles decreases, due to the aggregation of many small particles into relatively few larger ones, the interparticle distance increases and the collision grequency decreases. Therefore, it is necessary to allow increasingly more time, as the number of uncoagulated particles becomes fewer and fewer, to approach total removal of particles. The problem of the decreasing particle concentration is overcome today, in many applications, by providing increased solids contact. Before discussing solids contact techniques, mixing criteria (as they apply to coagulant and polyelectrolyte applications) will be discussed.

2. Mixing Techniques for Coagulation in Settling Operations
a. Mixing Devices
Agitation to achieve turbulent velocity gradients, in coagulation-flocculation-settling

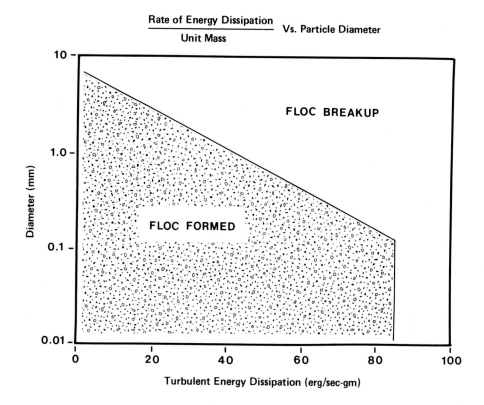

FIGURE 3. Various types of stirring blades used in coagulation-flocculation devices.

processes, is normally accomplished by either mechanical or hydraulic mixing devices. In the majority of cases mechanical stirring is used because of its "positive" operational aspects as well as being more controllable than hydraulic mixing in terms of turbulent energy applied evenly throughout a given space. Figure 3 shows a several types of commonly used stirring devices including paddle, turbine, and propeller type.

Historically, the paddle type stirrers have been used the most. Their configuration allows thorough but gentle (low shear) agitation at low shaft speeds. The anchor-type (1) is often a suitable design for vertically cylindrical-shaped tanks. The vertical shaft type (2) with a Christmas-tree shape is suitable for conical-shaped coagulation vessels. The reel-type (3) with its horizontal shaft is used in rectangular basins and other tanks with horizontally parallel walls.

The turbine type stirrers have come into use more recently for flocculation applications. They are suitable for vertical cylindrical shaped vessels where solids contact via sludge recirculation is employed. Presumably, the higher mechanical shear stress imparted on the floc, which tends to break it apart, is offset by the high solids concentration of recycled solids which tend to reaggregate it.

The propeller type stirrers are employed where high mechanical shear is not a significant problem; they may be used for initial coagulation because of their lower cost, high flow mixing capacity. They do tend to impart a substantial amount of mechanical shearing on flocculent suspended solids, especially when the suspension is detained in the zone of mixing for an appreciable length of time (i.e., more than 10 to 15 min).

An investigation of the various type mixers, in flocculation processes, was made by Hahn and Klute at the University of Karlsruhe, W. Germany.[15] In their paper they reported the following results: "propeller and turbine-type stirrers show a much higher turbulence in the vicinity of the stirrer as compared to the rest of the reactor while anchor-type and grid-type stirrers lead to a more uniform turbulence distribution (always at the same rotational speed and the same energy input)". They also concluded the following: "Differences in the turbulent flow field lead to different particle aggregation and floc distribution".

In comparing anchor-type, turbine type, and propeller-type stirrers on a "strongly destabilized system" (montmorillonite coagulated with calcium ions) the turbine type gave best performance at low rotational speed whereas the propeller gave best performance at higher rotational speed.

In comparing anchor-type, turbine-type, and propeller-type in a more "strongly flocculated system" (i.e., kaolinite flocculated with polyelectrolyte) "the more uniformly stirring anchor-type device appears to prove most efficient, particularly at higher rotational speeds".

It is, therefore, a matter of selecting the mixing device that best suits the particular solid suspension, coagulating agents, and tank configuration, rather than concluding any one type of device is best for all applications.

b. Mixing Stages

Following the initial stage of chemical addition and turbulent agitation to achieve particle destabilization and initial aggregation (generally termed coagulation), there are at least one or two further stages of particle growth that occur before the actual settling step can be effectively employed. For a practical approach, the various stages of coagulation-sedimentation can be described relative to the significant physical size and shape changes that are observed in the growth of the particles. These stages or changes may be expressed as, for both water and wastewater, the relative generalities shown in Table 7.

It is apparent from Table 7 that, if particle growth is to be sustained after initial coagulation, the degree of agitation (i.e., mixing speed) must be lowered. Otherwise, the higher hydraulic velocity gradients, required for initial coagulation, will shear flocs, over a certain diameter, into smaller particles. Thus, the size and settling rate of floc particles is to be controlled by the relative speed of the stirrers.

A relationship between floc particle diameter and turbulent energy dissipation (proportional to the degree of mixing) was developed, on theoretical and experimental considerations, by Delichatsios and Probstein.[16] Figure 4, which is based on this relationship for ferric hydroxide floc, indicates how floc size is limited to a certain maximum diameter at various degrees of turbulence. According to the authors, the floc break-up is due primarily to "unsteady turbulent pressure fluctuations" across the particle surface and secondly to "shear forces associated with the relative motion of the particle with respect to the continuous fluid".

The phenomenon of floc break-up in high turbulence is of course, well known. However, what this writer wishes to convey by reference to these studies is, in order to achieve efficient coagulation as well as sedimentation, it is imperative that the degree of applied mixing has to be controlled within both upper and lower limits. To achieve efficient coagulation as well as sedimentation prior to settling, the degree of mixing usually has to be gradually diminished if floc particle growth is to continue above a certain size. Table 7 illustrates this principle as it is generally applied in water and wastewater applications. The mixing velocities shown for coagulation, flocculation, and agglomeration, as separate stages, are applicable to standard type mixing paddles. The third mixing stage (agglomeration) is sometimes used in practice, in older water treatment plants that employ separate mixing tanks ahead of a settling basin. However, this step may be combined with the flocculation step in newer solids contact type clarifiers.

Table 7
SIZE AND SHAPE RELATIONSHIP OF FLOC PARTICLES DURING VARIOUS STAGES OF COAGULATION-SEDIMENTATION PROCESSES

Stage of treatment	Detention time (min)	Relative mix-ing[a] velocity (ft/sec)	Diameter (mm)/shape of particles	Ref.
Influent without chemicals	—	—	10^{-6}—10^{-3}/Colloids	1, 4, 7, 11
Influent without chemicals	—	—	10^{-3}—2×10^{-2}/Various particulates	7, 11
Chem. destabilization/coagulation	0.5—3	1.3—3.0	10^{-6}—10^{-1}/Discrete particles	1, 11, 15, 16
Floculation	20—40	0.5—1.0	10^{-6}—1/Amorph. flocs	1, 11, 15
Agglomeration	7—10	0.4—0.6	1—8/Amorph. masses	1, 11, 16
Sedimentation	30—90	<0.003	1—10/Amorph. masses	11, 16

[a]   Stirrer tip speed for mixing stages or liquid rise rate in settling.

A uniform addition and dispersal of high-molecular-weight polyelectrolyte (especially the weakly anionic and nonionic polyacrylamides at concentrations of about 0.01 to 0.2 mg/ℓ) to the flocculation step enables alum and iron flocs to become quite resistant to the degree of turbulent shear and pressure forces present in the flocculation step (i.e., 0.5 to 1.0 ft/sec stirrer tip speed). The floc growth rate can proceed to diameters as large as 2 to 5 mm with less entrained water so that the sedimentation step can directly follow flocculation. It is often observed that aggregation of flocculent particles at higher tip speeds results in less water entrained and, therefore, denser particles than at lower tip speeds.

It may not be advisable, however, to add the higher-molecular-weight polymer directly to a separate flocculation step if solids contact (sludge recirculation, for example) is not also employed. The polymer may tend to make the floc particles grow too rapidly minimizing the amount of contact opportunities the colloidal particles will have. The main purpose of the extended flocculation period (if no sludge recirculation) is to allow small floc particles ample time to contact and aggregate unflocculated and uncoagulated particles remaining after the coagulation step. If these small flocs were to grow too quickly, there would be less surface area available for effective removal of the existing colloidal solids. A plant that uses the three mixing basin approach and where high-molecular-weight polymer is successfully added to the agglomeration stage is discussed by H.G. Swope where ZP is used to control the chemical additions.[11]

Another method of polyelectrolyte application, generally useful in flocculation of many colloidal suspensions (either water or wastewater) is the use of the lower-molecular-weight cationics (i.e., highly charged polyamines, polyethyleneimines, etc.). These serve as primary coagulants and destabilizing agents either with or without inorganic coagulants. In this role, these polymers are added because they have the following advantages over inorganic agents:

1.   They are usually effective at one tenth (or less) of the inorganic coagulant dosage.
2.   In replacing inorganic salts, they produce less sludge volume.
3.   They add far less dissolved solids than inorganic salts and, therefore, are desirable in process water treatment, boiler water treatment and in recycling of wastewater for reuse.
4.   They are usually less pH dependent than inorganic coagulants.

Their main disadvantages are

1.   High specificity of activity in relation to different solid suspensions*
2.   Problems with biodegradeability of prepared polymer feed solutions
3.   Low resistance to chlorine and other oxidizing agents
4.   Relatively high cost per unit weight

In addition to the above it is usually found that, although cationic polymers will tend to form stronger flocs than alum or ferric salts (especially useful in filtration applications) at low dosages they do not always exhibit a noticeable improvement in particle settling rates. Therefore, their application should be carefully evaluated in jar tests, and possibly ZP tests.

When a cationic polymer does show a decided advantage, it normally should be added to the coagulation stage either in place of inorganic agents, with inorganic agents, or within a few minutes after. In all cases, allow at least 30 sec of rapid mix, 0.05 to 0.9 m/sec (1.5 to 3 ft/sec), after cationic polymer addition. The writer has found these time sequences to be generally applicable and in agreement with other investigators.[11,17] There have been no cases found, either within personal experience or work cited, where it has been advisable to add the cationic polymer before the inorganic coagulant. However, when dealing with wastewaters where variable pH values and other chemical characteristics may reverse the ZP of the colloids or the polymer, no possibility should be left unevaluated in a jar test study.

Many cationic polymers, especially the higher-molecular-weight varieties tend to react with suspended solids by both charge neutralization and bridging mechanisms. And, when a high-molecular-weight polyacrylamide (nonionic or anionic) is added to a later stage of mixing (i.e., flocculation or agglomeration) the resulting flocs tend to be particularly large, dense, and rapid settling. In such cases care must be exercised not to overdose with the high-molecular-weight polyacrylamide since both colloid flocculation efficiency may be adversely effected and significant residuals of polyacrylamide may pass through the settling stage to cause problems downstream.

An easily recognizable sign of overdosing with high-molecular-weight polymers (in settling and clarification operations) can be detected in jar tests. If glass beakers or jars are used, the overdosed flocs will exhibit a particularly strong affinity for the glass. When the treated liquid is poured from the beaker, the flocculated particles should be carried away with the liquid if it is not overdosed. If it is overdosed, floc particles will cling to the bottom and sides of the beaker after the water has been poured off.

3. The Settling Step

In keeping with the general objective of this chapter, a complete discussion of the basic theories applied to settling processes (such as Stokes' Law, Reynold's Number correlations, and other mathematical models) will be bypassed. Instead, a discussion of how polyelectrolytes may be applied to optimize the performance of existing settler designs, rather than how to design the optimum settling tank, will be undertaken. In doing so, the writer does not wish to underrate the importance of sound settler design. On the contrary, it is assumed that the basic settling tank design is satisfactory, but the objective of polymer application is to either increase throughput (modestly), improve economics of operation, or improve effluent quality (or maybe a little of all three). It would be a rare case where polyelectrolyte application alone could remedy

---

*   Some cationic polymers have been found to react primarily with the organic colloids and anions present, before they will coagulate suspended inorganics.[18]

PADDLE-TYPE STIRRERS

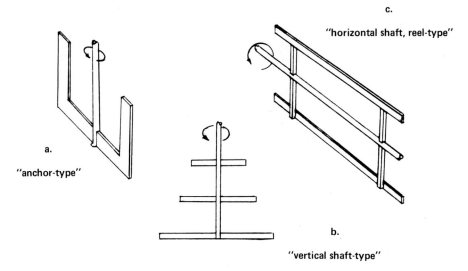

c.

"horizontal shaft, reel-type"

a.

"anchor-type"

b.

"vertical shaft-type"

TURBINE-TYPE STIRRER

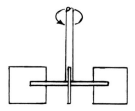

PROPELLER-TYPE STIRRER

FIGURE 4. Dependence of floc size on turbulent energy dissipated into a system of ferric hydroxide in water.

poor settling tank design. For discussions dealing with theoretical design concepts the reader is referred to the principles of sedimentation as explained by Camp,[12] Metcalf and Eddy,[13] Rich,[14] or Kynch,[19] as well as others.

In essence, a good basic design incorporates: *even distribution* of the solid-liquid suspension into the settling zone, *laminar flow* and adequate detention time through the settling zone, and *equal collection* of clarified liquid from the settling zone surface. These parameters are necessary to avoid the short circuiting, high velocity currents that can both fracture and sweep flocculated solids through the settling zone to the collection system at which point they escape with the effluent. It is to be understood that in this discussion of settling for clarification, *free settling* is the mode of solids separation from the liquid. This is to be distinguished from *hindered settling* and *compression* which are characteristic of more concentrated slurries or sludges.

In order for settling devices, such as continuous flow settling basins, reactor-clarifiers, and the like to operate properly, the essential parameters described above for good basic design must be adhered to first. Then, effective coagulation-flocculation and agglomeration of the influent suspension must be maintained during operation. Finally, there remains one last, but critically important criterion that must always be maintained. That is, the solid phase of the coagulated influent slurry must be free settling under quiescent or laminar flow conditions when introduced into the settling zone. Furthermore, the solids must settle freely to occupy a settled solids volume that is a low percentage of the influent volume in a shorter period than the detention time of the total flow through the settling zone. This, however, is not to say that the concentration of particles that accumulates in the settling zone cannot approach hindered settling conditions. Actually, this does occur in some of the solids contact type reactor-clarifiers.

a. Principles of Sedimentation

The rudiments of sedimentation principles have been studied and understood at least since about the turn of the century in that it was shown by Hazen in 1904,[20] particle removal depended on tank surface area rather than volume. The factors controlling the settling of discrete particles have been adequately defined mathematically. The factors controlling settling of flocculated slurries are more complex and not as readily determined by mathematical expressions. It is, therefore, usually necessary to determine settling rates of flocculent suspensions experimentally.[14,21]

In the classic representation of sedimentation, it is common to classify the various types of sedimentation as follows:

Class 1 sedimentation — Refers to settling of discrete "nonflocculating" particles in dilute suspensions.

Class 2 sedimentation — Applies to the settling of suspensions of "flocculating" particles.

Type 1 zone settling — Refers to the behavior of dilute flocculent suspensions where removal is principally by free settling.

Type 2 phase settling — This occurs when a suspension of flocculent particles reaches a concentration that causes them to contact each other obstructing the free settling paths of individual particles. The result is that the solid phase as a whole, assumes a common settling rate as the individual particles within it move about at varying velocities and directions while the liquid "channels" upwardly through it. This type of sedimentation is also termed "channeling".[21]

Type 3 compression — The concentration of flocculent solids is too high for individual floc particles to exist. The subsidence of the solid phase may be thought of as due to the weight of the solid mass in upper layers creating pressure on the lower layers thereby consolidating these layers into a porous solid (higher mass) area as water "filters" up toward less concentrated layers where channelling exists.[21]

The class 1 sedimentation of nonflocculent particles (those that do not coalesce to form larger particles) is not usually a problem in water and wastewater treatment unless they are too small to be effectively settled in the removal process. In those cases where they are too small, they are coagulated with inorganic salts and/or polymers which, then, produce flocs and make it a Class 2 sedimentation process. For this reason, the behavior of flocculent suspensions will be the principle concern in this discussion.

Figure 5 illustrates both graphically and pictorially, an example of the behavior of the three different types of Class 2 sedimentation. Although these are thought of as

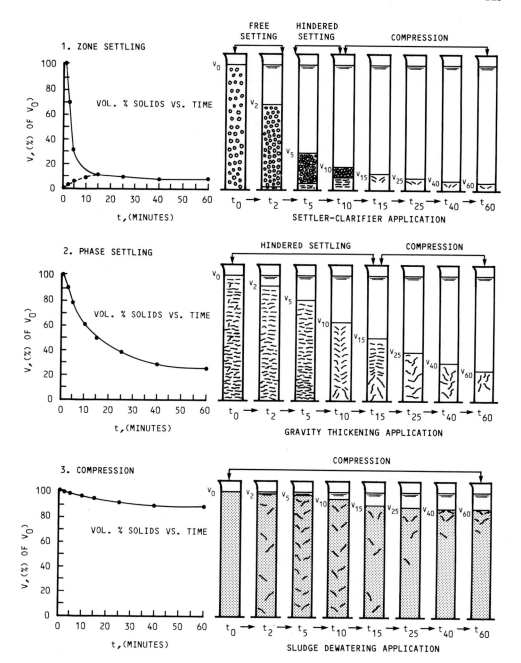

FIGURE 5. Graphical and pictorial representations of the three types of sedimentation that occur with flocculent suspensions. (The settling tests illustrated show a reduction in settled solids volume with time for three different initial solids concentrations.)

distinct cases of settling, in reality there are numerous cases where the behavior of the suspensions follows a pattern anywhere in-between or outside of the settling rates in this example.

In this example, a graduated cylinder (usually 1 to 2 ℓ) is filled to the mark with a particular suspension of flocculent particles. Then, the rate of subsidence of the solid phase is observed by recording the volume of solid phase (distance or height is also typically used) at times 0, 2 min, 5 min, 10 min, and other intervals up to 60 min or longer. Normally, the test is completed when it is apparent the solids mass will not settle much more even after a long period of time. (Settling times of 30 to 60 min are usually used for Type 1 and Type 2 sedimentation, whereas periods up to 24 to 48 hr may be used in evaluating Type 3 compression.) When the settling test is complete, the volume of the solid phase, at each time interval, is plotted as a percent of the initial sample volume against the time required to achieve that volume. The shape of the curve is then observed for an indication of the type of process that is required for solids removal from this sample.

If a curve as shown in Type 1 (zone settling) results, the removal of solids from the liquid is primarily by free settling due to the large distances between particles in the initial suspension. In this example, about 70% of the liquid volume is treated by free settling of the flocs and an additional 20% or so by hindered settling. The remaining 2% may be clarified by compression of the sludge mass if ample time is given to do so (i.e., 40 to 60 min). In this case, about 85 to 90% of the slurry is treatable by a settling tank or clarifier and another 2 to 7% by gravity thickening of the underflow.

If a curve more like the shape of Type 2 (phase settling) results, a settling tank or clarifier will not be very effective. In this case a gravity thickener is applicable because the solids do not settle freely at the start of the sedimentation period. In order to allow the sedimentation to proceed, upward flow through the solids mass must be eliminated so that the primary force exerted on the solids is the pull of gravity. Gravity thickeners are designed to allow this to occur, whereas clarification-settling tanks are not. With the slurry used in this example, the observed quantity of influent clarified is about 60 to 73% by volume. The sedimentation process proceeds primarily by channeling (i.e., 60 vol% of influent) and the rest of the way by compression (i.e., about 10 to 13% by volume). Channeling occurs due to the rise of water out of the sludge zone as the particles move downwardly into it. The solids mass becomes oriented into large irregular groups of solid agglomerates with many large channels running between and around them to allow the entrained water an escape route. Since the actual degree of sedimentation due to "phase settling" in full scale thickeners is a function of depth of the thickening zone, this type of test can only be used as a rough estimate of gravity thickener surface area requirements.

In the event that the settling test results in a curve similar to Type 3 (compression), it is an indication that even gravity thickening (on a continuous basis) will not effectively remove the suspended solids. The type of solid phase subsidence that occurs here is due to the consolidation of the individual floc particles that initially settled (first as individual particles and then as groups or agglomerates) into a denser matrix which becomes a continuous porous medium. The weight of the solid layers building upon each other forces water, (initially trapped in individual flocs) through the porous medium upward into the supernatant liquid on top. Depending on the nature of the solid material, its relative resistance to flow through it, and the volume (height) of the compressed mass, wide channels or narrow cracks throughout the mass can form to allow water to escape.

This brief illustration of the behavior of flocculent slurries in settling operations is meant only as a method for evaluating the most feasible type of settling operation to use. When it happens that the behavior of the slurry, upon settling, is an intermediate between either (1) zone settling and phase settling or (2) between phase settling and compression, at least two choices can be made. The first choice could be to employ the process that will handle the greater volume of solids which for (1) would mean going with a gravity thickener or for (2) with sludge dewatering. Both clarifiers and gravity thickeners work more efficiently with a lower concentration (i.e., vol% of in-

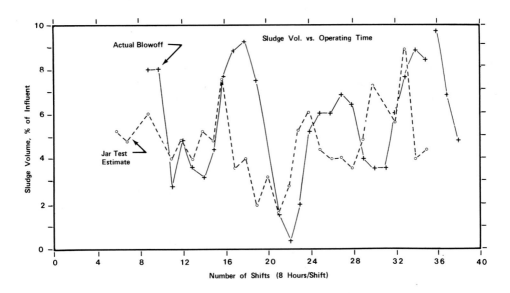

FIGURE 6. Comparison of blowoff volume from actual precipitator-settler operation with concurrent jar test results for the same chemically treated wastewater.

influent solids. This is not true in many cases for dewatering equipment. However, with the increase in effectiveness of solids handling in going from clarification settling, to gravity thickening, to sludge dewatering, there is usually a correspondingly lower efficiency in clarification of the liquid phase. There may also be increased equipment and operating costs involved with the higher solids handling method. This is related to the corresponding increase in surface area, power, or chemicals normally necessary for treating the higher solids concentrations.

For the second choice, it may be worthwhile to investigate polymers for conditioning the slurry into a more settleable material. Many times, a "border-line" slurry can be agglomerated with the proper polyelectrolytes (usually high-molecular-weight poly-acrylamides) and mixing techniques to produce settling behavior that can be accommodated by the higher hydraulic flow capacity, lower solids capacity process.

Sludge volume tests, such as the suggested wastewater jar test procedure, or the settling tests using graduated cylinders for measuring rate of subsidence, are valuable tools for predicting the rate of sludge volume withdrawal from the bottom of a settling unit. Although they may only give rough estimates for use in designing new settling basins, they are very useful in controlling the operation of existing units. As an example, Figure 6 shows a comparison of the sludge volumes predicted by jar tests with the simultaneous operation of a precipitation-coagulation-sedimentation type unit (i.e., a reactor/clarifier) for volume of sludge underflow on a wastewater treatment application. Since sludge withdrawal is periodic, the underflow is termed "blowoff". In both the jar tests and full scale system, the influent and chemical treatment are the same. The volume of sludge produced is given in terms of volume percent of influent. It is observed that the shapes of the jar test and full scale unit sludge volume curves are very similar about 80% of the time, but a change in jar test sludge volume occurs approximately 2 to 3 shifts (i.e., 16 to 24 hr) in advance of the actual change in full scale blowoff rate. This lead time in change of sludge volume requirement is to be expected since the detention time of the solid phase entering the unit is held up in the mixing and settling zones for a period of 18 to 24 hr at the average influent and blowoff rates required for this application.

The lead time can, in predicting sludge blowoff volume, be very helpful to the operator since he knows whether or not he will have to adjust the blowoff rate well in advance. In this case, as is standard practice in this type of equipment, the blowoff rate was adjusted so as to keep the top of the sludge blanket, in the settling zone, within a specific range of distances from the surface of the liquid.

The fact that the influent solids are measured as volume percent solids means that it is more directly related to the blowoff volume that must be removed than is the solids concentration (as weight per volume). Trying to relate blowoff volume to influent solids on a weight basis (i.e., mg/ℓ, kg/m^2, lb/ft^2) is subject to greater error since all solids entering the unit (especially in wastewater treatment) do not always maintain the same specific gravity (i.e., weight to occupied volume ratio). Since flocs can incorporate up to 99% water, it is basically impossible to predict their settled volume ratio (i.e., as sludge) based on an analysis of influent solids measured on a dry weight basis.

The difference in predicted blowoff volume (i.e., jar test) and actual blowoff is normally about 10% for the example shown, with the jar test prediction being about 90% of the actual. This difference is explained by the nature of the sludge solids. In this wastewater, the final sludge is composed primarily of calcium phosphates, metal sulfides, and hydroxides. It is rather dense (i.e., about 2% by weight as dry solid) and tends to solidify quickly. In order to effectively fluidize the solids to the point where they can be conducted from the sludge concentrator of the settler through the piping system into a collection sump, about 10% by volume additional water was required. A relatively incompressible sludge such as this requires more water to help conduct it through piping restrictions than a compressible sludge such as aluminum or iron hydroxide sludges. The particular sludge in this example has a coefficient of compressibility of about 0.71 as measured from specific resistance tests. The higher the coefficient of compressibility, the more compressible a thick slurry of the material tends to be. Table 8 shows a list of various materials with their corresponding coefficients of compressibility. Some of these were determined experimentally and others found cited in the literature.

b. Types of Settling Equipment for Clarification

There are two general types of settling tanks for clarification by sedimentation, batch settling and continuous flow.

The first type is batch settling. This requires the simplest type of tank. The dilute slurry is either pumped or drained into a tank, which may have a sloped or hopper type bottom, and is permitted to settle. Following a required settling period a valve may be opened at the bottom of the tank to withdraw the settled sludge. Then, the clarified liquid remaining may be drained through the same outlet to an effluent receiver or sewer or it may be drained or pumped out by way of an independent effluent collection system. Another mode of operation is to withdraw only the clarified liquid after the settling period and allow the settled solids to accumulate in a sludge zone or hopper. This method is more efficient in terms of reducing the sludge volume, when the influent suspended solids concentration is low, because the sludge can be allowed to compact for longer periods. Batch settling is usually limited to flow rates of 2 to 3 m^3/hr (i.e., less than 10 to 15 gal/min) in water and wastewater treatment because of the operational problems and extra tankage needed with larger flows.

One advantage of this method is that coagulation, flocculation, agglomeration, and sedimentation can all take place in the same tank with the use of variable speed agitators. Another advantage is that the treatment chemicals can be conveniently added during the mixing stages when the time is most appropriate. Also, chemical feed pumps

Table 8
COMPARATIVE COMPRESSIBILITIES OF VARIOUS MATERIALS

Material	Coefficient of compressibility	Degree of compressibility	Ref.
Sand	Approx. 0	Low	—
Petroleum industry (gravity separation)	0.5—0.7		23
Primary wastewater sludge	0.6—0.8		—
Activated (conditioned) digested	1.10		22
Activated (unconditioned) waste	1.11		—
Paper pulp (unconditioned) waste	1.15		—
Coal (froth flotation)	1.6		23
Acid mine drainage (lime treated)	10.5		23
Food waste (high BOD)	5—15		—
Alum sludge (water works)	14.5	High	23

and related flow proportioning equipment are not needed in small installations. The chemicals can be weighed or measured and added manually at the proper time. In reality, batch coagulation and settling is very close to reproducing actual jar test conditions.

The second type is continuous flow clarification equipment. There are two basic operational modes for settling in continuous flow equipment, horizontal flow and upflow. Horizontal flow settling basins may be either circular or rectangular. In the circular type basins the dilute slurry is pumped or drained into an influent distribution system located at the center of the basin. This enables equal delivery of the influent at the central surface, to the settling zone. The flow radiates outward until it is collected equally along the periphery of the tank again at the surface of the liquid. The radial flow is slow enough that the settleable solids can reach the bottom of the tank where the sludge is usually mechanically scraped into a pit or sump for periodic removal. Circular units are usually provided where large surface area is needed and they range from about 6 to 46 m (i.e., 20 to 150 ft) in diameter.

Rectangular units operate according to the same principle of horizontal flow, but the influent is admitted into one end of the tank and collected at the surface of the opposite end. A distributor or flume provides equal distribution across the width of the tank. On the opposite end of the tank the effluent is also evenly collected by flumes or weirs at the surface. Solids that settle to form sludge deposits are mechanically moved by sludge scraping or collection devices that move along the length of the basin, usually driving the sludge toward the inlet end of the tank where the deposits are the deepest. A sump or trough located along the same end of the tank (i.e., the influent) collects and concentrates the sludge for periodic blowoff.

A critical feature of the horizontal flow basins is that, as the flow continues to move across the surface of the tank toward the effluent collector, the solid suspension becomes more and more dilute. There is, in effect, no additional solids contact during the settling process. Therefore, any colloids or fractured flocs, that enter the settling basin with the influent, are likely to exit with the effluent. This type of settling step is highly dependent upon the effectiveness of the coagulation-flocculation step. Most rectangular flow basins are from 2.4 to 3 m (i.e., 8 to 10 ft) deep and have gently sloped bottoms (about 1%) to aid mechanical sludge withdrawal.

Upflow basins have come into general use for both water and wastewater treatment applications. There are several reasons for this. First, they take less surface area than horizontal flow basins because the coagulation-flocculation equipment is usually incorporated internally. Also, with an upflow settling zone, a fluidized blanket of flocculent material can be maintained that recirculates back through the mixing zone (by density currents) for increased particle contacts and improved coagulation efficiency.

Upflow basins are used for small flows (i.e., $3m^3/hr$) as well as for higher flows (i.e., 300 m^3/hr). The design of many upflow basins is readily scaled up and down so that they can be manufactured and sold as package units. The larger upflow units are circular (i.e., greater than about 50 to 60 m^2 surface area), but many smaller types are rectangular and designed to incorporate high rate settling tubes or inclined plates for improved clarification efficiency. For many of the reasons mentioned, upflow basins can operate at surface rates substantially higher than the horizontal flow basins (i.e., up to 3 to 5 m^3/hr instead of 1.2 to 2.5 m^3/hr).

Many types of upflow basins do not require mechanical scrapers or rakes for sludge removal. However, most all types require a periodic hydraulic flush-back to keep the sludge blowoff lines unclogged.

Solids contact in wastewater treatment is an effective means of removal for dissolved metals and other materials by addition of chemical precipitants or other agents to the coagulation step. Phosphates can be removed by addition of excess lime. Colloidal and dissolved organics are removed by alum in combination with powdered activated carbon added during the coagulation stage. Grease and oil in low concentrations (i.e., <1% by wt) can be removed by adsorption onto suspended diatomaceous earth.

The use of polyelectrolytes in these waste treatment processes by upflow solids contact is almost essential. Since wastewaters are usually variable in pH and suspended solids concentration, the use of inorganic coagulants alone requires considerable operational control to effect good coagulation of the suspended and precipitating solids that enter the settling stage. Polyelectrolytes are usually less sensitive to pH and suspended solids concentrations, especially the high-molecular-weight nonionic and anionic polyacrylamides. Addition of powdered carbon and/or diatomaceous earth adsorbents also require the addition of the same type of polymers to make them settle effectively.

Establishing a concentrated, fluidized sludge blanket (of hindered settling concentration range) provides a large reservoir of solids that can help take up residual coagulants, including polyelectrolyte, at times when there is an excess. Likewise, when the influent colloidal or dissolved solids range to peak levels and there is insufficient coagulant added to the coagulation or reaction step, the excess reactant in the blanket is normally sufficient to remove them. Therefore, upflow solids contact units can normally tolerate a higher range of fluctuations in influent solids than units without solids contact. Some units without sludge blankets, including horizontal flow basins, use a recirculation of sludge from the sludge concentrator back into the coagulation tank by mechanically pumping it. This seems to work satisfactorily with biological sludges which produce their own polymeric coagulants as secretions of the bacterial cells. However, with other types of sludges, especially alum and iron flocs and those coagulated with high-molecular-weight synthetic polymers, the pumping action can shear the flocs and either require the addition of more polymer or allow more suspended solids to pass out with the effluent if no sludge blanket is present in the settling zone to help capture the broken floc fragments.

The advantages of using solids contact are now well enough recognized that most clarifiers and reactor settling units incorporate the solids contact concept into the design in one way or another. Experience with this type of unit is an illustration of the theoretical concept that increased particle collision frequency enhances coagulation efficiency. An example of an upflow solids-contact unit (i.e., Precipitator) is shown in Figure 7. The sludge blanket extends throughout the mixing zone and into the settling zone. Due to the action of the agitator, the sludge blanket is fluidized in the

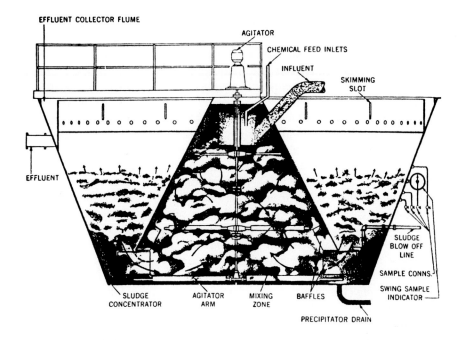

FIGURE 7. Section of Precipitator-Settler. The water flow pattern and thorough mixing of chemicals with influent and previously formed sludge is illustrated. (Courtesy The Permutit Company, Paramus, N.J.)

mixing zone and intermixes with the influent after an initial fast mix stage with chemicals. Flocculation proceeds rapidly in the high solids, turbulent flow mixing zone. As the flocs pass into the bottom of the settling zone with the flow of water, they intermix with a high concentration of floc aggregates existing in the settling zone blanket. As the flow moves upward, the turbulence gradually decreases as the surface area for settling increases. These conditions, as already discussed, are necessary for causing agglomeration of flocs into large rapidly settling particles. Since the settling zone becomes nonturbulent as the flow moves upward, the density of the sludge blanket increases and it acts as a fluidized upflow filter capturing and adsorbing fine colloidal solids.

The sludge blanket establishes a level at the height in the settling zone where its settling velocity equals the rise rate of the liquid. As the density of the settling zone blanket increases, a density gradient between the settling zone and mixing zone is established. A migration of denser suspended solids currents then moves downward, partly back into the mixing zone (i.e., sludge recirculation), and partly into the sludge concentrator where it is periodically drawn off as blowoff sludge. With addition of high-molecular-weight polymer (low concentration, approximately 0.01 to 0.2 mg/ℓ) directly to the mixing zone, by way of a dilution water distributor, it is sometimes possible to increase the density of light sludge blankets (i.e., alum) approximately two-fold for higher surface rates and decreased sludge blowoff volume.

Many types of upflow solids-contact units are designed with at least one sloping wall in the settling zone. If only straight vertical walls are used the established sludge blanket height is subject to abrupt upset by sudden changes in hydraulic and solids loading rates. This occurs because the upward liquid velocity through a concentrated slurry will establish a specific ratio of distances between particles (which can be thought of as pore spaces through a filter). As the velocity increases it increases the area of the pore spaces as well as the upward drag on individual particles. As a result, the sludge blanket rapidly expands. The sludge blanket rapidly shrinks if a sudden decrease in upflow velocity or solids loading occurs. With straight vertical walls the drag effect is not compensated for and the blanket will move upward or downward in proportion to the change in liquid velocity. Additionally, the change in pore space area due to change in velocity tends to compound the problem, making the blanket either expand or shrink even more rapidly. Under such conditions the top of the blanket can rise up to the collection system and exit with the effluent, or drop down out of sight.

Sloped wall units, tend to compensate for both of these problems caused by sudden hydraulic and solids loading fluctuations. The upwardly increasing area of the settling zone allows an increase in velocity (within design limits) to become zero as the blanket level moves upward a relatively small distance, therefore cancelling any increase in upward drag on the particles. Furthermore, the pore area becomes unchanged as the liquid velocity increase is cancelled at the higher settling zone level. The analagous effect occurs with a sudden decrease in influent hydraulic or solids loadings. Therefore, upwardly increasing area-sloped wall units appear to have a clear advantage over vertically straight wall constant area sludge blanket units.

4. Sludge Thickening Operations
a. Applications

Gravity thickening may be desirable for a dilute blowoff sludge from a large clarifier prior to final dewatering. If a settling rate curve similar to that described as Type 2 phase settling exists and the sludge flow is in excess of, say, 5 m³/hr, gravity thickening may be warranted. On the other hand, some wastewater treatment applications produce such voluminous amounts of sludge directly upon chemical neutralization and/ or precipitation that direct gravity thickening may be called for. Examples of these applications might be treatment of acid mine drainage or boiler fireside wash from the power utilities. Both of these wastes are highly acidic, containing high concentrations of dissolved and suspended iron, and yield several thousand mg/ℓ suspended solids, or more, upon neutralization with lime. Settling tests of such slurries may reduce the solids volume to 40 or 50% after 30 to 60 min settling time. Flocculating with a polyelectrolyte (e.g., high-molecular-weight polyacrylamide) may reduce this volume to 20 or 30% during the same settling period.

Gravity thickeners are also used for thickening water treatment plant sludges, as well as biological wastes from domestic sewage treatment. Below is a list of typical applications:

> Water treatment
>> Lime softening sludges
>> Alum sludges
>> Ferric hydroxide
> Wastewater treatment
>> Waste activated sludge
>> Primary sludge
>> Trickling filter (secondary settling)
>> Anaerobic digested sludge
>> Scrubber waste sludge

If it is determined that the sludge volume cannot be reduced to less than about 15 vol% after 30 to 40 min of settling, there is (most likely) too much sludge for an ordinary clarifier type settler to handle, even if a portion of the settling curve is in the free settling region. The reason for this is that most clarifier-settlers have sludge concentrator zones that are small in comparison to the surface area of the settling zone. If the underflow volume exceeds about 15% of the influent volume, the blowoff may

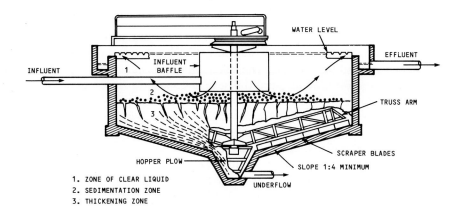

FIGURE 8. Typical concentration profile in a continuous gravity thickener.

be continuous and very little sludge concentration will be able to take place. Therefore, a settling device with a larger solids handling capacity, such as a gravity thickener, is required.

b. Design of Gravity Thickeners

A gravity thickener may be a simple tank (rectangular or circular) or a lagoon that sludge is periodically run into at one end, while the clarified supernatant overflows from the other. After a long period of time, and a dense sludge layer builds up, the basin is drained of all supernatant water and the settled sludge is scooped or drained out. This type of gravity thickener is satisfactory for small sludge flows (i.e., less than approximately 0.2 to 1.0 m³/hr), but for larger flows, continuous thickeners are usually more economical because tank size and land area can be substantially reduced.

Continuous gravity thickeners can be utilized favorably for both small and large sludge flows. They are usually circular vertically straight wall tanks approximately 3 to 4 m high, as illustrated in Figure 8. In a typical unit the influent slurry is introduced centrally about half way up from the bottom. Clarified effluent is collected from the liquid surface around the periphery of the tank. Some units are also equipped with a skimmer trough for scum, oil, and grease at the liquid surface. In such cases a submerged orifice effluent collector is required.

As the influent slurry enters the influent distribution system (baffled feed well) it may be mixed by mechanical stirring with polymer treatment, or may pass directly into the settling zone. As the flow enters the settling zone its initial downward motion gradually assumes a horizontal vector and then proceeds to gradually rise toward the surface as it approaches the walls of the basin. As hydraulic flow slowly progresses through the settling zone (the level occupying approximately the mid one third of the basin height) the solids fall out and settle toward the thickening zone (the bottom third, or so, of the basin).

To keep the thickened sludge at the bottom of the basin in a steady fluidized state, as it is drawn out as underflow, a mechanically driven rake slowly stirs the sludge mass. The bottom is typically sloped toward a central sludge hopper to promote complete movement of sludge from the periphery to the underflow outlet. Sometimes the rake arms are provided with vertical "pickets" to aid in separating the solids from water that has been entrapped in pockets or channels within the sludge mass.

In order to size a gravity thickener there are two important criteria for determining the surface area for a given hydraulic and solids loading rate. One criterion is the hindered settling rate of the suspended solids in the influent. This is determined by the straight line portion of the phase settling curve (Type 2) in Figure 5. The settling velocity can be estimated by running laboratory settling tests on the given sludge. The surface rate (rise rate) of the thickener would then be limited to a value equal to, or less than, the hindered settling velocity. The other criterion that has to be considered is the sludge consolidation time to reach the desired concentration of underflow solids. This is also a function of thickener surface area and a minimum surface area for the amount of thickening time required must be determined. The larger of the two indicated surface area estimates (i.e., surface area required for clarification or area required for thickening) must be used to size the thickener.

Several methods have been used for evaluation of thickening area required. Usually, these methods involve sludge settling tests in 2-ℓ graduated cylinders based on Kynch's "Theory of Sedimentation", presented in 1952.[19] Several applications of the "Kynch Unit Area Analysis" are to be found in the literature.[13,14,24] Although these tests are still widely used for sizing thickener area, evidence has been presented by Kos (1976)[21] that shows the Kynch analysis, while it holds true for discrete particulate suspensions, does not properly represent flocculent sludges (which most sludges are). Therefore, the Kynch approach should only be regarded as a rough approximation of the required unit area. In order to properly size thickener area, Kos recommends the following procedures:

1.   Find an actual, properly operating thickener used on the same kind of material, either in the literature or at hand. Determine or collect the operating data needed.
2.   Construct a pilot plant continuous thickener and perform a study subjecting it to various operating conditions to develop the relationship between underflow and possible suspended solids loading.

c. Polyelectrolytes in Gravity Thickener Operations

Sludges to be treated in gravity thickeners are usually the result of previous chemical coagulation operations. As such, it is usually difficult to predict how they will respond to any type of polyelectrolyte since residual effects of other coagulants can be quite predominant. Therefore, jar tests or other coagulation tests should always be performed to select the polymers to be used in gravity thickeners.

If jar tests are conducted to evaluate various polyelectrolytes, it may be determined that there is an optimum effective dosage for reducing the sludge volume by sedimentation. At higher or lower concentrations, a polymer may allow a larger settled sludge volume than the optimum dosage. In some cases it is found that increasing the polymer dose above optimum will actually increase the settled sludge volume to a level even greater than the untreated sample. In this case it is better to underdose than it is to overdose. It can be seen that when concentrated sludges or slurries are treated with polymers and allowed to settle along with untreated slurries for several days the untreated slurries will usually have the lowest sludge volume. One concept that explains this is that the polymer not only makes the sludge particles stronger and faster settling, but also, the integrity of the individual particles to keep entrapped water inside the floc is increased. As a result they are not readily consolidated by gravitational forces. The smaller individual particles of the untreated slurry gradually become more compressed, nested into the structure of the sludge matrix under the weight of the sludge layers above. Ideally, the correct concentration of polymer would be that compromise that produces a comparatively low solids volume both initially and after a long settling period. Figure 9 illustrates this concept of polymer overdosing in a simplified form. Figure 10 shows in more detail how one type of floc particle appears under a compound light microscope. Instead of actual branches as shown in Figure 9, the floc particle

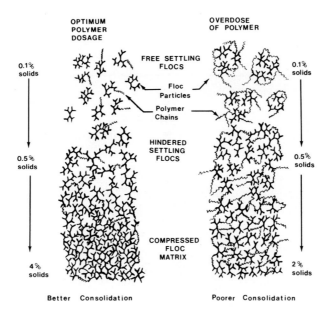

FIGURE 9. Effect of polyelectrolyte overdose on gravity thickening. The sketch illustrates how an excess of polyelectrolyte on floc surfaces can hinder the integration of floc branches into higher density, lower volume, sludge masses.

illustrated in Figure 10 shows continuous chains of smaller particles folded and wrapped many times so as to trap a large volume of water. Under pressure of other floc layers, these chains can join those of other flocs, making one continuous matrix.

Optimum dosages of polyelectrolyte can decrease the time it takes the sludge to reach the compression portion of the curve (Type 3, Figure 5). If this can be accomplished without significantly increasing the final sludge volume (over what would result with no polymer over a long period of time), the underflow solids can very often be doubled over that which is possible without use of polymer. When adding polymer to the thickener it may suffice to install mixers in the influent well to create sufficient agitation to thoroughly contact the individual sludge particles with a dilute polyelectrolyte stream before the flow passes into the settling zone. If the stirring is not thorough and polymer contacts the sludge masses as highly localized pockets, the agglomeration will be nonuniform and will result in either unnecessary consumption of polymer, poor sedimentation of a portion of the solids, or both.

Propeller or turbine type mixers with shaft speeds of about 300 rpm have been used for this type of application. Mixing should not be so severe as to cause turbulence in the settling or thickening zones. Mixing of the influent slurry just prior to entering the thickener is another alternative. Thick sludges such as these can be effectively flocculated with higher speed mixing because of the great collision frequency that causes floc formation to exceed floc breakup.

Some very thick influent slurries (i.e., greater than 1.5 to 2% by weight) have been found to be very difficult to effectively coagulate with polymer, even with high speed mixing. In this case it is sometimes advisable to add the polymer to a dilution water stream that will effectively reduce the influent concentration to less than 1% by weight. Studies are described in the literature where dilution of the influent solids concentration to less than 1% by weight has been beneficial. Influent dilution without use of polyelectrolytes has also been shown to be beneficial in some cases.[25]

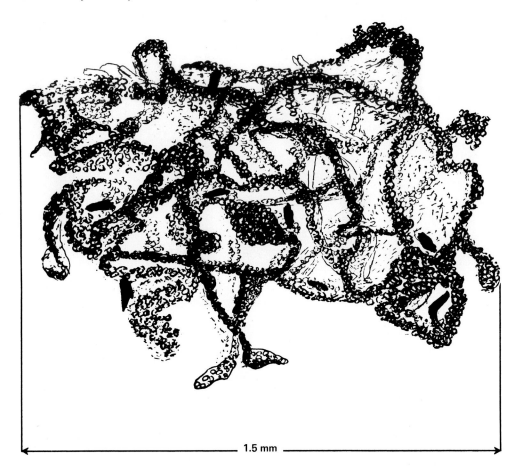

1.5 mm

FIGURE 10.   Appearance of a flocculent hydroxyapatite particle through compound microscope. This hydrated floc is composed of chains of small calcium phosphate crystals of about 10 microns diameter. The particle is stained with alizarin red [at pH 11 to 11.2] to impart a deep purple color for contrast. Darker portions of the chains are in foreground of the three dimensional field of view. The large dark particles, entrapped in the net-like matrix, appear to be firmly held and move with the floc particle. (Magnification × 100; shown at 74%.)

Table 9 shows typical values of important operational parameters for various thickener applications. These are the reported results of various studies cited in the literature. Unfortunately, the lack of proper mixing techniques for polymer application in some gravity thickening studies, may have resulted in some misinterpretation of the true effectiveness of the polymers for these applications. There is, in all probability, a need for continued study for improvement of mixing techniques in conditioning concentrated slurries (i.e., greater than 1.0 wt% suspended solids) prior to gravity thickening. The term conditioning is used for these concentrated flocculent slurries, instead of coagulation or flocculation, since the solids have already undergone the primary destabilization and particle growth steps.

In clarification operations, flocculation steps involve a slow rate of particle growth with large volumes of entrapped water due to slow agitation and low available particle concentration (i.e., number of particles per unit volume). When a resulting flocculent suspension of light particles (such as a metallic hydroxide) is settled to form a sludge of about 1.0% by weight the solid phase forms a continuous matrix with a large volume of pore space. Intertwined chains of primary coagulated particles (similar to that

Table 9
TYPICAL GRAVITY THICKENER OPERATING PARAMETERS FOR VARIOUS APPLICATIONS WITH AND/OR WITHOUT POLYMER CONDITIONING

Type of sludge	Surface rate (m³/hr/m²)	Polymer used	Solids loadings (kg/m²/day)	Underflow wt % total solids	Ref.
Water plant, by case					
Alum and powdered activated carbon	0.12—0.2	Without	15—24	2.5—3.0	21
Alum	0.2—0.25	Without	25	1.5—2.0	—[a]
Alum	0.2—0.25	Nonionic to anionic hmw[b] polyacrylamide	25	3—4	—[a]
Biological waste					
Primary	1.0—1.4[a]	Without	100	8—10	26
Primary and trickling filter	1.0—1.4[a]	Without	60	6—8	26
Primary and waste activated (60 to 40 wt ratio)	1.0—1.4[a]	Without	40	3—6	26
Waste activated	0.12—0.25[a]	Without	9—11	2.4—2.6	26
Waste activated	0.5[a]	Cationic polymer	22—35	2.6	26
Utility Waste					
Mixed air preheater and fireside washes, etc. (metallic hydroxides)	1.9	Low anionic hmw polyacrylamide	68	10	27

[a]   Values estimated by author.
[b]   hmw — high molecular weight.

shown in Figure 10) form the solid matrix with the supporting liquid dispersed among the voids between them so that both phases occupy the same total space or effective volume. The real volume of the solid phase is, however, only a fraction of the effective volume.

Conditioning of these slurries with high degrees of agitation and addition of bridging polymers involves a fracturing and restructuring of the solid matrix into aggregates of a more granular, spherical (cluster-like) shape (i.e., Figure 9). These aggregates have a higher particle density and hence, occupy less total space per unit weight of solid. As a result, large spaces between the aggregates are opened up which enable the solid phase to further consolidate and occupy substantially less volume upon sedimentation.

D. Conclusions

In retrospect, of all gravity sedimentation processes, the general practices of coagulation, flocculation, and sludge conditioning operations are carried out in order to maximize the rate of particle growth and sedimentation while minimizing the volume of the settled mass. In effect, two intermixed phases consisting of a dispersed solid and a liquid, occupying the same total volume (i.e., space), must be "unmixed" so that the phases can be separated into two component volumes, each containing as little intermixing of the separate phases as possible. This can be accomplished through careful selection of coagulants and polyelectrolytes as well as properly controlled degrees of agitation and rates of suspended solids and hydraulic loadings. Each of these parameters has maximum and minimum limits that must be maintained in a proper balance if efficient performance is to prevail. If, in the practice of bench tests, we can produce a slurry in which the solids will settle to the proportional volume available for sludge concentration in an actual settling unit, and in the time period available, at least nature is on our side. Efficient unit performance then depends on sound equipment design principles and attentive monitoring by the plant operator.

Operator attention in gravity sedimentation applications is an extremely important factor in achieving good performance. It must be constant and thorough and requires a greater measure of good judgement than many other water and wastewater treatment operations. Jar tests and similar methods discussed in this chapter can be a great deal of help to the operator in controlling the day to day variations in the operation.

III. TERMINOLOGY

Agglomeration — The bringing together of visibly sized flocs into larger masses of randomly arranged (i.e., noncrystalline) particles, mainly through the action of slow mixing and bridging mechanisms.

Aggregation — In general, the bringing together of small particles into larger ones. Applies to coagulation, flocculation and agglomeration, as well as the orderly growth arrangement of crystalline structures.

Agitation — In water and wastewater treatment; the general act of forcing the liquid into a state of hydraulic turbulence.

Clarification — The general process of removing suspended solids by sedimentation, in water and wastewater, usually following a coagulation-flocculation step.

Coagulation — In water and wastewater treatment; the process of combining dissolved substances (such as ions) or colloids (such as macromolecules) into larger particles by destabilization of the primary particles.

Coagulant — A substance added during water or wastewater treatment to bring about destabilization of primary particles.

Consolidation — The reduction in volume of the solid phase of a suspension during the hindered settling and compression stages of gravity sedimentation.

Destabilization — The neutralization or destruction of the physical and/or chemical forces holding a solid phase suspended in a liquid.

Flocculant — A chemical agent added to a suspension of solids in a liquid to flocculate the small particles into larger ones.

Flocculation — The growth of unstable, microscopic particles in water, into small visible amorphous masses (of indefinite shape and arrangement) brought about by a low degree of turbulent agitation.

Flocculent — The characteristic physical and chemical properties of amorphous, gelatinous, hydrous (i.e., floc-like) solid particles.

Mixing — General term for the bringing together of different substances into a uniform combination forming a blend where the original substances are not easily distinguishable from each other. In water and wastewater this can apply to the uniform distribution of solids within a liquid or of liquids within a liquid.

Precipitation — In water and wastewater treatment; the chemical formation of solid masses, separable from the liquid, out of dissolved or ionic species in water. An example is removal of calcium ion by softening processes. This can also be referred to as coagulation, but it is not the destabilization of colloidal or molecular particles, which is also called coagulation.

Sedimentation — The removal of particles, that are heavier than the suspending liquid, by gravity settling, usually under quiescent or laminar flow conditions.

Sludge Blowoff — The portion of concentrated solids removed from a settling unit which is required to keep the sludge from building up and overflowing the unit. Usually, the blowoff is taken off at periodic intervals from a quiescent sludge concentration zone at the bottom of the settling zone.

Sludge Blanket — A fluidized mass of solids in the settling zone (and mixing zones of some units) of a sedimentation unit that is suspended by either turbulent mixing and/or upward hydraulic flow to maintain a relatively stable position with respect to the tank even though the liquid is moving through it.

Sludge Recycle — The movement of fluidized solids, of settleable size, from a zone of high concentration to the initial mixing (i.e., coagulation) zone to promote increased solids contact (i.e., collision frequency).

Solids Contact — The process of precipitation or coagulation of primary particles (either dissolved or colloidal) with chemical reagents, in the presence of a relatively high concentration of previously formed solid particles. This term is usually used synonymously with sludge blanket contact and sludge recycle.

Stirring — This term is commonly used, in water and wastewater treatment, to indicate the action of putting the water in motion so as to "mix" all the constituents or components within the liquid.

Subsidence — A general term used to indicate the sinking of a solid phase within a liquid phase. In settling operations, it is commonly used to denote the state of volume or height reduction of the solid phase during hindered settling and compression as opposed to the free settling of individual particles.

Thickening — In water and wastewater sedimentation; the process of making concentrated solid slurries (in a state of hindered settling) more dense by providing proper conditions for gravity to drag the solids down into an increasingly smaller volume while liquid is forced up and out.

Turbulence — The state of fluid motion in which the flow of individual shear planes are not in a laminar condition as defined by Reynolds' correlations. Rather the movement of individual shear planes are in a state of disarray with respect to each other.

REFERENCES

1. **La Mer, V.K.,** Coagulation versus the flocculation of colloidal dispersions by high polymers (polyelectrolytes), in *Principles and Applications of Water Chemistry,* Faust, S.D. and Hunter, J.V., Eds., John Wiley & Sons, New York, 1967, 246.
2. **Gutcho, S.,** Waste Treatment With Polyelectrolytes And Other Flocculants, Noyes Data Corporation, Park Ridge, New Jersey, 1977.
3. **O'Melia, C.R.,** Coagulation in water and wastewater treatment, in *Water Quality Improvement By Physical And Chemical Processes,* Gloyna, E.F. and Eckenfelder, W.W., Jr., Eds., University of Texas Press, Austin, 1970, 219.
4. **Stumm, W.,** Chemical interaction in particle separation, *Environ. Sci. Technol.,* 11(12), 1066, 1977.
5. Water and Waste Treatment Data Book, 10th printing, The Permutit Company, a division of Sybron Corporation, Paramus, N.J., 1961.
6. **Black, A.P. and Harris, R.H.,** New dimensions for the old jar test, *Water Wastes Eng.,* 6(12), 49, 1969.
7. **Riddick, T.M.,** Zeta potential: new tool for water treatment, *Chem. Eng.,* 68(13), 121, 1961; 68(14), 141, 1961.
8. Waters Associates, Inc., SCD - Streaming current detector, 61 Fountain Street, Framingham, Mass.
9. ASTM, *1974 Annual Book of ASTM Standards,* American Society For Testing and Materials, Philadelphia, 1974, 818.
10. **Lai, R.J., Hudson, H.E., and Singley, J.E.,** Velocity gradient calibration of jar-test equipment, *J. Am. Water World Assoc.* 67(10), 553, 1975.
11. **Swope, H.G.,** Zeta potential measurement, *Water & Sewage Works,* R-64, 1977.
12. **Camp, T.R.,** Sedimentation and the design of settling tanks (Paper No. 2285), *Trans. Am. Soc. Civ. Eng.,* Vol. 111, 895, 1946.
13. **Metcalf & Eddy, Inc.,** *Wastewater Engineering — Collection, Treatment, Disposal,* McGraw-Hill, New York, 1972, 283.
14. **Rich, T.G.,** *Unit Operations of Sanitary Engineering,* John Wiley & Sons, New York, 1961, 81.
15. **Hahn, H.H. and Klute, R.,** Effect of turbulent flow conditions upon the process of coagulation, presented at the Division of Environmental Chemistry, American Chemical Society, Chicago, August 24-29, 1975, 1.
16. **Delichatsios, M.A. and Probstein, R.F.,** Scaling laws for coagulation and sedimentation, *J. Water Pollut. Control Fed.,* 47(5), 941, 1975.
17. **Duff, B.F., Barkley, W.A., and Wu, Shen-Chih,** Time parameter evaluation in dual flocculant destabilization. Reproducibility evaluation of the flocculation of a synthetic latex colloidal system, Water Resources Res. Inst., WRRI-062, W76-12211, New Mexico State University, University Park, 1976, 106.
18. **Narkis, N., Rebhun, M., and Sperber, H.,** Flocculation of clay suspensions in the presence of humic and fulvic acids, *Isr. J. Chem.,* 6, 295, 1968.
19. **Kynch, G.J.,** A theory of sedimentation, *Trans. Faraday Soc.,* 48, 166, 1952.
20. **Hazen, A.,** On sedimentation, *Trans. Am. Soc. Civ. Eng.,* L111, 63, 1904.
21. **Kos, P.,** Fundamental principles of gravity thickening, presented at Annual Meeting American Institute of Chemical Engineers, Chicago, November 26 to December 2, 1976.
22. **Eckenfelder, W.W. and O'Connor, D.J.,** *Biological Waste Treatment,* Pergamon Press, Oxford, 1961.
23. Water Pollution Research Laboratory, Industrial Waste Sludges, Water Pollution Research, Her Majesty's Stationary Office, London, 1962, 1964, 1966.
24. **Perry, J.H., Ed.,** *Chemical Engineers' Handbook,* 4th ed., McGraw-Hill, New York, 1963, 19.
25. Process Design Manual for Upgrading Existing Wastewater Treatment Plants, U.S. Environmental Protection Agency Technology Transfer, October 1974.
26. **Torpey, W.N.,** Concentration of combined primary and activated sludge in separate thickening tanks, *J. Sanit. Eng. Div. Am. Soc. Civ. Eng.,* 80(1), 1, 1954.
27. **Carvallo, D.A. and Patel, D.K.,** Design and Operation of a Wastewater Treatment Facility At An Existing Fossil Power Plant, presented at American Institute of Chemical Engineers 71st Annu. Meet., Miami, November 12 to 16, 1978.

Chapter 5

POLYELECTROLYTES AND COAGULANTS FOR THE FLOTATION PROCESS

James G. Walzer

TABLE OF CONTENTS

I. INTRODUCTION

Chemical coagulants and coagulant aids are employed extensively in the flotation process. Chemical treatment affects not only the quality of the product, but also the capital and operating costs of the process equipment employed.

The flotation process is used to separate suspended solid or liquid particles from a liquid phase. In the flotation process gas (usually air) bubbles are introduced into the suspension. The air bubbles become attached to suspended particles making them buoyant and available for separation. Applications include clarification, sludge thickening, and fractionation.

There are three basic flotation processes commercially available: Induced Air Flotation (IAF), Vacuum Flotation (VF), and Dissolved Air Flotation (DAF).

In the IAF process air bubbles are introduced into the suspension by mechanical shear. The process is employed for selective separation of particles from a suspension. This is normally accomplished by pretreatment with surface active agents. The process is also used to accelerate separation of particles having a lower specific gravity than water, e.g., separation of demulsified oils from wastewater. Due to the inherent problems with mechanical shear the process is seldom employed where coagulants are used.

In the VF process the water to be treated is first saturated with air at atmospheric air pressure then deaerated with vacuum. The process is seldom employed due to the complexity of the equipment, the capital cost, and the oxygen deficient effluent. The DAF process is employed in the vast majority of applications, and will be dealt with most extensively in this chapter.

II. THE DISSOLVED AIR FLOTATION PROCESS

The DAF process employs the principle of increased solubility of gas in solution at elevated pressures (Henry's Law). In the flotation process the stream to be treated is saturated with air at several times atmospheric pressure. When the pressure is released, air in excess of atmospheric saturation comes out of solution in the form of tiny air bubbles which attach to the suspended solids and float them to the surface. Figure 1 shows the relationship of the solubility of air in water to pressure applied at various temperatures. Flotation units typically operate at 30 to 70 lb/in.^{2}g.

The released air bubbles become attached to the suspended particles by one of the following mechanisms: (1) condensation, (2) collision, or (3) entrapment.

In the condensation mechanism, air in excess of atmospheric saturation comes out of solution by formation on the surface of the suspended particle. This is not the dominant mechanism, however, since nonturbulent depressurization of the suspension results in a degree of super saturation;[1] hence, it is necessary to first achieve near complete air release by turbulence then attach to the particle.

The collision mechanism is perhaps the most significant in the flotation process. In this mechanism the air-to-solids bond is created by collision during random motion. Air bubble and particle size must be controlled to some extent to insure that there is a sufficient radius of attachment to maintain the bond until separation. Particle size is controlled by chemical coagulation.

The entrapment mechanism, which provides a "permanent" air-to-solids bond, is the predominant mechanism when the final step of chemical flocculation occurs after air release. Air bubbles become embedded in the floc mass.

After the air-to-solids bond is complete, flotation will occur if the net combined specific gravity of the air-solid agglomerate is less than 1.0. Rise rate of the undisturbed

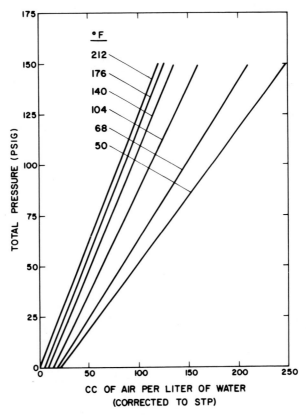

FIGURE 1. Solubility of air in water at various temperatures and pressures.

agglomerate is governed by Stokes' Law. The revised Stokes' Equation for the flotation process is:

$$V_t = \frac{(\rho \text{ liquid} - \rho a) \, Da^2}{\mu \, 18} \tag{1}$$

where V_t = terminal velocity agglomerate, ϱ = density of liquid, ϱa = density of agglomerate, Da = diameter of the agglomerate, μ = viscosity of the liquid and g = acceleration due to gravity. Actual separation of the suspension in the flotation unit will also be governed by the solids and air concentration, and the degree of turbulence.[2]

The flotation process is employed where separation of particles having specific gravities close to that of water is desired. Where applicable the flotation process will provide faster separation and higher ultimate solids concentration.[3] Figure 2 compares separation of waste activated sludge solids by flotation and sedimentation. Note that while the final sludge volumes are nearly equal, the flotation sludge will have much interstitial air which displaces water.

A. Equipment Used in the Flotation Process

The DAF unit consists of two major components, the retention tank and the separation vat. Figure 3 depicts a typical DAF unit. The retention tank is a pressure vessel designed to provide sufficient time for dissolution of air into the stream to be treated. There are a variety of air introduction systems available, most employing a sparger or eductor.[4]

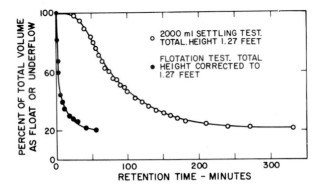

FIGURE 2.   Separation of waste activated sludge solids by flotation and sedimentation.

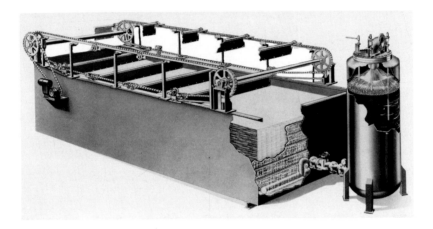

FIGURE 3.   Typical dissolved air flotation unit with air dissolution tank. (Courtesy The Permutit Company, Inc., Paramus, N.J.)

From the retention tank, the stream is released back to atmospheric pressure in the separation vat. The system is usually designed so that most of the pressure drop occurs in the transfer line between the retention tank and the separation vat so that the effects of turbulence are minimized. In the separation vat air in excess of atmospheric saturation comes out of solution in the form of tiny air bubbles which become attached to the particles in the suspension resulting in flotation. The separation vat is equipped with a flight scraper mechanism which removes the floated material to a recovered solids compartment. Clarified effluent is drawn off from the bottom of the vat. There are a number of different vat configurations and process application methods available to satisfy the conditions of a particular application.[5,6]

B. Economics of Chemical Treatment
In general, coagulants and coagulant aids are used in the flotation process to increase the particle size so that bubble-solid attachment can occur. Colloidal solids and dispersed oils do not respond to flotation, and are discharged with the treated effluent if not coagulated.

Beyond simple coagulation, which produces a particle that can be separated, coagulants perform two other major functions which establish the overall economy of the flotation process.

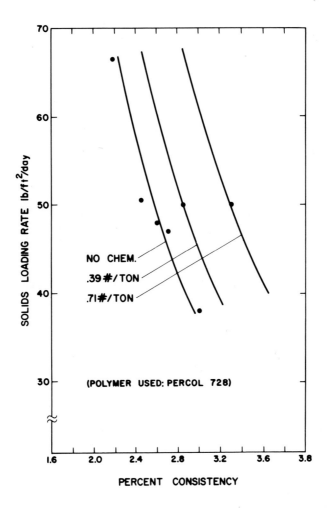

FIGURE 4. Float solids percent consistency at various solids
loading rates with various amounts of chemical conditioning.

1. Flocculants are added to increase the volume of the air-solid agglomerate which
    will result in a faster separation rate. Since the separation (particle rise) rate es-
    tablishes the size of the flotation unit, flocculation has an important economic
    impact on the capital cost of the system.
2. Flocculants are added to increase the solid particle density to produce a sludge
    of higher consistency (percent solids), or inversely a decrease in water content.
    The sludge concentration establishes the size of the sludge dewatering equipment
    downstream of the flotation unit.

   The importance of these two functions is realized in reviewing Figure 4. This graph
shows the size of a particular flotation unit required to produce a particular sludge
concentration at various chemical dosages.
   If we assume that a 2.8% sludge consistency is desired, we can see operation without
polyelectrolyte pretreatment will require a flotation unit having 25% more surface area
than operation with 0.39#/ton polyectrolyte to condition the sludge.
   Also we can see that operation at a constant solids loading rate, while varying the
polymer treatment, will produce different sludge consistencies, which again establishes
the size and capital cost of the dewatering equipment.

III. CHEMICAL TREATMENT PROCEDURES

Chemical pretreatment procedures employed in the flotation process usually fall into one of three categories.

Precipitation/flocculation — This process involves precipitation of a previously dissolved material. In most cases the precipitated material is ready for flocculation before removal, as in precipitation of a heavy metal followed by flocculation with an organic polyelectrolyte, e.g. removal of a zinc as the hydroxide. In some cases the precipitate is too fine for direct flocculation. In these cases a metallic salt coagulant such as alum or copperas is used in small quantity. Coagulation of precipitated calcium carbonate in the cold lime-soda softening process is an example.

Charge neutralization/flocculation — This process involves neutralization of a high surface charge on an insoluble dispersed material. The first step of the procedure would require addition of a metallic salt coagulant, or highly charged (usually low-molecular-weight) organic polyelectrolyte. Neutralization of the surface charge on the colloid or dispersion results in "pinpoint flocculation" of the suspension. This procedure is followed by flocculation of the pinpoints in the second operational step. Whenever applicable high charge polymers in the coagulation step are preferred, because the polymer will not produce additional sludge for disposal. Metallic coagulants contribute difficult to dewater metallic hydroxides.

Demulsification/charge neutralization/flocculation — In the first step of this procedure chemicals are added to "break" emulsified fats and oils from solution. Acidification or treatment with ionic surfactants is usually employed. The final two steps are similar to the procedures outlined above. In addition to the chemicals described in the procedure outlined above, acids and alkalis are often used to control pH. Almost all of the coagulants require optimization of pH for peak efficiency.

A. Flotation Process Application Methods and Chemical Additives

Since there are many steps to the flotation process, and several modifications available, due consideration must be given to the point of chemical addition and method of injection.

In general one must insure that the proper mixing of chemical with waste occurs to allow sufficient contact. In addition to this the effects of shear action on floc particles must be investigated. For example, in a charge neutralization step, rapid mix is usually necessary to insure optimum utilization, but shear is not of importance since the product is a pinpoint floc too small for flotation. In flocculation, the objective is to form a voluminous structure, which would be broken down during a rapid mix procedure. Proper mixing for flocculation is normally provided by a mechanical paddle flocculator which tumbles the entire suspension, rather than divides it. Mixing may also be achieved by gentle hydraulic motion.

There are three DAF process application methods in commercial use today. Chemical injection procedures will depend on the waste being treated, and the method of treatment. The three methods are (1) direct aeration, (2) partial aeration, and (3) effluent recycle.

Figure 5 presents schematic diagrams of the three process application methods. Please note that the point of chemical addition is an oversimplification and is included to indicate that chemicals are normally used. The actual procedure is developed after investigation of points of addition and mixing requirements.[7]

In the direct aeration or full flow pressurization method, the entire stream to be treated is pressurized in the retention tank. This process is usually employed where the suspended solids exceed 800 mg/ℓ. The process is first considered for applications where floc shear is not a major problem. Many wastes will simply reflocculate after

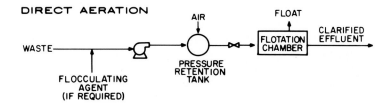

DIRECT AERATION

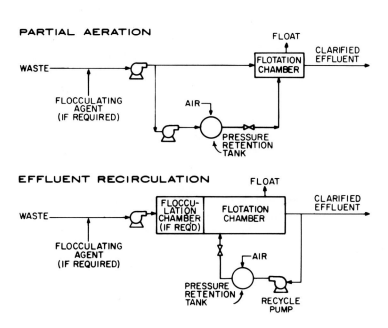

PARTIAL AERATION

EFFLUENT RECIRCULATION

FIGURE 5. DAF methods for combining air with waste stream suspended solids. (Courtesy The Permutit Company, Inc., Paramus, N.J.)

being sheared several times. In cases like this, chemical flocculants are normally added to the suction side of the pressurization pump, or the inlet to the retention tank to insure proper mixing with waste. In this procedure the floc particles reform in the inlet compartment of the separation vat. Note that the final point of shear is the pressure release valve in the transfer line between the retention tank and separation vat.

The direct aeration method may also be employed for shear sensitive floc situations, where flocculation occurs rapidly.[8] In these cases the final point of chemical addition is the inlet compartment of the flotation unit. In this case air bubbles are entrapped in the flocculation of the pinpoints previously formed. The net result is a highly efficient removal process due to the formation of the permanent air-solids bond.

In the partial aeration process, only a portion (usually 30 to 50%) of the stream to be treated is pressurized in the retention tank. The remaining portion of the waste flows by gravity or low pressure pump to the inlet compartment of the DAF where it mixes with the aerated stream. This method is limited to applications where lower suspended solids concentration are anticipated, since the air available for flotation is less than that available in the direct aeration process. However, for low solids streams this process is more economical since a lower percentage of wastewater is pumped at high pressure. The design considerations for direct aeration hold true for partial aeration.

In the effluent recycle process, a portion of the clarified effluent from the DAF is pressurized in the retention tank. From the retention tank the recycle is depressurized and injected into the inlet compartment of the flotation unit where it admixes with the incoming wastewater. This process eliminates the problems with mechanical shear since only clear effluent passes through the pressure release valve. Since most flotation flocculation procedures occur at relatively slow rates, and since most of the flocs produced are delicate in nature, the effluent recycle method is the most common in use today. It is also employed where a higher applied air rate is necessary, since the recycle flow can be several times the raw flow.

Before designing the effluent recycle process, investigations are centered around the degree of mixing and mixing time. The coagulation step normally entails rapid mix by injection into a pump suction, inline injection usually ahead of a static mixer or serpentine section of pipe, or rapid mix tank. The proper procedure will depend on the actual contact time required. The flocculation step is normally carried out in a mechanical paddle wheel flocculator which is either external to, or integral with, the DAF unit.[9]

B. Selection of Chemical Treatment Scheme

Before selection of the ultimate treatment scheme one must first consider the disposition of the final products, i.e., whether recovery or disposal is desirable. Often it is desirable to recover a solid constituent for reuse even though its presence will enhance flocculation.[11] If the constituents are not recoverable, the evaluation becomes a study of the cost effects comparing capital and operating costs. Since chemical pretreatment is an important contributor to capital cost, detailed investigations are normally conducted in the predesign stages.

C. Laboratory Techniques

While there is no substitute for pilot plant study information to establish design criteria, most often this is impractical. Hence, it is necessary to employ a reliable laboratory procedure for developing design criteria. The flotation "Bomb Test" procedure described below is a valuable tool in predicting performance of the full scale equipment, operating on a stream having the characteristics of the sample analyzed.

The basic test equipment consists of a small pressure vessel, graduated separatory funnel, tubing, stopwatch, and auxillary laboratory equipment. The pressure vessel or laboratory retention tank is equipped with pressure release valve, pressure gauge, air inlet, water outlet, and discharge connection. This equipment enables prediction of retention tank operating parameters.

The separatory funnel is essentially the separation vat. The funnel is graduated so that particle rise rate and final sludge volume can be recorded. In addition sample connections are provided so final product quality may be established. Depending on how the equipment is used, all three process application methods can be evaluated. Figure 6 depicts the laboratory apparatus set up for evaluation of all three methods.

Since the effluent recycle process is the most complex, we will deal with this laboratory technique exclusively. Direct and partial aeration techniques become evident. Coagulation and flocculation jar tests are run on a gang stir (same as sedimentation) to establish optimum chemical dosages for a particular floc quality. The flocculated sample is then filtered in a Buchner funnel. The filtrate is saved for use as the pressurized recycle in the test (plant-clarified effluent). If the particular waste under study is difficult to filter, tap water may be used as the initial source of recycle. However, several subsequent bomb tests must be performed, each time saving the subnatant for use as recycle in the following test. This enriches the recycle source with the minerals contained in the waste. This is important since the waste characteristics influence the bubble formation.

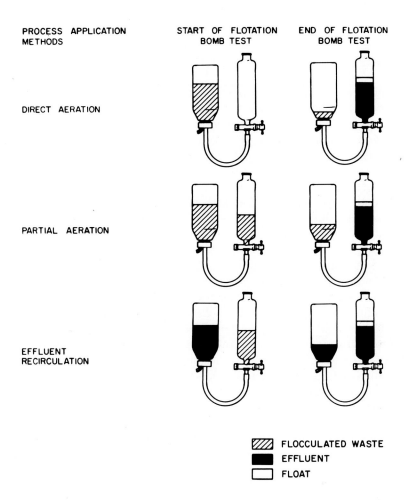

FIGURE 6. Laboratory "Bomb Test" method of simulating aeration modes. (Courtesy The Permutit Company, Inc., Paramus, N.J.)

Clarified recycle water (to be used in laboratory retention tank [bomb])[12] and air is added until the operating pressure under study is reached. The bomb is then agitated for several seconds to insure proper dissolution of the air. Pressure is then increased to operating pressure since part of initial air charge goes into solution resulting in a pressure decrease. Sufficient recycle water must be placed in the bomb to insure that no free air above the water level will be discharged into separatory funnel. This will create shear which will destroy the floc. Furthermore, if there is a high pressure drop in the bomb from initial pressurization to final discharge, one must question the validity of the test.

The chemically flocculated waste sample is placed in the separatory funnel, and the volume of the waste and floc appearance are noted. The bomb is inverted and connected to the stopcock fitting on the separatory funnel.

The stopcock and pressure release valves are opened and a portion of pressurized recycle is released into the separation funnel. After the desired volume of pressurized recycle is released into the funnel, the valves are closed and the flotation observed. Particle rise rate is measured by timing the distance traveled with a stopwatch; effluent (subnatant) and sludge (supernatant) are withdrawn from the cylinder.

From the laboratory bomb test several things can be determined. This information is valuable as a tool for selection of equipment for a new application, as well as a tool for evaluating full scale plant performance. The following information is learned from the bomb test.

1.   The particle rise rate and its changes are noted. The particle rise rate establishes the maximum overflow rate the equipment may be operated at and maintain particulate removal. In troubleshooting chemical dosages can be adjusted to determine optimum levels for elimination of solids carryover. A detailed analysis of rise rate data[11] is normally made before equipment design.[13]
2.   The recorded volumes of waste and recycle will parallel the ratios of influent required on the full scale equipment.
3.   The bomb pressure will predict the retention tank pressure.
4.   The sampled sludge consistency and effluent suspended solids will be similar to that achievable in the plant.

In evaluating bomb tests one must realize that the bomb test is a static operation performed under ideal conditions. Some scale-up considerations must be given.

When one is involved in the design of a system for a new application, pilot plant tests are a must. This is particularly true for wastewaters exhibiting constituent mix changes. The reliable plant will include equipment having detention times and hydraulic characteristics similar to the full scale equipment. Otherwise, validity is questionable. In order to stress the need for testing, Figure 7 has been included. This graph shows the different quality products produced from various plant influents at several applied air-solids ratios. Note that at an air-to-solids ratio of 0.02, sewage sludge can be three times the concentration of a particular chemical sludge.

IV. INDUSTRIAL APPLICATIONS AND COMMON CHEMICAL PROCEDURES

A. Pulp and Paper

The flotation process is used extensively in the pulp and paper industry for both recovery and waste control. Since fiber is present in most streams to be treated, polyelectrolytes can normally be used alone. Metallic salts such as alum are employed when a side benefit is realized, e.g., where alum is added to the pulp slurry for sizing or coloring.

The most common application is recovery of fiber lost in "white water" discharged from the paper machines. In this application the unit is referred to as a "save-all". Polymers are commonly used on save-alls to enhance the capture of fibers which are recovered to the pulp supply. Production of a clear effluent is also desirable since the treated water can be reused for washing operations. This recovered water is a valuable resource since it bears the same chemical analysis as transport water used in the mill. The presence of the polyelectrolyte in the sludge increases the probability of retention during the second pass through the system.

Another application is removal of fines from fractionation/washer filtrates. In this application the ultimate objective is recovery of wash water as described above. Floated solids are normally disposed of.

Flotation units have also been used as primary[11] and secondary[14] clarifiers treating final mill effluent. Since paper fibers are close to water in specific gravity, flotation is normally a more effective treatment alternative. Since papermaking fibers are not susceptible to hydraulic shear at the operating condition in the flotation process, direct aeration is most frequently employed. Polymers are typically injected into the retention tank and inlet compartment.

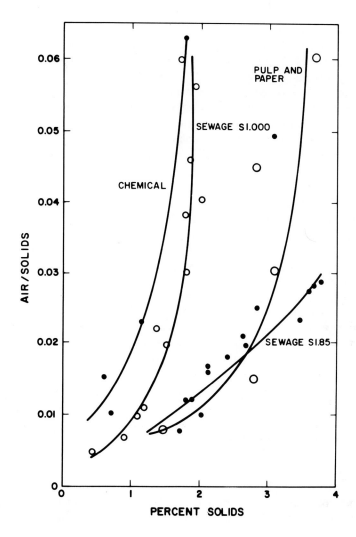

FIGURE 7.   Float solids concentration as a function of air solids ratio for various wastes.

B. Food Processing

The flotation process is used in several areas of the food processing industry. The primary function of the unit is recovery of oil, and greases which can be recovered are converted into usable by-products. A secondary function is removal of foulants ahead of a biological waste treatment process. In this process, emulsified oils and greases are normally demulsified chemically. Whenever possible food grade organics are used in lieu of metallic salts, since they will have a negligible effect on the recovered material. After demulsification, pinpoint flocs are flocculated with polymer for flotation. Since the flocs formed in this operation are delicate in nature, the effluent recycle process is normally employed.[15] Furthermore, the pressure drop at the release valve in the direct aeration process will enhance re-emulsion of the oils, defeating the purpose of the process.

C. Oily Waste Treatment

The flotation process application considerations for oily waste are similar to those outlined in section IV.B.,[16] the only exception being that the floated sludge is disposed of more frequently. Since this material has a negligible value, chemical treatment alternatives are governed by dewatering economy and effluent quality.

D. Latex and Synthetic Rubber

The DAF process has been employed for both primary and secondary clarification in this industry.[17] Primary treatment normally consists of coagulating latex emulsion with a metal salt to form "crumb rubber". This "crumb rubber" is then flocculated with a polymer. The recycle process is normally employed because latex emulsion will coagulate when subjected to shear, which would cause internal fouling of direct aeration equipment.

E. Biological Waste Treatment

DAF has been employed extensively for thickening of waste activated sludge. Flotation produces a higher consistency sludge than sedimentation, which provides more economical sludge dewatering. The DAF process has also had limited application in secondary clarification, but is the subject of extensive investigation as evidenced by the titles in the References of this chapter.

Since biological suspensions normally coagulate naturally in the process, yet remain near the specific gravity of water, flotation is a prime candidate for separation. The effluent recycle technique is normally employed to take full advantage of the delicate but natural floc formation. Polyelectrolytes are normally added to increase final sludge consistency. Metallic salts and high-charge polymers are used for coagulation, followed by high-molecular-weight flocculation.

In more recent years DAF has been applied for algae removal from oxidation ponds and tertiary lagoons.[17] Not only is the algae particle close to water in specific gravity, but during daylight hours, impounded algae floats to the water surface to receive more sunlight for the photosynthesis process.

F. Other Applications

The flotation process has also been employed for a variety of other applications including chemical plant waste, petroleum and petrochemical, metal finishing, and raw water coagulation. These applications methods and chemical treatment procedures employed are similar to those described previously. It should be understood that the flotation process can be employed for most applications where sedimentation can be employed. Economics should be the basis for process selection.

V. FUTURE TRENDS

At the present time significant experience in using the flotation process has been accumulated and reliable guidelines for process evaluation and design have been established. In order to improve the process efficiency, energies have been directed toward bettering automatic process control. In some of the more modern installations instrumentation has facilitated smoother operation by buffering changes in the stream characteristics. Equalization, pH control, flow proportionate chemical feed, and turbidity proportionate chemical feed are techniques frequently considered. More sophisticated systems provide for cascade control using a variety of monitoring techniques. Even more exotic monitoring procedures such as zeta potential have been utilized.[18] As physical conditions observed indicate performance trends, internal process modifications are investigated for optimization. For example, since the effect of solids concentration has been observed,[18] sludge recirculation has become a subject for study; sludge recirculation concepts are of interest for ultimate chemical conservation.

REFERENCES

1. Friend, W., The flotation of papermaking fibres, *Tappi,* 39(6), 172A, 1956.
2. Lundgren, H., Recent advances in air flotation technology, *Tappi,* 53(2), 287, 1970.
3. Oil-Water Separation Process Design, Manual on Disposal of Refinery Waste, American Petroleum Institute, New York, 1969.
4. Vrablik, E. R., Fundamental principles of dissolved air flotation of industrial wastes, *Proc. 14th Ind. Waste Conf.,* Purdue University Press, West Lafayette, Ind., 1959, 743.
5. Bratby, J. and Marais, G., *Saturator Performance In Dissolved Air (Pressure) Flotation,* Vol. 1, Pergamon Press, Oxford, 1975, 929.
6. Masterson, E. M. and Pratt, J. W., Applications of pressure flotation principles to process equipment design, 2nd Waste Treatment Conf., Manhattan College, New York, April 1957, 232.
7. Groves, S. E. and Lundgren, H., Recovery of undissolved solids from rubber plant wastewater, *Ind. Wastes,* p. 40, 1974.
8. Mennel, M., Merrill, D. T., and Jorden, R.M., Treatment of primary effluent by time precipitation and dissolved air flotation, *J. Water Pollut. Control Fed.,* 46(11), 2471, 1974.
9. Hutchenson, O. F., Flotation processes use and results in paper mill waste water clarification, *Tappi,* 41(7), 158A, 1958.
10. Balden, A. R., Flexibility key to design of machining plants treatment facilities, *Ind. Wastes,* p. 6, 1970.
11. Lundgren, H. and Walzer, J. G., Primary treatment by a flotation system, Proc. 9th Annu. Tappi Environ. Conf., Houston, 1972, 171.
12. Eckenfelder, W. W., Jr., Rooney, T. F., Burger, T. B., and Gruspier, J. T., Studies on Dissolved Air Flotation of Biological Wastes, 2nd Waste Treatment Conf., Manhattan College, New York, April 1957.
13. Talmadge, W. P. and Fitch, E. B., Determining thickener unit areas, *Ind. Eng. Chem.,* 47(1), 208, 1955.
14. Jackson, M. L., Shen, Chia-Chouh, and Plopper, C., Deep Tank Flotation Biological Treatment, presented at Pacific Northwestern Pollut. Control Assoc. Water Pollut. Control Fed., Seattle, 1976.
15. Reed, S. W. and Woodard, F. E., Dissolved air flotation of poultry processing waste, *J. Water Pollut. Control Fed.,* 48(1), 107, 1976.
16. Rohlich, G. A., Application of air flotation to refinery waste waters, *Ind. Eng. Chem.,* 46(2), 304, 1954.
17. Bare, W. F. R., Jones, N. B., and Middlebrooks, E. J., Algae removal using dissolved air flotation, *J. Water Pollut. Control Fed.,* 47(1), 153, 1975.
18. Riddick, T. J., *Control of Colloid Stability Through Zeta Potential,* Creative Press, Claremont, Calif., 1968. 1.
19. Springer, A. M., Flotation Clarification and Thickening Trials, NCASI Field Study Report, National Council for Air and Stream Improvement, Kalamazoo, 1975.

Chapter 6

SLUDGE DEWATERING

William L. K. Schwoyer

TABLE OF CONTENTS

I. INTRODUCTION

Probably no other liquid-solid separation operation has been as dramatically affected by the advent of polyelectrolytes as has dewatering. Polyelectrolytes, especially the synthetic polymers, have made mechanical gravity dewatering practical. The dewatering effectiveness afforded by polyelectrolyte flocculation, coupled with the desirable operating characteristics of mechanical gravity dewatering devices (as described more fully later in the chapter) have given this relatively new method of continuous mechanical dewatering a full share in the sludge dewatering equipment marketplace.

Signaling continuous mechanical gravity dewatering as the most profoundly affected equipment does not mean to diminish the effect polyelectrolytes have on the efficiency of other dewatering methods, whether they be gravity drying beds, vacuum filters, centrifuges, or in some instances, filter presses. The latter represents the least definite case of polyelectrolyte utility. In some cases polyelectrolytes are helpful, in others, a detriment. This, too, will be covered in more detail later when filter presses are discussed.

In keeping with the theme of this book, water and wastewater treatment, dewatering will be couched in terms of sludges generated in the processes of raw water, sanitary sewage, and industrial residue treatments.

A. Sequence and Continuity

Dewatering is the last mechanical liquid-solids separation unit operation to which a waste stream is subjected prior to disposal of the separated liquid and solid phases. In most instances, the sludge feed contains a relatively small proportion of the total initial raw water (liquid) content, and most of the solids. The cake discharged by the dewatering device will go to its final disposition, whether it be landfill, land application, incineration, or reuse.

The separated filtrate, centrate, or whatever it might be designated (consisting of a predominately liquid phase containing varying proportions of solids), is usually returned to some point downstream of the dewatering operation where, and in such a manner that, the added hydraulic and solids loadings will not upset the system equilibrium. This same principle is used in handling filter backwash.

1. Continuity

While this chapter is not meant to philosophize, there are certain concepts about liquid-solids separation that can be helpful in understanding how systems operate and in determining why they malfunction. One such idea is that of compatible contiguous operations.

Ideally, a waste treatment system will consist of unit operations that are integrated with regard to hydraulic and solids handling capacities, from the solids generation step to ultimate sludge disposal. That is, the solids-rich discharge from one unit is of the proper volume, concentration, and physical consistency to be economically handled by the succeeding piece of operating equipment. There should be some overlap or redundancy built into the system, preferably proportioned over the various unit operations, in order to give the system the flexibility to accommodate the normal operation variations, with periodic excursions.

This sounds too simple to be worthy of mention, or too idealistic to be practical. Many existing treatment plants, both old and recently designed and built, have mismatched sequential unit operations. Mismatching can occur in two directions: too much overlap, where a given piece of equipment is not capable of sufficiently reducing the volume of the influent stream to justify its capital and operating costs, or not enough overlap to properly process the influent as it is received.

Designers, purchasers, and operators are faced with a tradeoff between increased capital costs for the insurance against over-capacity, and the unknown increased future operating costs of under-capacity. For one reason or another, usually unanticipated growth, or more stringent treatment requirements, most plants become overloaded or reach capacity sooner than anticipated. Operating an overloaded plant invariably results in excessive operating costs. These costs usually take the form of greater chemical consumption and increased volumes of wetter sludges to be disposed.

Under some circumstances, polyelectrolytes have been found to be helpful in coping with overloading by increasing the solids settling rate in a clarifier or thickener, and increasing the capacity of the dewatering system. These increases are realized when a polyelectrolyte can be added where one is not being used, or by increasing the dosage where one is being employed below its maximum effective concentration. The improvements are incremental, at best, and costly. At times, the most cost-effective polyelectrolyte dosage is below that amount that will give the greatest unit capacity.

2. Sequence

Continuity was discussed before sequence because it can be considered the factor necessary to make a treatment system work successfully. Sequence, while it may be argued to be the same as continuity, in this chapter refers to the relationship of one liquid-solids separation unit operation with respect to another in the treatment plant flow scheme.

Historically, the waste treatment section of an industrial installation usually diluted its earnings. Waste treatment equipment was, therefore, installed only when really necessary.

The evolution of waste treatment systems usually started by adding the waste treatment step adjacent to the main plant operation. Successive treatment steps were added as required.

It has been almost axiomatic that the flow of operating adjustments is one directional, emanating from the primary or production operation through the various waste treatment unit operations. When operating upsets such as production spills or process changes occur, each successive waste treatment operation has to adjust to accept the feed from the previous step. The magnitude and difficulty of the adjustments often increase with each successive step.

Dewatering, being the last liquid-solids separation step, can be called on to handle larger hydraulic loads, higher solids loads, or poorer quality sludge than it was sized to do.

Only recently have investigators taken a different approach in designing waste treatment facilities.[1] Impetus for the radical departure from the traditional method of designing a waste treatment system, which generally centered on primary and secondary treatment, has been supplied by two factors. Sludge handling accounts for approximately half the construction and operating costs of a waste treatment plant. Also, dewatering characteristics of a sludge, particularly a biological sludge, are the most difficult unknowns to predict in a waste treatment system design.

New concepts of system design start by determining the dewatered cake characteristics required for the chosen ultimate disposal method, whether it be zero discharge, incineration, composting, land application, or hauling to a regulated landfill. Once these criteria are established, the dewatering unit capable of meeting the criteria is chosen. After the dewatering unit is chosen, the secondary treatment and thickening system capable of meeting the influent characteristics requirements of the dewatering system is chosen. This tack is diametrically opposed to the traditional method of designing biological waste treatment plants which centered design emphasis on a particular biological process chosen to best accommodate the waste as received, paying little attention to the sludge handling portion of the plant.

Public environmental concern has had such far-reaching effects as to implement changes in manufacturing processes in order to simplify waste treatment. The current and future rising costs of sludge management make it a significant factor in plant profitability.

Figure 1 shows the approximate influent and discharge suspended solids contents for the more common liquid-solids separation devices above the abscissa, and the discharge solids requirements for various sludge disposal methods below. A biological sludge system was chosen because it represents one of the most commonly used treatment methods. This graphical method of expressing unit operation feed and discharge suspended solids levels allows a quick visual grasp of the following: (1) comparison of solids concentration for various methods of performing a given unit operation, (2) the degree of continuity between various equipment types in progressing from one concentration operation to the next, and (3) the equipment capable of producing the solids concentration required for a given sludge disposal method.

A raw waste stream usually has a suspended solids loading range of 100 to 1000 mg/ℓ. Clarification will raise the level to 15,000 to 25,000 mg/ℓ (1.5 to 2.5 wt/vol%) in the sludge discharge (blowoff). If thickening is used, the underflow from a gravity unit, or the float from a flotation thickener, will be in the 2.5 to 5.0 wt/vol% range. The next solids concentration step, broadly called dewatering, results in discharge cake solids, ranging from approximately 12 to 45 wt/vol%.

It is important for anyone reading and using the above-mentioned solids content numbers to recognize that the actual values obtained for any given "biological sludge", treated by any given combination of liquid-solids separation steps or devices, can vary considerably from the numbers stated. They probably will skew toward lower values. A few of the causes of deviation between reported values and those obtained for a specific installation will be covered later in this chapter.

Again referring to Figure 1, when the appropriate dewatering system capable of delivering the cake solids required for sludge disposal is considered, one factor to weigh is the number of subsequent unit operations above the line needed to meet the cake solids requirements below the line.

A value analysis of the various alternatives will be an important factor in choosing a system. Economic factors equal, operating personnel usually favor the fewest pieces of equipment necessary to perform the work.

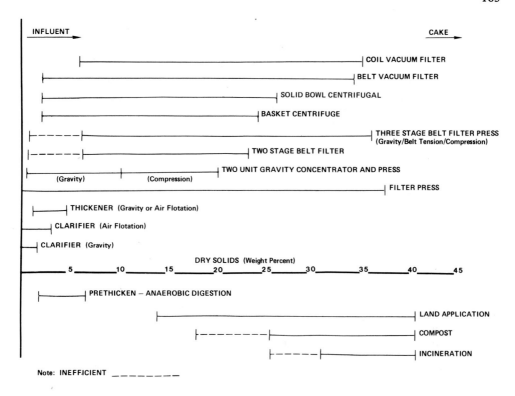

FIGURE 1. Influent and discharge solids ranges for liquid-solids separation operations, and acceptable solids content range requirements for subsequent processing of a biological sludge.

In the past, and even today, volume reduction obtained during dewatering received considerably less attention in sludge handling discussions than did the influent and cake solids weight percents. However, when the final discharge cake is moved to the ultimate disposal site, whether it be landfill or land application, the volume of material to be accommodated during transportation and disposal is one of the prime factors, if not the prime factor, to be considered in an economic evaluation.

Figure 2 shows the relationship between the sludge weight percent, the weight and volume of water present, and the solids weight as the weight percent of the sludge increases. It is a graphic reminder that in dewatering, solids contents are increased by removing water.

If we start with a 2% sludge we have 2 g of solids in 98 g of water. Raising the solids content of that sludge requires removing half of the weight of the water present. That represents approximately halving the original sludge volume for most biological wastes which have specific gravities in the 0.94 to 1.02 range. It becomes apparent that the greatest volume reductions occur during the first three doublings of the solids contents, going from 2 to 16%. The cost of reducing sludge volume becomes increasingly greater as the solids content increases.

Each of the three broad liquid-solids separation unit operations shown, clarification, thickening, and dewatering, employ devices that process the waste in two different ways. They are either batch or continuous operations. Polyelectrolytes are important in batch-type operations, but it is in continuous processing where they have made the greatest contributions to process efficiency.

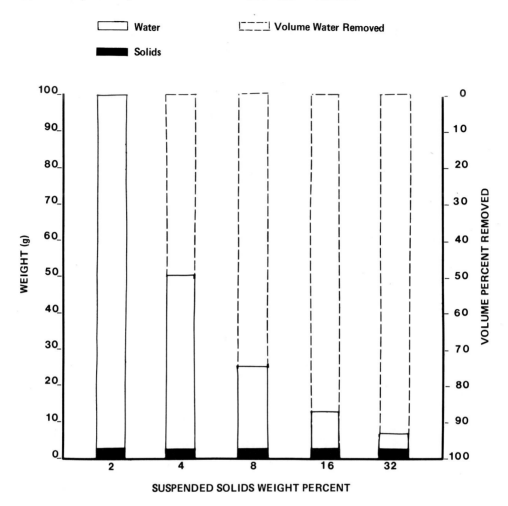

FIGURE 2.   Volume reduction as a function of solids content for a 2% influent biological sludge.

II. FLOCCULATION

The authors of several of the other chapters, particularly Chapters 5 and 7, give clear, detailed discussions of the mechanisms by which polyelectrolytes operate with regard to their specific liquid-solids separation step. There is no need to cover again the structure, types, and basic operating mechanism of polyelectrolytes when used to gather solids. This chapter will look at the relationship of the two properties of polyelectrolytes, charge and molecular weight, with their ability to further agglomerate the particles that were destabilized and removed from suspension during the coagulation process required for each of the unit operations, such as clarification or filtration (where low concentrations and/or small particle-size solids had to be removed from the bulk of the liquid stream).

A. Differences Between Flocculation and Coagulation

Prior to dewatering, the sludge obtained from a clarifier, thickener, or filter backwash usually again undergoes chemical treatment. This treatment is termed flocculation and can consist of the addition of a single chemical, or a series of chemicals. It is not at all uncommon for the chemicals added in the flocculation process to be the same ones used for the initial coagulation charge neutralization, although this is not always the case. During flocculation, the coagulated small solids clusters are further gathered into flocculum 1/8 to 3/8 in. diameter, depending on the type of equipment being used.

1. Dosage

One practical difference between coagulation and flocculation is the amount of polyelectrolyte added in the various processes as noted below:

Equipment	Process	Polyelectrolyte dosage (mg/l)
Clarifier	Coagulation	0.25—1.50
Filter	Coagulation	0.05—1.00
Thickener	Coagulation-flocculation	2—10
Dewatering	Flocculation	15—150

It should be noted that the aforementioned dosages represent only the amount *added* at that particular step, not the cumulative total for each step up to and including that step.

2. Charge Density Versus Molecular Weight

A second practical difference between coagulation and flocculation is the manner in which the two main characteristics of a coagulant or polyelectrolyte, electric charge density and molecular weight, are employed. Electric charge density is employed in neutralization of the electric charge built up on the double layer surrounding the solids particles surfaces. Molecular weight is associated with the bridging ability of the conditioner. Bridging is usually associated with medium- and high-molecular-weight polyelectrolytes, and refers to the ability of the polyelectrolyte to gather and hold together the charge neutralized fine flocs.

Coagulation of fine, dispersed solids in suspension requires neutralization of the electric charge before the suspension can be destabilized and the particles agglomerated. Conditioning chemicals with high electric charge densities are required. Most of the particles found in nature have a negative surface charge and require positively charged conditioning chemicals. The hydronium ion (H_3O^+), metal cations (particularly Ca^{+2}, Fe^{+2}, Fe^{+3}, and Al^{+3}), and cationic polyelectrolytes fulfill this service.

The closer the surface charge of a particle approaches to zero, with or without prior conditioning, the more bridging becomes the important operator. Schwoyer[2] reported laboratory work in which a ferric hydroxide-based sludge was flocculated and dewatered equally well by cationic, nonionic, or anionic polyelectrolytes of the same chemical base and molecular weight, indicating that bridging was the only operator of consequence in that particular instance.

The hydronium ion is a strong charge neutralizer and a very weak bridger. Hydroxide flocs formed by di- and trivalent cations are very effective charge stabilizers and mildly effective bridgers. Cationic polyelectrolytes run from mild to strong charge neutralizers and strong to mild bridgers. That is, usually the charge neutralization and bridging abilities of organic cationic polyelectrolytes run counter to one another. Most cationic polyelectrolytes are in the 1×10^4 to 1×10^6 molecular weight range, whereas most nonionic and anionic (polyacrylamide based) are in the 3×10^6 to 1.2×10^7 range.

There are considerably more cationic polyelectrolytes, both natural and synthetic, than there are nonionic or anionic. Mention is made in Chapter 7 of the consideration that cationic polymers* tend to interact more with the particles on which they are working than do nonionics or anionics. This characteristic of cationic polymers makes

---

* Polymer, a generic term for polyelectrolytes, is often used interchangeably with polyelectrolytes when discussing chemical conditioning for *liquid-solids separation*.

prediction of the correct polymer based on prior experience much more difficult than with nonionics or anionics. More of this aspect of polyelectrolytes is covered in the laboratory testing section of this chapter.

B. Chemical Dosage Versus Performance

Clarification is usually done by one of two methods, gravity or flotation. On an overall basis, the chemical conditioning requirements for both methods are the same. Flotation requirements might be at least a bit more for the same degree of effluent clarity, since dissolved air flotation units operate at two to four times the hydraulic loading rate of gravity settling clarifiers (see Chapter 5).

Thickening is again usually done by gravity settling basins (picket fence paddles) or dissolved air flotation. Again, the chemical dosages for dissolved air flotation can be higher than those for gravity settling because of higher hydraulic loading rates. Flotation thickeners can also be run without any conditioning chemicals. Centrifugation is also used for thickening, employing polyelectrolytes at various dosage levels as a way of controlling solids capture and final cake solids.[3]

Dewatering is performed by sand or mechanical drying beds, several types of continuous gravity/press combinations, vacuum filters, filter presses, centrifugals, bag filters, and other type units. The use of polyelectrolytes as sludge conditioners for this vast variety of units ranges from absolutely necessary (for continuous gravity/press type units), through being required for increased solids capture or higher capacity (for centrifugals, vacuum filters, or drying beds), to questionable (for filter presses). The wide variety of dewatering equipment available, with much of the equipment having individual dewatering characteristics, gives polyelectrolytes a wider application spectrum in this area than in the other liquid-solids separation processes.

III. TESTING

The technology of manufacturing the polyelectrolytes used for water and wastewater liquid-solids separation processes can be explained and predicted for the most part, as is evident in Chapter 1. In short, it can be considered a science. Application of these well defined chemicals to conditioning waters and wastewaters for the various liquid-solids separation unit operations is still limited enough in technical basis to be considered "art". It is not a lack of understanding of the properties of polyelectrolytes that limits predictability in their application; it is the poor knowledge of the components of the stream to be treated, and thereafter, the manner in which they individually or collectively will react with the polyelectrolytes. The above is particularly true of wastewaters where it can be argued that complete categorization of constituents is not possible. Most people involved with the application of chemicals and equipment in wastewater treatment come to the conclusion that each waste is different enough to prevent predicting the best chemical treatment with any degree of certainty. The inability to consistently predict qualitatively, let alone quantitatively, the correct conditioners gives reason to believe that there can, at times, be significant components in the waste of which we are not always aware. There is also reason to believe that we are not fully aware of the effect the physical and chemical properties of the wastes have on the conditioning and separation processes.[4] This will be discussed in more detail later in the chapter.

The amount of capital and operating monies committed when a treatment system is chosen makes surveying, characterization, and testing prior to selection more than a matter of prudence for all parties involved.

A. Testing Scale

As was mentioned earlier, dewatering characteristics are the least understood aspect of waste treatment plant design and operation. Rapidly shifting demands for dewatered cake characteristics further compound this problem. It is important, therefore, that the proper perspective is obtained with regard to the capabilities of the various scales of testing available, i.e., bench test, pilot unit trial, and full scale equipment.

The greatest purview is usually obtained by employing all three, or at least two of the three test scales when possible. At times physical or financial constraints will minimize the amount of testing in developing design and performance data. The waste stream may not be available, or accessible in very limited quantitites, making pilot or full scale testing impractical, if not impossible. Pilot and full scale tests are considerably more expensive than bench tests, especially laboratory bench tests, and frequently will be bypassed due to costs involved — at times a "penny wise and pound foolish" approach for all parties concerned.

The interdependence or relationship of the three testing scales (which really differ in more ways than just quantity of sample treated) might better be seen by roughly categorizing their abilities to supply information needed as good, fair, or poor. These comparisons are general and arguments might be made for any of the classifications based on a specific case. From Table 1 it can be seen that sludge characterization is best done in the laboratory where the analytical equipment is available to quantitatively measure the chemical and physical characteristics of the waste, and records are available to compare it with other wastes previously characterized and categorized. Frequently this is difficult to do in the field, if for no other reason than the cost involved in having personnel and equipment in the field.

It is extremely important to recognize that waste characterization can be done well only if the sample or samples tested are representative of the waste to be treated, and preserved in such a manner that the characteristics do not change during shipment and storage. This point cannot be emphasized enough. A sample or samples taken for laboratory (bench) evaluation will not profile the changes in waste stream chemical and hydraulic characteristics to the degree that can be done by full scale or pilot tests. Laboratory tests, however, are most efficient in choosing the chemical treatment system. The time and money that would be required to choose the best chemical conditioning, make-up, and feed systems on a full size equipment or even pilot scale, without previous information, would most probably be prohibitive. In many actual instances, even if a sample did not fully duplicate the waste stream characteristics, the chemical conditioning system chosen was sufficiently correct to be optimized with minor adjustments when field tested on a larger scale.

Scale-up, the application of laboratory (bench) and pilot plant data in a full size installation, is a factor many people frequently assume automatically happens, and will be discussed in more detail later.

B. Laboratory (Bench) Testing

To date, the smallest scale testing has been referred to as laboratory (bench) tests. This type test might practically be geographically subdivided into two subcategories. Laboratory tests would denote the test work being performed in a location having laboratory facilities such as balances, ovens, mixers, vacuum, etc. available for running controlled, quantitative tests.

Bench tests might also be conducted in the field, making qualitative determinations that have a lesser degree of quantitative accuracy than when run in a laboratory, since little equipment is available. The most advantageous aspect of field bench testing over laboratory is that the sample is usually fresh. For some readily degradable sludges,

Table 1
TEST SCALE UTILITY

	Laboratory bench	Pilot	Full scale
Waste characterization	Good	Poor	Poor
Waste consistency	Poor	Good	Good
Chemical treatment choice	Good	Fair	Poor
Hydraulic considerations	Poor	Fair	Good
Scale-up and design	Poor	Fair	Good
Economic analysis	Poor	Fair	Good

both organic and inorganic, field testing is a necessity. Better sampling is generally an additional advantage of field testing because the bench testing is frequently combined with a process survey.

Practically all equipment used to mechanically dewater sludge or separate liquids and solids, does so by physical means. Their performance depends on the physical, rather than chemical, characteristics of the phases they are separating. Of course, the phases must not chemically attack the separation device. Drying beds, dryers, and incinerators can be included with the above equipment.

Conditioning, chemical or otherwise, then might be considered as a means of imparting to the two phases (usually the solids) the physical characteristics required for efficient processing in the particular device chosen. This will vary from drying bed to belt press, to centrifugal, to vacuum filter, etc. In fact, the requirements can change from one zone to another of a piece of equipment, such as in the case of a belt press. The solids or collection of solids need enough integrity to withstand the mechanical and hydraulic forces encountered during separation of the phases.

Laboratory testing, then, consists of three steps: (1) waste characterization, (2) conditioning (usually chemical), followed by (3) a procedure developed to simulate the dewatering characteristics of the particular mechanical device to be used. The second phase is often referred to as "jar testing" (a misnomer carried over from settling tests). The third phase methodology may consist of determining relatively universal parameters such as specific resistance or compressibility coefficient of the conditioned sludge. Also, it may be more specific for a given dewatering device, such as the filter leaf test for vacuum filtration, bench centrifugation for centrifugals, and roll or press tests for continuous gravity/belt press type units. The chemical conditioning also will generally vary for the different type units.

1. Characterization

Characterization consists of determining both the chemical and physical properties of the waste. When combined with information obtained in a process survey pertaining to how the sludge was generated and what will be required of the dewatered cake and filtrate in subsequent processing operations, it can give preliminary insight into the conditioning system required. Good waste characterization and process survey can greatly reduce the amount of conditioning testing required for the desired result.

Table 2 lists measurements that might be made during a complete waste characterization work-up. Characterization should be done prior to any testing for any type of waste treatment, not only dewatering. It serves as a reference point on which performance guarantees can be based and against which the waste can be compared once the system is placed in operation.

The absence of a sufficiently complete characterization of the waste to be treated is often the main cause for disagreement between all concerned parties with respect to

Table 2
WASTE CHARACTERIZATION PARAMETERS

1. pH
2. Solids texture
3. Solids color
4. Liquid color
5. Odor
6. Settleable matter (volume/volume percent)[a]
   Settling Time

   ---

   1 hr
   24 hr
7. Sludge volume index
8. Total solids % (weight/volume percent)
9. Total volatile solids (weight/volume percent)
10. Total suspended solids (weight/volume percent)
11. Total volatile suspended solids (weight/volume percent)
12. Acidity[b]
13. Alkalinity[b]
14. Total organic carbon (liquid phase)[b]
15. Specific conductance[b]
16. Specific resistance
17. Zeta potential[c]
18. Turbidity[c]

[a] This measurement includes the total combined settleable and floatable layers when there is floatable material.
[b] These parameters are useful for completely characterizing a dewatering application candidate waste, but not of primary importance in all instances.
[c] Not usually measurable for dewatering sludges.

how successfully a piece of equipment or a process is meeting the expectations of the designer and the operator. Not all of the parameters listed in Table 2 affect the performance of the sludge dewatering unit to the same degree. Settleable matter, total solids, total suspended solids, total volatile solids, pH, and specific resistance are the minimum determinations for a good "fingerprint" of the sludge. While color and odor do not directly bear on the dewaterability of a sludge, changes in these easily discernible characteristics can be good indicators of changes in more significant properties. Parameters 1 through 6 can be determined in the field when the waste stream is sampled, with a minimum of equipment. "Hydrion" paper may be used for a rough pH determination and a graduated cylinder (100 mℓ minimum) may be used for determining settleable matter, that portion of the total waste volume occupied by the solids.

Settleable matter, or solids volume percent, is not often used as a waste stream descriptor. Sludge Volume Index (SVI), a measurement derived for biological treatment systems, is more common. The 1- and 24-hr measuring time periods for settleable solids are arbitrarily chosen for high and low density solids particles, respectively.

Many of the other tests are completely detailed in "Standard Methods for the Examination of Water and Wastewater".[5]

2. Conditioning
The first insight into choosing a conditioning system, as mentioned earlier, is gained during the waste characterization. Broad categories, or certain types of sludges historically respond to various charge polyelectrolytes i.e., cationic polymers for sanitary sludges, and anionic polyacrylamide based for inorganic metal hydroxides. A considerable amount of effort, both documented and undocumented, has been spent in at-

tempting correlations between wastes and conditioning systems which will lead toward moving coagulation and flocculation from the realm of practiced art to that of technology. La Mer was one of the pioneers in this area.[6] Vostrcil and Juracka,[7] O'Brien,[8] and Bradley,[9] have published more recent work along these lines.

Direct, reliable correlations between wastes and flocculation systems are still some time in the future, because of the almost infinite variety of possible components present in waste streams. At this time we have no economical way of detecting the components, and insufficient knowledge of what levels of various components have significant effects on the conditioning process.

Few wastes, especially sludges, are single components. Municipal wastes are frequently comprised of sanitary and industrial streams. Wastes having undergone clarification, thickening, neutralization, or other operations prior to dewatering, most likely have gained some new components along the way which can significantly change the conditioning requirements.

Properties usually considered belonging to the liquid phase, such as pH, can significantly influence the conditioning reagent choice or its effectiveness. This situation frequently occurs if an organic sample is shipped or stored without proper preservation, one of the greatest disadvantages of laboratory testing some distance from the waste source. The problem can also occur in pilot plant scale tests when drawing waste from a storage supply that represents holding conditions different from those encountered under normal operations.

A lowering of pH usually accompanies biological decomposition of organic wastes under septic conditions. The hydronium ion (H_3O^+) generated during the decomposition is available to neutralize the negatively charged double layer associated with the surfaces of the organic particles. Availability of hydronium ion usually changes the conditioning chemical requirement. Readjusting the sludge pH to its original point will most likely again shift the conditioning requirement. In most instances the writer observed, the shift was toward the direction of the requirements for fresh sludges, but not exactly duplicating them.

Since cationic polyelectrolytes or combinations of metal salt coagulants and polyelectrolytes are usually used for organic sludges, directly testing a septic sludge could result in choosing a polyelectrolyte that might not function at all under normal conditions. A second probability would be underestimating the treatment cost because the hydronium ion neutralized part of the surface charge normally acted upon by the polyelectrolyte or metal salt coagulant.

Sodium or potassium hydroxides should be used when attempting to restore original properties to a septic sludge sample for testing. They have the lesser effect on the sludge with regard to its dewaterability than do other bases available, due to their lower positive charge density and soluble salts. The calcium cation (Ca^{++}) introduced when lime is used for neutralization is itself reasonably effective in neutralizing the negative surface charge on the organics, aiding in their coagulation. In addition, calcium forms precipitates with many of the anions present in wastes, both inorganic and organic. These precipitates can give erroneous results with regard to dewaterability, cake solids, and volume and filtrate clarity.

Table 3 lists the U.S. Environmental Protection Agency recommended storage procedures for biological waste samples.[10] Refrigeration at 4°C should be used for storing and shipping sludges. Freezing sludges for storage or shipment would in itself dewater them and render them useless for subsequent testing.

Presently there are no other "standard" methods for preserving organic wastes for shipping or storage. Addition of 2 mℓ/ℓ of a 40% formaldehyde solution can be used with caution. Tests should first be run to determine that the formaldehyde does not

Table 3
RECOMMENDED STORAGE
PROCEDURE

	Sample storage	
Analysis	Refrigeration @ 4°C	Frozen
Total solids	Acceptable	Acceptable
Suspended solids	Up to several days	No
Volatile solids	Up to several days	No

exert any coagulation or flocculation effect on the sludge that would cause erroneous results in the conditioning tests. The total organic carbon content would also be increased and of questionable value in a characterization.

In spite of the above-mentioned drawbacks to the formaldehyde preservation, it is a better choice than no preservation at all, and more easily accomplished than refrigeration at 4°C. Therefore, under some circumstances, it might be the most practical choice.

a. pH Adjustment

Both the dewatered cake and the filtrate (or centrate) frequently have some pH range requirement for disposal. Usually adjusting the pH to meet the discharge requirement is advisable prior to scanning the conditioning chemicals, whether they be polyelectrolytes, metal salt coagulants, or combinations. A pH range of 6 to 10 is most commonly employed in an attempt to avoid costly equipment materials of construction, and to render the filtrate innocuous to its receiving body. Higher final pH ranges might result when ferric hydroxide and lime are the major or sole conditioning chemicals used, such as with vacuum filters and filter presses, which will be discussed later.

Polyelectrolytes are of the most interest; therefore, they will receive the most attention. Biological sludges will usually be in the pH range of 6 to 8 and will need little, if any, adjustment. The same is true of sludges that have already been subjected to a liquid-solids separation step such as clarification and/or thickening, since the adjustments were made at that time.

Waste streams fed directly to the dewatering stage from industrial processes usually show the wide pH deviations. Neutralization of these streams can radically change their dewatering characteristics by precipitating or, in some instances, dissolving solids.

When dissolved solids are precipitated as a result of pH adjustment, it is best to remove them as thoroughly as possible in the dewatering stage. The alternative is to have them precipitate when the filtrate is blended into a receiving body and the pH normalized. Even if the filtrate is returned to the head of the process, the dissolved solids will come in contact with a pH range of 6 to 9 in the final effluent of the plant. Post precipitation, the delayed precipitation of solids from an effluent (usually due to a pH shift), can cause interior plugging of a final effluent filter bed or deposition of suspended solids in the plant discharge. A closed loop system could conceivably operate in a manner that would not require optimum dissolved solids removal, but it would be the exception.

Sludges obtained from a water softening process using lime to precipitate calcium carbonate or a calcium carbonate/magnesium hydroxide mixture are generated at a pH range of 10.5 to 12.0 for complete precipitation. Dropping the pH to the 6 to 9 range resolubilizes a portion of the solids. These are then returned to the softening process to be reprecipitated, possibly overloading the system.

Table 4
pH EFFECTIVENESS OF
POLYELECTROLYTES

	pH range
Cationic (tertiary, mannich)	⩽8
Cationic (quaternary)	⩽8
Nonionic (polyacrylamide)	Any
Anionic (polyacrylamide)	⩾4

Table 4 lists the effective pH ranges for some of the more common type polyelectrolytes.[11] When operating outside these ranges the efficiency of the polyelectrolytes will decrease. There are isolated cases where a specific polymer will operate far outside its normal usage range, for some unexplainable reason. One such case is a laboratory test where a mildly charged, anionic polyacrylamide-based polymer was as effective as a cationic polymer of the same chemical base and molecular weight in flocculating the undissolved silt residue from an alum clarification sludge from which the alum was reclaimed, by adjusting the pH to 1.5 to 2.0 with sulfuric acid.[12]

Neutralization can involve the addition of either base or acid. In practice, neutralization prior to dewatering usually involves adjusting the pH upward after the addition of a metal coagulant such as alum, ferric chloride, ferric or ferrous sulfates, or in some cases, a mineral acid.

The bases most commonly used for pH adjustment of acidic sludges are lime and sodium hydroxide (caustic soda). Less commonly used are potassium hydroxide, sodium bicarbonate, and sodium carbonate. Each chemical has its particular advantages and disadvantages with regard to sludge dewatering.

Sodium hydroxide can be obtained as a concentrated solution which makes handling easy. When used, it should be diluted to the point where it can be accurately metered and mixed with the sludge, using the equipment chosen. Using too concentrated a feed solution can result in overshooting the desired pH range. Sodium hydroxide is also a single function chemical with regard to sludge dewatering. Its sole activity is pH adjustment. Charge neutralization obtained from the Na^+ cation is negligible for all practical purposes. Most of the common anions encountered in waste treatment form soluble sodium salts. If effluent-dissolved solids content is important, sodium hydroxide might not be the best neutralizer choice.

Lime (Ca [OH$_2$]) is usually the least expensive base available. It has limited solubility (1.6 g/ℓ at 20°C) and is not as convenient to handle as sodium hydroxide. Using a fully solubilized lime solution for pH adjustment can require an impractical volume to be added with regard to solution holding and metering facilities and sludge dilution. In practice, a trade-off is made, wherein the lime is usually fed as a 1 to 3 wt% suspension (10 to 30 g/ℓ). The acidity in the larger volume waste is utilized to solubilize the undissolved solid lime. This method works reasonably well as long as provisions are made to keep the particulate lime suspended during holding and through the metering system. A characteristic of a lime neutralization system is a gradual pH increase with time, as the suspended lime dissolves. Overshooting the desired neutralization point often happens, both in the laboratory and full scale installations, unless sufficient time is alloted for the full suspended lime solubilization.

Lime is a multi-functional chemical with regard to sludge dewatering. The hydroxyl anion (OH$^-$) neutralizes acidity as was just discussed. In addition, the divalent calcium cation (Ca^{++}) is an effective coagulant by virtue of its ability to neutralize the negative surface charge associated with most naturally occurring particles in suspension. Several calcium salts of anions found in waste streams are considerably less soluble than their sodium analogs, as mentioned earlier. When lime is used in conjunction with alum $(Al_2[SO_4]_3 \cdot 18\ H_2O)$, ferrifloc $(Fe_2[SO_4]_3)$, or copperas $(FeSO_4 \cdot 7\ H_2O)$, calcium sulfate $(CaSO_4)$ will precipitate depending on the concentration of the Ca^{++} and SO_4^{--} ions present. The solubility of calcium sulfate is approximately 2400 mg/ℓ. Solubilities below this level might be obtained by the mass action effect, depending on the concentrations of Ca^{++} and SO_4^{--} ions present.

Calcium sulfate, being a crystal, is itself a good filler material which can improve the dewaterability of very compressible wastes such as oils, latex, and metal hydroxides, to name a few. Its advantages frequently outweight its disadvantages, making lime the preferred base for pH adjustment in dewatering.

Existing use of a given base somewhere in the plant usually makes standardization a consideration. Only a close analysis of standardization vs. dewatering performance for the particular application can indicate which to choose. Usually the sludge dewatering demand for the specific material is such that dewatering results are not sacrificed for standardization. They should never be when running the laboratory tests.

Anhydrous ammonia is a base that is used in industry for various reasons. At times it is used to neutralize pickle liquor from steel mills. Limited laboratory work done on these sludges showed them difficult to flocculate with polyelectrolytes. The exact reasons why iron hydroxide formed by neutralization with ammonium hydroxide could not be flocculated as well as the same hydroxide formed by neutralization with lime or sodium hydroxide, was not determined. It was thought that the higher vapor pressure of ammonium hydroxide or its ability to complex both the ferrous and ferric irons which were present, were the causes. There is little work in the literature pertaining to this type system, and a considerable study would be required before any system could be considered prudently designed.

b. Screening Flocculants

As mentioned previously, waste streams have very individual characteristics and should be tested to determine which of the candidate flocculants best meet the needs. Best usually is evaluated in terms of cost effectiveness.

Effective chemical conditioning requires that there be maintained a certain ratio range, usually based on weight between the sludge solids concentration and the conditioning chemicals. The wider the range that is operative, the more effective the system will be in coping with the day-to-day variations that are encountered in normal operations.

Cassel and Johnson reported that the characteristics of a 2 to 1 ratio activated/primary municipal sludge varied so widely that no single polyelectrolyte was found which could adjust to the daily variations in the quality of sludge received during a pilot trial.[13] The activated sludge was generated from a modified (high-rate) aeration process, with ferric chloride (25 mg/ℓ) and an anionic polyelectrolyte added just ahead of secondary sedimentation to aid in solids capture. The sludge feed to the units ranged from 4 to 7% solids. The percent is assumed to be reported on a weight/weight (w/w) basis rather than on a weight/volume (w/v) or v/v basis. The importance of weight to volume relationships when discussing high-molecular-weight polyelectrolytes will be discussed later in this chapter under Mixing.

The above reference is made only to show that the sludge dewatering operation must face all the variations in sludge consistency and quality, usually at a greater magnitude, resulting from fluctuations in previous operations. It is not meant to imply that single polyelectrolyte systems are ineffective. On the contrary, they have been operating very successfully, for over 10 years, on a daily basis, on the same type sludges. Operating experience has shown that once the particular dewatering system has been incorporated into the overall plant operation, the chosen polyelectrolytes will accommodate the usual fluctuations encountered. Wide fluctuations or a significant shift in sludge quality are often indicative of process problems in one of the previous unit operations and should be investigated insofar as they could affect effluent quality.

Most flocculant suppliers are good sources of technical assistance in selecting the type of flocculant that will most likely flocculate a given type waste. Equipment manufacturers also can be of assistance in further delineating which type flocculants generally impart the particular characteristics to the sludge that are necessary for good equipment performance.

c. Single Component Systems

A good flocculant screening plan will include an initial rough test of single component systems. If adequate performance cannot be obtained, then more complex systems must be investigated.

Single component systems for dewatering are generally polyelectrolytes. A minimum of 9 to 12 candidate polymers, most of them of the general electric charge anticipated, will be evaluated in a prudent screening. For example, a biological sludge screening might include eight cationics (high, medium, and low charge), a nonionic, and three anionics of varying electric charges.

This initial phase is a qualitative test. It consists of placing 15 to 25 mℓ of sludge in a small beaker and adding the polyelectrolyte to the sludge in portions with gentle stirring, so that the scale runs from underdosing to overdosing. Separating the suspension into clustered solids and clear liquid indicates good flocculation. If floc clusters form and then fade with continued polymer addition, the polymer might be resuspending the floc. Resuspension of solids can result from overdosing, especially with cationics, which have relatively high charge densities and low molecular weights. Floc degradation can be caused by factors other than polymer overdose. Attrition of the flocs due to too violent mixing is also common.

The nonionic and anionic polymers would not be expected to be operative in the case of biological sludges; however, unless the sludge is pure, there may be components that shift the net surface charge toward neutrality or a positive charge. This can occur when combining a tertiary treatment sludge containing calcium, aluminum, or iron phosphate with primary and secondary sludges. It also could happen by adding iron or aluminum salts to the secondary clarifier, as mentioned by Cassel and Johnson.[13]

In laboratory tests studying the flocculation of biological sludges with combined metal salts (ferrifloc, alum)/polyelectrolyte systems,[14] the polyelectrolyte requirement ran from a high cationic charge through a nonionic to an anionic, as the metal salt content increased from zero. This system, at some point, becomes a three component system as base is added to raise the pH depressed by the free acidity in the metal salt coagulants.

d. Multi-Component Systems

Multi-component systems must be used when single component systems, primarily polyelectrolytes, will not adequately condition the sludge for the particular dewatering unit in which it is to be processed. This category encompasses a wide variety of materials, the most common of which are listed in Table 5. There are no fixed sequences in which the components are added. Experimentation is required in order to determine not only the combination, but also the sequence.

Qualitative initial screening tests can be run in much the same manner as are single component systems. However, attention must be paid to the quantitative aspects of

Table 5
MULTI-COMPONENT CONDITIONING SYSTEMS

Type	Example
Dual polyelectrolyte	Anionic/cationic polymers
Polyelectrolyte/filler	Paper pulp/cationic polymer
Metal coagulant/neutralizer/polymer	Alum/lime/anionic polymer
Acid/metal coagulant/neutralizer/polymer	H_2SO_4/alum/lime/anionic polymer
Filler/coagulant/neutralizer/polymer	Fly ash/ferrifloc/lime/anionic polymer

multi-component systems. In many instances, such as with a dual polyelectrolyte system and the alum/lime system, a balance must be maintained between certain, if not all, of the components present.

Dual polyelectrolytes are usually used when the sludge particles have a net electrical surface charge near zero. The first polymer is added in an amount sufficient to impart a net charge to the small floc so that the second polyelectrolyte (oppositely charged) will be able to form larger and more stable floc. In this type system, at least one of the polymers is of a high molecular weight in order to utilize the bridging effect it imparts.

A metal salt/pH adjuster/polyelectrolyte system requires balances between the net electrical charge on the particle surface and the metal hydroxide which coagulates it. Before the coagulation can take place, a pH balance must be maintained in order to precipitate the metal hydroxide, as was explained earlier. This range is usually between pH 6.5 to 8.5 for aluminum and iron hydroxides.

The above range can be better defined for the separate metal hydroxides. Aluminum hydroxide is amphoteric, that is, it can be dissolved by either acid or base. It is least soluble in the pH 7 to 7.5 range. At pH 8.3 and greater, it goes into solution as the aluminate ion. Sodium aluminate may be considered the basic analog of filter alum (aluminum sulfate). In solution it contains free basicity just as alum contains free acidity. In certain instances where an acidic waste must be both neutralized and coagulated, sodium aluminate might be used. Rarely is its use economically justified, however, due to the great price difference between it and the alum/lime combination.

Iron hydroxide can exist in both the ferric and ferrous forms. Ferric hydroxide is essentially fully precipitated over a pH range of 6 to 11, whereas ferrous hydroxide does not fully precipitate until near pH 8.4. Black-brown ferrous hydroxide when exposed to air under neutral or alkaline pH conditions will gradually oxidize to rust-colored ferric hydroxide.

Once the sludge/metal hydroxide balance required for coagulation has been reached, a polyelectrolyte is added. The latter balance is not as critical as the initial charge neutralization or pH neutralization balances, in that the particle bridging afforded by the high molecular weight of the polymer is of equal importance to its electrical charge. Overdosing anionic polyelectrolytes rarely produces particle resuspension (as do the cationic polymers) due to electrical charge repulsion. The greatest practical problem associated with anionics and nonionics is filter medium blinding by the mobile polymer that has not securely bound to the solids surfaces. This possibility becomes more probable with increased molecular weight. Once attached to a filter medium surface, the polymer is not easily dislodged. This is one reason why ferric sulfate or ferric chloride and lime are frequently used with vacuum filters and filter presses.

Fillers are employed when the resultant floc formed by polyelectrolytes or polyelectrolytes in conjunction with metal salt coagulants is too compressible to be processed in a particular dewatering unit. Liquids such as oil, fine colloidal dispersions such as latex and certain paper mill secondary fines, and organic surface water color coagulum are examples of wastes which might require fillers. The amount of filler (sometimes called body feed, sweetener, or a variety of other terms depending upon the industry and the dewatering equipment) will be a function of the initial compressibility of the floc to which it is being added; its own overall porosity in terms of particle size distribution and compressibility; and, the forces exerted upon the floc during the dewatering process. Typical fillers are diatomaceous earth, clay, paper pulp, sawdust, fly ash, incinerator ash, and lime, to name a few.

The great disadvantage associated with adding fillers is the increased sludge volume that must be handled from that point through ultimate disposal. Sludge disposal costs can be expected to become a larger portion of the total waste treatment plant operating costs in the future.[1] Therefore, a filler evaluation must take into account both the chosen dewatering and sludge disposal methods. For example, incineration can represent a significant operation cost unless the sludge can be dewatered to the dryness required for autothermal combustion. Swanwick[15] suggests that polyelectrolyte conditioned cake solids of 24.8 wt% will be comparable to a copperas (Fe SO_4) and lime conditioned cake of 34.8 wt%, using 50% excess air and obtaining an exit temperature of 850°C.

e. Chemical Dosage Determination

Once the candidate flocculants or flocculating systems have been determined by screening, they must be examined quantitatively in order to determine equipment sizing and chemical costs. A chemical cost determination is quite important since for many types of dewatering equipment it represents a significant, if not the major, operating cost. The quantitative testing also more clearly defines the operating latitude of a given flocculant or flocculating system than does the screening test, since it is run under more controlled conditions, over a wider and more defined flocculant range.

The quantity of sludge used varies with the specific test being used and the parameter being measured. For example, determining the filtrate suspended solids content of high capture dewatering unit will require more sample than when determining only the cake solids weight percent. The volume of solids present in the sludge is also a controlling factor. Usually, the sample size is in the 100 to 500 mℓ range.

While the laboratory test equipment must not be a direct scaled-down model of the full size plant, there must be some means by which the full scale equipment critical requirements can be correlated with the laboratory test. Each type of dewatering device has its own special test or modification of a standard test which attempts to satisfy this need. The tests might lend themselves to mathematical expression, such as the filter leaf test or specific resistance and compressibility coefficient, or might be more in the realm of art as some of the tests empirically developed for belt presses. It is important that some method exist by which the results obtained in the laboratory test can be extrapolated to operating equipment requirements. Some of the specific tests will be discussed later; however, Vesilind covers many of the more common tests in good detail.[16] Metcalf and Eddy cover most of the same tests on a less fundamental basis.[17]

When running flocculation tests, especially with polyelectrolytes, it is imperative that fresh chemicals be used. In some cases, fresh means more than freshly prepared. A polyelectrolyte as received from the manufacturer has a shelf life; the container will usually carry an expiration or manufacturing date. Before mixing the polymer solution, the manufacturer's technical bulletins pertaining to that product should be studied. They often are most helpful in gaining insight into how the particular polymer will have to be handled on a full scale installation.

Lime and most of the other inorganic reagents discussed should not be exposed to atmospheric conditions for any period of time. Frequently the salts pick up atmos-

Table 6
CHEMICAL MAKE-UP CONCENTRATIONS

Chemical	Concentration (wt%)
Polyacrylamide-based polymers	0.1—0.3
Mannich polymers (liquid)	2
Lime	1—3
Sodium hydroxide	5
Alum, ferric chloride, ferrifloc, copperas	5
Mineral acids	5
Calcium chloride	10

pheric moisture which causes actual reagent weight employed to be decreased by the proportion of adsorbed moisture, unless this is compensated for during weighing. Lime reacts with the carbon dioxide in the atmosphere, forming calcium carbonate, in addition to taking on moisture. This can happen in significant quantities.

The above-mentioned aspects of using old chemicals results in poor chemical performance, which translates into higher chemical costs than would actually be encountered under normal operating conditions where stock is regularly rotated. The problem can be even more pronounced for high-molecular-weight polyelectrolytes where the bridging action associated with high molecular weight is required. Molecular weight tends to break down with time, especially when the polymer is in solution. For this reason, manufacturers generally recommend that polymer solutions be freshly prepared on a daily basis. After several days of storage in dark bottles a decrease in viscosity of a 0.1% polyacrylamide solution can be detected. Exposure to light tends to increase the breakdown rate.

Chemicals are generally mixed at concentrations that are practical to handle in both the laboratory and a full scale installation. A great disparity between the two could lead to problems in equipment sizing or system performance. Table 6 lists some suggested chemical make-up concentrations that should meet laboratory and operating requirements. They are intended to serve only as a guide. The best concentration range for any particular chemical should be obtained from the chemical supplier. Many of the inorganic chemicals can be prepared in significantly higher concentrations which would reduce storage tank size. In practice it is usually difficult to adequately mix concentrated reagents with sludges so that their full effectiveness is obtained. The concentrations listed in the table might still undergo a dilution as the reagent is being introduced into the waste stream. This is usually the case where organic polyelectrolytes are used.

Determining chemical dosages is nothing more than a titration of the sludge with the conditioning chemical to a desired end point, or a series of end points. In a multi-component system, each component has an end point. An example of the latter would be treating an oily waste with alum, lime, and an anionic polyelectrolyte. A series of beakers containing equal quantities of sludge would be given increasing amounts of alum. Lime would then be added to adjust the pH to a 6.5 to 7.5 range, if necessary, at which time the samples should be evaluated with regard to how well the oil has been tied-up by the aluminum hydroxide floc that was formed. The initial desired end point is complete scavenging of the oil and other turbidity by the aluminum hydroxide so that the sludge consists of two distinct phases, clear water and small flocs. The beakers containing the higher alum dosages make the end point difficult to discern. Usually the clarity can be seen at the meniscus when the beaker is tilted slightly.

The sample requiring the least amount of alum for turbidity removal is the sample to be further tested with the polyelectrolyte. Two titrations were involved in getting to this stage, the alum against the turbidity and the lime against the total acidity of the sludge, plus that associated with the alum. This floc is next titrated with the anionic polyelectrolyte until an end point of solids clusters and clear water is obtained.

One method of performing this titration is to continually add aliquots of polymer, while stirring, until the solids clusters form. Once the correct dosage is determined, the test should be repeated, adding the entire polymer dosage at one time, in order to minimize any errors that might be introduced by the longer mixing time (associated with the batch test) than might be encountered during continuous operation. One of the requirements to be observed during this stage is the time required for complete flocculation under given mixing conditions. This will have to be related to the detention time built into the full scale equipment. The same observation applies to the alum coagulation and lime neutralization steps. Earlier, when discussing lime neutralization, it was pointed out that the pH can drift upwards with time when a lime slurry is used for acid neutralization, until all the particulate lime is dissolved.

This one set of titrations represents the chemical dosages required for complete turbidity flocculation; it may not represent the full requirements for processing in a given piece of equipment. Repeating the titrations, increasing the amount of alum added, with subsequent lime and polymer determinations, would be required to find a system that gave the resultant floc the integrity needed to withstand mechanical handling and applied filtration pressure.

In this particular system alum might not be as good a coagulant choice as ferrifloc $(Fe_2[SO_4]_3)$. The effective insolubility range of aluminum hydroxide is narrower than that of ferric hydroxide, pH 6.0 to 8.3 for aluminum hydroxide vs. 5.5 to 12.0 for ferric hydroxide. This necessitates tighter lime feed control when using alum. In addition, while aluminum hydroxide is a more effective charge neutralizer than ferric hydroxide on a weight-for-weight comparison, it does not dewater as well as ferric hydroxide.

The titration method just described does not take into account any dilution water added during actual operation. The reagents are generally employed at the concentrations listed in Table 6, 5.0 wt% for alum and ferrifloc, 3 wt% for lime, and 0.1 wt% for the polymer.

A stirring device capable of being correlated with the operating installation flocculation system mixer should be employed when running laboratory tests. At times, flocculation tests must be run in the field, removed from utilities. Under these conditions, the second or equipment requirements portion of the test procedure might be precluded. Experience with regard to specific dewatering equipment requirements, on the part of the person running the test, will be required to judge how much integrity must be built into the system over that required for floc formation.

Gravity/press or band-filter type dewatering units most easily lend themselves to reasonable correlation between these less refined test conditions and operating installations.

The equipment shown in Figure 3 was used to determine the polyelectrolyte requirements and predict process results for a gravity dewatering unit operating on a calcium carbonate sludge generated by a cold process lime softening operation.[2] Influent sludge was measured in the left-hand disposable 400-mℓ, polypropylene beaker (200 mℓ). The anionic polyacrylamide based polymer (0.1% concentration) was added by means of the 10-mℓ, disposable polypropylene syringe. Mixing was done by means of the wooden stick. A piece of monofilament nylon screen taken from the dewatering unit was used to determine the conditioned floc particle integrity requirements. The second polypropylene beaker was used to catch and measure the filtrate volume. Machine capacity was estimated, based on the drainage rate and cake formation. Influent and

Table 7

LABORATORY TEST VS. FULL SCALE
UNIT RESULTS COMPARISON FOR A
CaCO₃ LIME-SOFTENING SLUDGE

	Laboratory	Full-scale
Influent		
pH	9.5	9.4
Solids (wt%)	9.4	10.6
Chemical requirements		
Type polyelectrolyte	Anionic	Anionic
lb/ton solids treated	0.41	0.32
Dewatering		
Cake solids (wt%)	44.5	50.5
Capacity (gal/hr)	600	870

From Trans. 22nd Annual Conf. Sanitary Engineering, University of Kansas, Lawrence, February 2, 1972. With permission.

dewatered cake solids weight percent had to be determined using an oven, but the influent solids volume percent was determined using the volumetric beaker.

Table 7 compares the laboratory test results and sample received with the full-scale pilot unit testing the softener blow-down. The greatest difference is in the capacity, where the estimate was approximately two thirds of that actually obtained. Capacity was the most subjectively based parameter listed. It will be noted that all the laboratory results were more conservative than the operating results.

While this correlation can be considered good for practical purposes, the conservative laboratory results might have been due in part to the fact that the sludge was being conditioned to be treated by a press following the gravity dewatering unit, and that was not done on the full scale equipment. In any case, it should be recognized that the agreement was good, considering the many variables encountered in waste treatment.

f. Chemical Requirement Expression

The chemical requirement in Table 7 is reported as "lb/ton solids treated", one of the two methods commonly used to express the chemical dosage. It refers to the pounds of polyelectrolyte required to treat a ton of influent sludge solids, on a dry solids basis. The SI (International System of Units) analog is grams polyelectrolyte per kilogram dry solids treated, where 1 lb/ton = 0.5 g/kg. These units are independent of flow rate and are used primarily for evaluating chemical performance and cost.

The other method used to report chemical consumption is milligrams per liter, which refers to the milligrams of polyelectrolyte (dry basis) added to a liter of sludge. It is sometimes reported by the old convention, ppm (parts per million). The milligrams per liter connotation is an operating term used to designate proportional flow between the polyelectrolyte and influent waste. Waste volume replaces weight of dry solids because the weight of waste solids is not the only factor controlling the polyelectrolyte requirement. A shift in particle size distribution or sludge pH can have as great an effect as a change in weight. In addition, changes in solids content cannot be determined rapidly enough to be a practical operating control. Proper sludge conditioning depends upon maintaining the polyelectrolyte-sludge solids ratio within a reasonable range. This is usually done by fixing volume ratios by means of positive displacement polyelectrolyte and sludge feed pumps. The polyelectrolyte volume is directly relatable

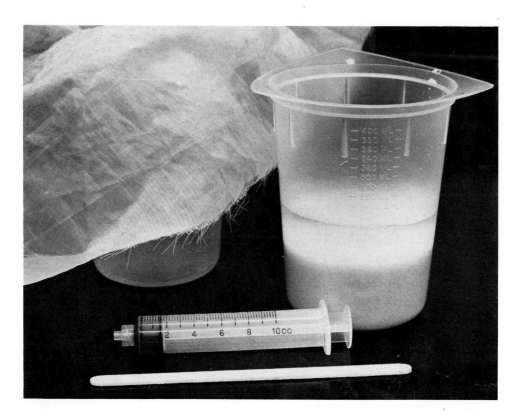

FIGURE 3.   Field test equipment used for determining polyelectrolyte requirements for gravity dewatering. (Courtesy University of Kansas, Lawrence.)

to its weight since the solutions or slurries are prepared at known concentrations. Any change in polyelectrolyte demand by the influent sludge can be, and usually is, assessed by recording the polyelectrolyte and sludge pumping rates and taking a sludge sample for suspended solids content. Dividing the product of the polyelectrolyte concentration and pumping rate by the product of the sludge-suspended solids concentration and its pumping rate, produces the first expression discussed, grams polyelectrolyte per kilograms dry solids treated.

Figure 4 shows the polyelectrolyte requirement for dewatering an extended aeration sludge, expressed in both terms. The curves represent average dosages of a high cationic charge density, medium-molecular-weight polymer, Primafloc C-7, manufactured by Rohm and Haas Co., required for processing the waste in a gravity type dewatering unit. Average dosages were obtained by determining and averaging the polymer dosages required for (1) the first sign of destabilization of the solids suspension (small floc formation), and (2) the maximum floc size and integrity obtainable (maximum dewatering rate). The correct or optimal polymer dosage for dewatering falls into a wider range than do the optimum requirements for other liquid-solids separation operations such as clarification, thickening, and filtration. Usually, the floc size and strength resulting from neutralizing the electric charge on the surfaces of the sludge particles are not large enough for processing in the variety of dewatering units employed, which can subject the flocs to significant ranges of force magnitudes and types. Many of the tests or parameters commonly used for comparing polyelectrolytes, i.e., electrophoretic mobility (zeta potential), capillary suction test, and specific resistance, to name a few, assess the activity of a polymer with regard to only a portion of

181

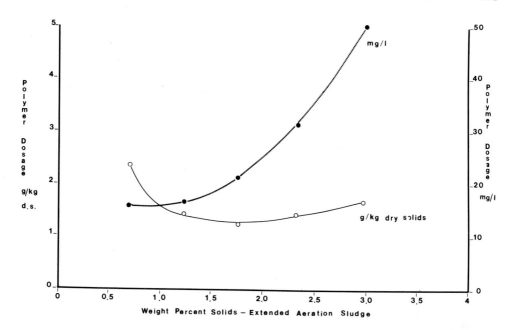

FIGURE 4. Polyelectrolyte requirements vs. total suspended solids weight percent for an extended aeration biological waste.

the significant factors involved in applying the polymer in a specific dewatering situation. They are very good, basic investigatory tools when viewed in the proper perspective.

In Figure 4, the left ordinate shows the weight per weight relationship (grams polyelectrolyte per kilogram dry solids); the right ordinate shows the weight per volume relationship (milligrams per liter). The solid line represents the weight per weight relationship; the broken line, the weight per volume. Both curves can be considered typical when testing a wide range of influent total suspended solids contents. The milligram-per-liter line (broken) frequently is nearly flat at low concentrations, then increases in slope with increased total suspended solids content, the slope varying from case to case. It indicates that at low concentrations the mass of particles present is not the only factor influencing the polyelectrolyte dosage requirement. A threshold for polyelectrolyte dosage is frequently encountered when working with various concentration sludges, particularly biological sludges. That is, a 0.5 wt%-activated sludge might require the same total amount of polyelectrolyte as a 0.8 or 1.0 wt% sludge.

The grams polymer per kilogram solids line (solid) is frequently seen only as the left side of the parabolic curve, ending somewhere around 1.8 to 2 wt% in this case. This shows a relationship frequently encountered, that, at low sludge solids concentrations, the efficiency of the polyelectrolyte (cost per weight solids treated) increases as the influent solids increases. Discussion of the right side of the parabolic curve, beyond 2 wt% will be covered under Section IV.C. of this chapter.

Roberts and Olsson,[18] reported on laboratory studies in which they attempted to determine the relative contributions of macroscopic (>10 μm) and colloidal (<10 μm) particles, respectively, to dewatering of low concentration ($\leqslant 1$ wt%) activated sludge. Centrifugation was used to separate the two particle-size fractions so that they could be recombined in various ratios. The sludge mixtures were then treated with various dosages of a high-molecular-weight (15×10^6 by viscosity) cationic polyacrylamide derivative. Dewaterability was determined from capillary suction time measurements. The electrophoretic mobilities (zeta potential) of the colloidal phases as a function of added polyelectrolyte were determined. They concluded from their tests that the optimal polyelectrolyte dosage for dewatering is independent of the content of macroscopic (>10 μm) particles in the sludge, and therefore dependent on the concentration and type of colloidal particles. They further postulated that the correlation between optimal polyelectrolyte dosage for dewatering with that for neutralization of the initial negative charge of the dissolved and colloidal particles, suggests that dissolved and colloidal particles play a decisive role in dewatering of sludges of the type studied, by reacting with the monomeric and polymeric anionic colloidal material present to build three-dimensional gel-like aggregates, which trap out the macroscopic particles present and to give a readily dewatered floc structure.

In a second delivered paper covering an extension of their initial test work, this time using a combination of the original cationic polyelectrolyte and aluminum hydroxide, Roberts and Olsson reiterated their original conclusions,[19] adding the observation that the polyelectrolyte requirement can be substantially reduced by addition of aluminum hydroxide. Electrophoretic mobility measurements were suggested as a method of determining the aluminum hydroxide range. The authors mentioned that preliminary investigations (to be reported later) indicate the essential correctness of this view.

Another facet of the effect of colloidal particles on filterability, as postulated by Karr and Keinath,[4] will be discussed in Section III.B.3.f.

3. Testing For Dewatering Characteristics

Once the chemical conditioning system to be used has been chosen for a given waste sludge, the quantities of the conditioning chemicals required to process the sludge in a particular piece of equipment or with regard to a particular parameter must be established. The tests available to do this can be classified in many ways. We shall divide them into two categories: (1) general tests that measure a given parameter, i.e., capillary suction time and specific resistance, and (2) tests which attempt correlation with a specific type or piece of dewatering equipment, i.e., filter leaf test for vacuum filters and laboratory centrifuge test for centrifugals. We will touch upon several of these tests and comment generally about their applicability to evaluating polyelectrolytes for dewatering.

a. Buchner Funnel Test

This test and its many modifications are the dewatering analogue of the jar test for clarification. The utility of the test depends upon how clearly one defines his objectives and how astutely he manipulates the elements of the procedure to measure the effects desired. This test procedure lends itself to reproducibility, and the problems encountered in applying it to full-scale equipment are usually centered in correlation. While its accuracy can be questioned, it must be evaluated in the proper perspective. It is a simple test using simple equipment, reproducible, and working with wastes whose characteristics are not as well defined or stable as those of the test procedure. Probably for these reasons, the test is accepted by those who evaluate the application of polyelectrolytes to sludge flocculation.[9,20]

b. Specific Resistance

This is the most frequently used and reported form of the Buchner Funnel Test. The apparatus employed in Figure 5 is the same as employed for the test variation discussed in Section III.B.3.a. Specific resistance is used primarily for measuring and comparing the resistance of the solids in the sludge to the passage of the liquid phase under a given pressure, usually applied by vacuum. Test conditions are more closely monitored than when using the apparatus for evaluating polyelectrolyte applicability and dosage.

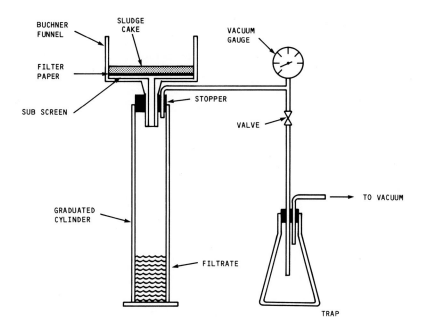

FIGURE 5. Schematic diagram of a specific resistance test apparatus.

Specific resistance and coefficient of compressibility, determined by plotting the results of specific resistances run at various pressures, frequently appear in mathematical expressions of filtration theory. The specific resistance and coefficient of compressibility test procedures can be found in several of the references listed in this chapter,[4,16,17,20] so they will not be covered in detail. Specific resistance is determined by plotting the elapsed filtering time (θ) per corrected filtrate volume (v) vs. the corrected filtrate volume (v) at a given temperature and vacuum pressure. Figure 6 is a typical plot for specific resistance of an extended aeration sludge. The slope of the line is the constant "b" which is used in the following "specific resistance" formula:

$$r = 6.91 \times 10^6 \times \frac{bA^2p}{uc} \tag{1}$$

in which r = specific resistance in s^2/g, u = absolute viscosity in centipoises, c = initial suspended solids of the sludge in mg/mℓ, p = filtration vacuum in psi, A = filter area in cm^2, and b = slope of θ/V vs. V curve in s/mℓ^2.

When comparing specific resistance values, better filterability occurs at lower values. Care must be exercised when using values, in that specific resistance can be reported in various units. Vesilind gives a conversion table for the various terms employed in the equation.[21] The most commonly reported unit for r is s^2/g. It can be changed to a m/kg value by the following conversion:

$$s^2/g \times 9.81 \times 10^3 = m/kg \tag{2}$$

Coefficient of compressibility (Sc)* is determined by running the specific resistance (r) at various pressures and plotting them as functions of pressure on log paper. The

* Sc is used to differentiate coefficient of compressibility from s, the SI abbreviation for second.

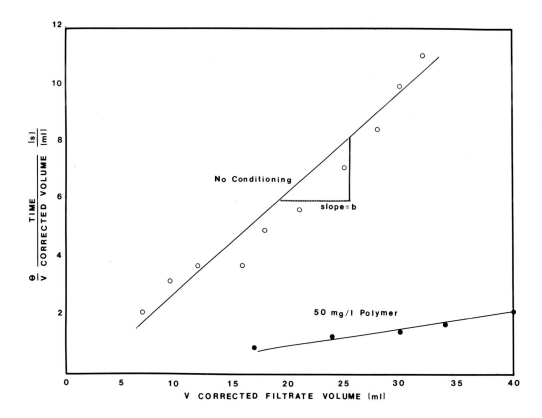

FIGURE 6.   Specific resistance plot for an extended aeration sludge.

slope of the line is the coefficient of compressibility (Sc), an indication of how much the solids are compressed. See Figure 7. Sc and r are related in the emperically derived equation:

$$r = r_o P^{Sc} \qquad\qquad (3)$$

where r_o is the specific resistance at unit pressure. Incompressible solids, such as sand, have Sc values close to zero. Coackley reports values of 0.70 to 0.86 for unconditioned digested sludges and 0.60 to 0.80 for unconditioned activated sludges.[22]

The particle size distribution is of importance in these measurements for reasons other than their surface electrical charge. Karr and Keinath[4] use specific resistance to evaluate the influence of particle size on sludge dewaterability in their paper which postulates that particle size is the single most important factor influencing sludge dewaterability. They employed the segregation of sludge suspended solids into various particle-size fractions as did Roberts and Olsson,[18,19] but to a greater degree — four classifications as opposed to two for the latter. The various particle-size fractions were then blended with different sludges in the amounts necessary to give a desired particle size distribution for the suspended solids phase of the sludge. They concluded that the supracolloidal particle sizes (1 to 100 μm) are responsible for poor sludge dewaterability in that they have the ability to blind both the filter cake and the filter medium. In support of their postulation, they report that an unconditioned anaerobically digested sludge with a high supracolloidal solids concentration had a specific resistance value of 99.5×10^{13} m/kg as opposed to a value of 7.0×10^{13} m/kg for a similar sludge from

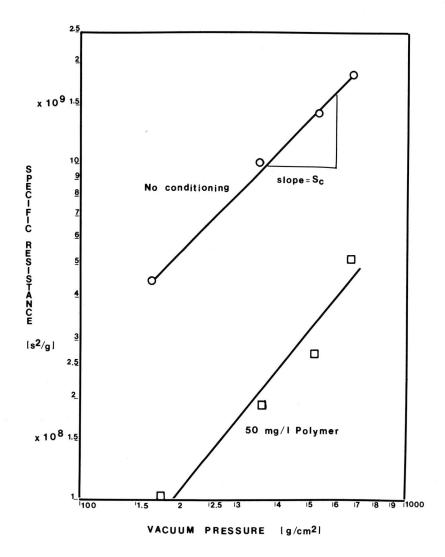

FIGURE 7. Coefficient of Compressibility plot for an extended aeration sludge.

which most of the supracolloidal particles had been removed. They assumed the lower 7.0×10^{13} m/kg value (better filterability) was caused primarily by the fluid flowing through an unblinded cake, and therefore, the difference between the two specific resistance values (92.5×10^{13} m/kg) is attributable to blinding. In summation of a portion of their tests, they assigned the influencing factor for the total specific resistance results as: 7% for filtrate flow through the unblinded cake, 23% to filter medium blinding, and 70% to cake blinding.

One set of data does show that the dewaterability of the activated sludge improved as pH decreased, and that the converse was also true. The pH also appeared to alter the particle size distribution insomuch as the quantity of supracolloidal solids was found to decrease with pH. This agrees with the recognized ability of the hydronium ion to neutralize the negative electrical charges on the surfaces of the organic solids, allowing coagulation to take place, as mentioned earlier in Section III. B. 2. of this chapter.

Test data are not frequently seen for polyelectrolyte conditioned sludges because polyelectrolyte conditioning appears to cause an abrupt flocculation of the fine solids

Table 8
SPECIFIC RESISTANCE AND COEFFICIENT OF
COMPRESSIBILITY OF AN EXTENDED AERATION
SLUDGE AS A FUNCTION OF SLUDGE CONDITIONING

Solids (wt%)	Conditioning	Specific resistance[a] (sec²/g)	Coefficient of compressibility
1.4	None	1.81×10^8	1.00
1.4	Indirect freezing	1.63×10^8	0.96
1.4	20 mg/ℓ Prim. C-7	1.34×10^8	1.14
1.4	20 mg/ℓ Prim. C-7 + indirect freezing	8.73×10^7	1.17

[a]   Vacuum pressure @ 49.3 cm Hg.

particles, which results in too rapid a filtration rate to be reliably recorded. Karr and Keinath in their above-mentioned paper reported only one test in which a polyelectrolyte treated sludge was tested.[4] They reported that the addition of 100 mg/ℓ of Dow Purifloc® C-31, a cationic polymer, caused the sludge to dewater so well that it was impossible to measure its specific resistance. A change in structural rigidity of the particles, by conditioning, was thought to have an influence on the dewaterability. Table 8 compares specific resistance and coefficient of compressibility values for a 1.4 wt% extended aeration sludge as a function of conditioning by freezing, addition of Primafloc® C-7 (Rohm and Haas) cationic polymer, and a combination of freezing and Primafloc® C-7.

Because specific resistance is a relatively basic parameter, investigators used it to predict performance for various methods of dewatering, or in the case of the Capillary Suction Time (CST) test,[23] as a method of correlating more readily performed tests to equipment performance. Coackley,[22] in his discussion of the development of a mathematical theory of the filtration process mentions that Jones employed specific resistance to calculate vacuum filter yields.[24] Comparisons of predicted and actual yields in four cases showed the predictions to be +11%, +14%, −3%, and −18%. Jones also employed specific resistance to calculate the time required to press sludge in a filter press.

In more recent work, Wilhelm concludes that filtration data for recessed plate or plate and frame pressure filters can be correlated by plotting cycle time vs. a correlating factor that includes the K-value (related to specific resistance), filter area, cake and feed solids concentration, and cake thickness.[25]

Nebiker et al.,[26] as a result of an investigation concerning correlating specific resistance to gravity dewatering on sand drying beds, concluded that laboratory tests can predict dewatering rates on drying beds in the same manner that vacuum filter performance can be predicted, with the aid of specific resistance determinations. They found the important factors to be initial solids content, depth, specific resistance, coefficient of compressibility, filtrate viscosity, and a dimensionless media factor "m" which relates the sludge dewaterability on sand and other media to its dewaterability in the Buchner funnel.

The application of specific resistance to so many varied situations resulted in various investigators finding it not completely suitable in every instance. Karr and Keinath in their earlier mentioned work with supracolloids encountered two major limitations of the specific resistance test with regard to their work.[4] The specific resistance was concluded to be dependent on the solids concentration of a sludge when the concentration of supracolloidal solids (1 to 100 μm) were sufficient to cause blinding, contrary to

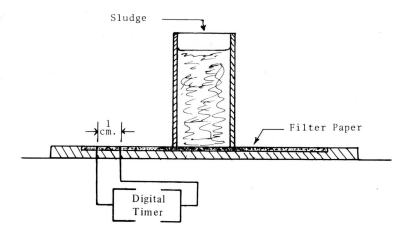

FIGURE 8.   Schematic diagram of a capillary suction time (CST) test apparatus.

specific resistance theory. Secondly, the test does not account for all the properties of a sludge, such as pick-up and release characteristics, scrolling or channeling properties, and resistance to shear.

Kos and Adrian in a recent examination of cake filtration theory based on new developments in continuum and soil mechanics, suggest that use of average specific resistance values, those usually obtained, should be limited to evaluations of dewatering operations similar to the Buchner funnel test.[27] They further state that it is not applicable to analysis of the full dewatering cycle of a rotary vacuum filter because it is valid only when the cake formation is unrestricted.

The shortcomings of the specific resistance test should be viewed in relation to the use for which it is being employed. Its inadequacies arise when it is being used as a basis for mathematical analysis, where one strives for greater and greater precision in definition. When used for predicting the performance of a polyelectrolyte with a waste stream which is usually poorly defined and seldom constant in character, it may be considered an adequate test.

c. Capillary Suction Time

Capillary Suction Time (CST) is the last of the general conditioning tests we will discuss. Gale and Baskerville,[23] in 1967, reported on a new method which was rapid and required little experience, utilizing the capillary suction of a porous medium to effect filtration. To convert the suction time into specific resistance to filtration, a calibration curve must be prepared for the suspended solids concentration concerned. This test has gained wide acceptance in testing polyelectrolytes for dewatering, especially as a first screening step prior to running specific resistance.[4,16,18,19]

The CST is the time required for the liquid fraction of a sludge to travel a given distance in a porous or adsorbant paper. Figure 8 is a schematic of a CST test apparatus which consists of a cylindrical sludge reservoir setting on a piece of porous paper (usually filter paper). The water travel through the paper is measured by the two sensing electrodes which start and stop a digital timer.

Care must be exercised in running the test and interpreting the results. Mangravite et al.[28] point out that the test loses differentiation when the sludge solids are low or the sample well flocculated under which conditions the CST value approaches that of water. Gale and Baskerville report that capillary suction pressure of the paper will determine the pressure difference available to cause filtration, and that the filtrate-adsorbing capacity of the paper will affect the volume of filtrate, thus the depth of

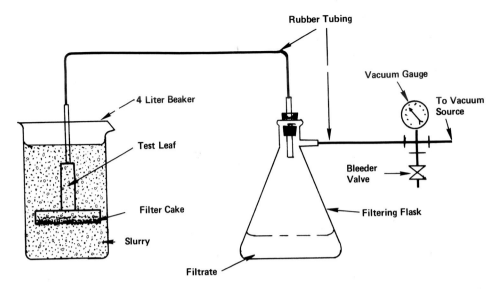

FIGURE 9.   Schematic diagram of a Leaf Test apparatus used for laboratory prediction of vacuum filter performance. (Courtesy Komline-Sanderson, Peapack, N.J.)

the cake built up when a given area of paper has become wetted.[23] They also reported that any machine-made paper has a "grain", and that resistance to flow is approximately 20% less along the grain than against. This means that the measurement should always be made in the same relationship to the grain. Mangravite et al. caution that the specific resistance test must be run at a vacuum equivalent to that of the capillary pressure of the CST filter paper in order to minimize error in correlation between the two tests.

The CST test cannot give information about the dewatered cake or the filtrate suspended solids, whereas the Specific Resistance Test and other modifications of the Buchner Funnel Test can. Karr and Keinath observed that the CST test seemed to be more sensitive to changes in concentration of fragile settleable solids than was the Specific Resistance Test, even though the total settleable solids concentration was maintained constant.

d. Filter Leaf Test

The Filter Leaf Test (FLT) is a modification of the Buchner Funnel Test used specifically for vacuum filters. It differs from the Buchner Funnel Test in that a leaf, consisting of a support grid covered by a piece of the filter cloth used on the full scale machine, replaces the Buchner Funnel. The leaf is constructed so that a vacuum can be drawn in it. In this test, the circular leaf (92.9 cm² area) is immersed in a beaker of suspended solids, rather than having the suspension poured onto the cloth in the funnel, although the latter method can be run by fitting the filter leaf with a cylindrical wall which will act as a reservoir to hold the sample. The leaf test can be used for simulating precoat filtration conditions, as well as straight cloth filtration, by first immersing the leaf in a precoat suspension and building the precoat layer. The leaf is then immersed in the suspended solids to be dewatered.

Figure 9 is a schematic drawing of an FLT apparatus. There should be no valves located in the lines between the test leaf and the filtrate receiver or between the receiver and the vacuum pump. Valves could cause hydraulic restrictions which result in measurement of the flow through the valve rather than the true flow through the filter media and cake. A detailed description of this test procedure is given by Dahlstrom and Silverblatt.[29]

The test allows prediction of yield, cake moisture, filtrate characteristics, vacuum requirements (pressure and air flow volume per unit surface area of filter cloth), and media selection to the degree of accuracy of the sample submitted.[30] Most of the data found in the literature pertains to unconditioned sludges, or sludges conditioned with inorganics such as ferric chloride/lime for coagulation or diatomaceous earth for body feed. The test is not frequently used with polyelectrolytes, because polyelectrolytes are not frequently used as sludge conditioners for most type vacuum filters. Coil filters are the exception to the rule because filter media blinding is not a big problem. This is covered in more detail under the section pertaining to full scale equipment.

e. Centrifuge Test

There has been considerably less universality in devising and adapting a laboratory test procedure that can be correlated with full size centrifugal performance than has been done for vacuum filters or filter presses. The task is complicated by the number of centrifugal types, with operating differences and the complexity of centrifugation itself, in contrast to vacuum and pressure filtration. Vesilind[31] mentioned the latter in his discussion of centrifugation, one of the more complete references on laboratory testing for centrifugation scale up in circulation. His emphasis is placed on correlating the handling characteristics of the sludge between laboratory centrifuges and prototype machines, rather than on methods of conditioning the sludge so that its physical characteristics will be within the acceptable range for processing in any given type centrifugal. The importance of polyelectrolytes to successful, economical centrifugal operation was pointed out by him with mention of the first successful installation of a centrifugal in the U.S. at San Mateo Company in 1960.

Vesilind mentions thirteen operating and machine variables which affect centrifugal performance. Of these he signals two having the greatest influence on machine performance for correlation in laboratory testing. They are solids retention time (or flow rate to the machine) and centrifugal force. These two properties can be evaluated in a variable speed laboratory centrifuge using 15-ml tubes. A centrifuge of this type can develop as much as 15,000 gravities.

Good approximation of yield results can be obtained using the equation:

$$\text{Percent Recovery} = \frac{(C_o - C_i)}{C_o} \times 100 \qquad (4)$$

where C_i is the centrate solids concentration and C_o is the feed solids concentration.

Another factor Vesilind mentions as necessary for correlation is a measure of the scrolling ability of sludge, its ability to be removed from the centrifugal. This is done by means of a penetrometer which measures the firmness of the solids layer as it is packed in the tube after the supernatant has been decanted. The penetrometer reading is designated P, a percent value. A prediction accuracy of approximately 10% between the laboratory and fullscale centrifugal performance is obtained using the following equation:

$$\text{Estimated Percent Recovery} = \left(\frac{C_o - C_i}{C_o}\right) \left(\frac{P}{100}\right)^{0.1} \times 100 \qquad (5)$$

where C_o is the feed concentration, C_i is the centrate clarity measured at the same centrifugal force and residence time as the full size unit, and P is the penetration percent value.

Chemical conditioning, by polyelectrolytes, is mentioned by Records,[32] who utilizes the CST to determine the correct polymer and dosage. The correct dosage is the minimum amount capable of almost instantly producing large flocs (5 to 10 mm) and clear supernatant. The floc must not degrade after vigorous shaking. He suggests that a good flocculant application will give a dosage level below 2.5 kg/ton of dry solids. Laboratory tests are run in either a "bottle spinner" or small tubular bowl centrifuge. The tubular bowl unit lends itself to correlation with larger units.

f. Filter Press Tests

Considerable differences of opinion exist among manufacturers and users of filter presses as to whether polyelectrolyte conditioning should be used with them. The main reasons cited against polyelectrolytes are that the pressures and hydraulics encountered inside the press degrade the floc particles which are initially more compressible than those formed with inorganic coagulants, and that the long chain polymers tend to blind the filter media. Both arguments have credence in that body feeds (incompressible solids added to the sludge prior to entering the filter cavity) are frequently added to many classes of wastes, and once a filter cloth or mat is polymer blinded, it is very difficult to clean. Polyelectrolytes find their way into filter press dewatering as part of the sludge to be dewatered, being added as a conditioner prior to an earlier liquid-solids separation step, such as clarification. Under these circumstances, much the same as with belt vacuum filters, the polymer conditioned sludge usually performs as well as or better than the same type sludge without it.

Since this book is concerned primarily with polyelectrolytes, and this chapter with laboratory tests used to evaluate their performance in various types of equipment, we will limit our discussion to the more unsophisticated tests and apparatus already discussed, which have been used to predict filter press performance (Buchner Funnel, Specific Resistance, and CST Tests).

Those desirous of more information concerning laboratory testing at various levels of sophistication should refer to the AIChE Equipment Testing Procedure for Batch Pressure Filters,[34] publications by Tiller et al. concerning the role of porosity in filtration,[35-37] publications by Shirato et al. concerning expression under various pressure conditions,[38-42] Schwartzenberg et al. on removal of water by expression,[43] and Thomas and Purdy on batch filters.[44]

g. Gravity Dewatering/Pressing

Continuous mechanical gravity dewatering units are relatively new developments as compared to vacuum filters, filter presses, and centrifuges. They are unique in that they were developed specifically to dewater wastewater sludges, primarily biological sludges, whereas dewatering these sludges was just another application for the older liquid-solids separation devices. The newness and diversity in design of these units, in addition to the relatively recent interest in sludge dewatering, might be considered the more probable reasons why little has been published in the way of laboratory test procedures, especially mathematical modeling. Kos and Adrian point out that the continuous or step-wise increased pressure applied make the actual cake formation and concentration changes (occurring in the cake with time) so complicated that it is impossible to describe these operations by one value characterizing the average filtration properties.[27]

Flocculation by polyelectrolytes plays a far more significant role in the operability of these units, aside from their cost effectiveness, than it does for any other type of dewatering equipment mentioned, including centrifugals. Because of this, the general plan followed in laboratory testing for gravity dewatering units consists of considerable emphasis on polymer selection followed by empirical tests developed to determine the suitability of the conditioning system to meet the special requirements of a specific machine. Seldom are the tests well described by mathematical expressions, and they should at this time be considered more in the realm of art.

Vesilind gives a laboratory exercise pertaining to the flocculation test used for the Dual Cell Gravity Solids Concentrator (DCG),[33] a primary dewatering unit employing only a gravity head for initial water drainage, followed by the weight of the rolling cake to express additional water from itself prior to discharge. This unit will be covered in more detail later in the chapter.

The equipment required to run the "DCG Roll Rest" is shown in Figure 3. Table 7 shows the type of correlation obtainable from this simple test and apparatus. The test is a straight titration of the sludge with the polyelectrolyte to an end point signified by the separation of the two phases into floc particles and a clear liquid phase. The suspension is then poured onto a piece of screening used on the full-scale equipment and opposite ends of the cloth lifted and lowered alternately to cause the solids on the screen to form a roll. If the solids do not roll, the conditioning is not adequate to withstand the forces to which the floc particles will be subjected in the dewatering unit. Usually this situation can be improved by adding more polyelectrolyte.

Requirements for other gravity dewatering units might be more closely simulated by some other manipulation of the floc once it is formed. Drainage and floc integrity are checked for one unit by pouring the floc over a stainless steel screen and covering it with a second screen. The screen-sludge-screen sandwich is then arched to simulate the action encountered in the full scale unit. The sludge which was first placed on a moving belt for initial gravity drainage, undergoes compression from the tension in the supporting and cover screen belts as they reverse direction while going over a roller.

While these test procedures seem crude and unreproducible, they have proved adequate for predicting chemical conditioning and operating performance for these structurally simple, but hydrodynamically complex machines. Most people associated with the application of polyelectrolytes and these new machines for dewatering are aware of the need for better mathematical modeling of the processes. The sheer diversity of equipment designs employing various types and permutations of filtration driving forces suggests that any modeling will have to be done by the equipment manufacturers. This type of equipment is in its infancy in comparison to the other types of liquid-solids separation equipment that have been employed for some time in manufacturing a product. The manufacturers have had to place a higher priority on obtaining a large enough share of the suddenly materializing, legislated market caused by the worldwide concern in ecology. Therefore, many versions of gravity/press type units have not been operating long enough to build a data base of sufficient breadth and reliability for good mathematical modeling.

Several other factors pertinent to the application of gravity/press type dewatering units are worth mentioning. First, the properties of the polyelectrolytes and the equipment employed are already better defined, or are capable of definition, than are the waste streams they treat. It was mentioned earlier, but is of sufficient importance to be reiterated — characterization of and variation in waste streams will keep correlation between laboratory and operating units in the realm of art for some time. Secondly, the filter media employed in these units is considerably more porous than that used for vacuum or pressure filtration. Its primary function is support, not solids capture. Solids capture is performed initially by the bridging capabilities of the polyelectrolyte, and later during pressing by the close packing of the solids. In some instances the optimum polyelectrolyte dosage for a gravity dewatering unit could result in a slight polyelectrolyte residual in the liquid phase. The filter papers used for both the Buchner Funnel and CST tests are very susceptible to plugging by polyelectrolytes, especially high-molecular-weight polyacrylamide-based ones. Whether meaningful results could be obtained using a more porous paper in the CST procedure is questionable, in view of the other variable already associated with the test.

IV. SCALE-UP

Scale-up, in this chapter, will be limited to discussion of several factors related to chemical conditioning that frequently receive little thought when translating laboratory findings into full scale equipment operation. Some of them might seem obvious, but they are frequent sources of problems for designers and operating personnel. Others are less discernible and difficult, if at all possible, to deal with under normal design considerations. A good source for scale-up considerations, referenced by equipment type, is *Solid/Liquid Separation Equipment Scale-Up.*[45]

A. Conditioning System Complexity

When running laboratory tests, every effort should be made to determine the simplest conditioning system possible that will perform reliably. One reason polyelectrolytes gained popularity is that they frequently were able to do the work previously done by two chemicals. Each additional chemical used in the laboratory means an additional chemical makeup and feed assembly, usually a separate reaction or holding tank, possibly a transfer pump, design considerations for reaction times in each step, and chemical storage capital considerations. Operating considerations include added operator time requirements for mixing, setting, and adjusting each additional feed. Since good equipment performance depends upon proportioned feeds of the waste influent and all conditioning chemicals, the chances for sub-par performance in a system of "n" conditioning chemicals is (n + 1) factorial.

Most types of dewatering equipment include provisions for adding one chemical in their standard package. Two or more chemicals involve specially designed flocculation systems, or less controllable utilization of in-line chemical addition and mixing for one of the stages. The latter, in some instances, can be effectively used, and will be covered in more detail later in this section.

Primary consideration in chemical choice in laboratory testing usually rests in how well a given chemical performs. In full scale installations other considerations can be of varying significance, depending on the circumstances. Chemical storage space and handling conditions might make sodium hydroxide more practical to use than lime for pH adjustment. Too wide a demand range for a polyelectrolyte that can be mixed and fed only in low concentrations might exceed the chemical feed system delivery range.

B. Sequence and Timing

One of the more common problems encountered in translating laboratory flocculation test results to full scale installations has to do with sequence and timing. The need for sequence continuity is fairly well observed. Timing is far less obvious in its importance.

Every step in the laboratory procedure has its counterpart in the full size system. Sufficient detention time must be allowed between chemical feed points to insure complete mixing and reaction if the operating chemical dosages and floc characteristics are to be in agreement with the laboratory results. Reaction times can vary for different wastes and conditioning chemicals, in addition to physical variables such as temperature, solids concentration, and mixing speed.

When investigating poor equipment performance, a good practice is to run laboratory tests on the same influent waste to be sure that the conditioning chemical or chemicals used can adequately flocculate the waste. Once the dosages have been determined on the laboratory scale they should be separated into consecutive discrete stages. The full size system should be viewed as consisting of corresponding stages and checked to see that the same degree of conditioning is obtained at the corresponding stages. As obvious as this seems, it is one of the most commonly overlooked aspects of poor equipment performance.

C. Mixing

No other aspect of sludge dewatering is overlooked to as great a degree as is mixing, both in theory and practice. Much work has been done in understanding the physical aspects of filtration or expression in various types of equipment. Considerable efforts have been expended also toward understanding the chemistry of flocculation. Mixing, on the other hand, appears to be taken for granted. It is of equal importance to chemical and equipment choice. Too little mixing does not bring all the components together in the homogeneity needed for uniform floc integrity. Too strong or too long mixing can cause floc degradation.

One of the earlier investigations recognizing the importance of mixing was that of Hahn and Klute.[46] It dealt with the effects the different turbulent flows (set up by three types of mixers) had upon the floc formation and floc destruction of two different types of suspended solids systems. Propeller, turbine, and anchor type stirrers were tested at low (34 rpm) and high (84.5 rpm) rotational speeds, maintaining constant energy input. Operating on a "strongly flocculating system" (54.5 mg/ℓ kaolinite and 0.1 mg/ℓ polymer), the uniformly stirring anchor mixer proved more efficient than the propeller and turbine mixers, especially at the higher rotational speeds. In contrast, when acting on a "strongly destabilized system" (28.2 mg/ℓ of montmorillonite and 0.01 mol/ℓ of Ca^{+2}), the turbine type mixer showed the greatest "coagulating efficiency" at low rotational speeds, and the propeller type at the high speeds. The latter two mixer types are hydrodynamically typified by much higher turbulence in the vicinity of the stirrer.

This work suggests there is no one correct mixer for all the waste and conditioning systems that can possibly be encountered, particularly when another variable, stirrer speed, is introduced. The least fluctuation was encountered at the lower speeds where the turbine produced slightly better conditioning than the propeller. More point scatter was obtained for the "strongly flocculating system" (kaolinite and polymer). Many of the waste and conditioning systems encountered in everyday operations should be more viscous than the strongly flocculating system and should, therefore, be expected to be less predictable. It should be pointed out that the criterion for "coagulating efficiency" (changes in particle size distribution) might not be completely valid when making an evaluation with regard to a specific dewatering device. The results should be viewed much the same as those obtained for specific resistance, a relative value for comparing mixing options.

Recent work pertaining to the effects of mixing and polyelectrolytes on kaolin and quartz suspension was reported by Keys and Hogg.[47] Working with 1 to 3 wt% suspensions, a turbine mixer, and 50 to 1000 mg/ℓ stock nonionic and anionic polyacrylamide polyelectrolytes in the 11 to 15×10^6 molecular-weight range they concluded the following:

1. Under almost all conditions, polymers should be added continuously without additional mixing after polymer addition is complete.
2. Optimum incorporation of primary particles into floc is usually obtained by using vigorous agitation for relatively short time periods.

3.   Flocculation is not strongly affected by initial concentration of polymer stock solution as long as sufficient mixing accompanies the polymer addition.
4.   When scaling flocculation tests, the rate of agitation should be scaled so as to maintain constant impeller tip speed. The mixing time should be adjusted so that, at the optimum impeller speed, mixing (and polymer addition) takes place for a fixed number of revolutions of the impeller.
5.   The mixing effects appear to be quite general. Very similar trends were observed in experiments carried out using three different polymers and two different solids at various concentrations.

Their test procedure might be considered a version of the jar test procedure mentioned in Chapter 4. Most of their conclusions are in agreement with, or seem logical additions to, current thinking pertaining to coagulation for clarification or sludge thickening. Their conclusion pertaining to the minimal effect of polyelectrolyte stock solution strength does not disagree too much with manufacturers' recommendations in the polymer concentration range they employed. (See Chapter 8.) Operating experience indicates that the adverse effects of using a polymer feed concentration of 1000 mg/ℓ (0.1 wt%) tend to minimize as the influent waste suspended solids approach the optimum volume percent range for dewatering.

It should be noted that reference was made to the volume percent range of the influent solids. This measurement is of equal or greater importance than the weight percent because the dewatering devices sense volume, not mass. Coagulants and coagulant 4aids (polyelectrolytes) are usually added to gather the fine, bulky colloids into flocs that will be of the proper size and strength to allow rapid expression of the liquid phase by whatever force is to be applied regardless of the mass of the particles. Mass and volume are related by the particle density, or the packing density when dealing with more than one particle. Therefore a 3 wt% suspension of quartz (high density) would be expected to have a smaller volume of solids than an equal volume of a 3 wt% activated sludge (lower density).

D. High Solids Volume Suspensions

The volume percent of settleable solids in a sludge is taking on more importance than it has in the past, and more than it is being given recognition for at this time. This is especially true when high-molecular-weight polyelectrolytes are being used to flocculate thickened sludge prior to dewatering. Most investigators, such as those mentioned in the previous section, study systems in which relatively small amounts of a polyelectrolyte are added to suspensions containing small or modest volumes of suspended or settleable solids, situations normally encountered in clarification or sludge thickening, not dewatering. For most commonly employed dewatering devices, especially vacuum filters and the newer gravity belt presses or "bandfilters", there is a minimum solids content for cost-effective operation.

Thickeners are required, preceding the dewatering unit, in order to reduce the influent sludge volume to the point where the dewatering unit is not hydraulically overloaded. Polyelectrolytes may or may not be employed in the thickening stage to increase the solids capture and thickening rate (settling rate in gravity thickeners, rise rate in flotation thickeners). If polymers are added, they might or might not have some beneficial value in dewatering depending on many factors. When polymers must be added, there is an optimum solids volume percent range at which the polymer can be mixed with regard to machine performance and cost effectiveness. Mixer configuration and speed, polyelectrolyte viscosity, and system temperature are additional factors having significant impact on the mixing effectiveness in addition to the solids volume.

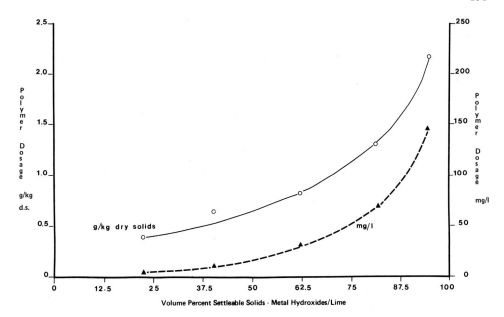

FIGURE 10.   Polyelectrolyte requirements vs. total suspended solids volume percent for a lime neutralized mixed-metal hydroxide sludge.

Figure 4 shows a parabolic relationship between the polyelectrolyte requirement in grams per kilogram and the influent solids weight percent for an extended aeration biological sludge, the optimum value being somewhere around 1.75 wt%. Increased polyelectrolyte demand at lower influent solids content was explained in terms of the amount of polyelectrolyte needed to neutralize the surface charges on the colloids. Nothing was said about the right side of the curve, beyond 1.5 to 2.0%.

Both operating experiences and laboratory tests on various types of wastes, ranging from calcium carbonate to extended aeration biological, strongly indicate that the increased polymer consumption is due to ineffective mixing of the waste particles and viscous polyelectrolyte. A suspended solids range of 40 to 60 vol% was found to be optimum for the specific mixer/dewatering device combination used, a three-blade propeller mixer and a gravity dewatering unit.

An interesting observation made on full scale equipment and then rechecked in the laboratory, on several wastes, was that the volume of polyelectrolyte solution added in excess of the optimum grams per kilogram (polymer per dry solids) ratio, in order to flocculate a high solids sludge, diluted the sludge so that it approached the upper limit of the effective solids volume percent range. This was checked in the laboratory by first determining the volume of a given concentration polyelectrolyte (0.1 wt/vol %) required to flocculate a high solids sludge. Various volume combinations of polyelectrolyte and make-up water, equaling the original volume of polyelectrolyte solution added, were tested for the same degree of floc formation. The polymer requirement (grams per kilogram) trended downward toward the optimum, as mentioned earlier. Inability of the polymer solution and the closely positioned solids to mix is clearly visible, as is the tendency toward complete mixing as dilution occurs.

These observations are not offered as a complete explanation of why the polymer consumption curve tends to swing upward after a given solids concentration range, nor does it infer that a decrease in polymer cost effectiveness will occur for all combinations of mixers and dewatering unit requirements.

Figure 10 is a plot of the average polyelectrolyte requirement for flocculation of a lime neutralized-mixed metal hydroxide clarifier blow-down sludge as a function of the settleable solids volume percent. The plotted parameters vary from those in Figure 4 only in that Figure 4 used weight percent, rather than volume percent settleable solids, for the abscissa. Weight percent was used because the extended aeration solids were present both as settleable and floating phases, something that occasionally occurs with aerobically digested sludges. The fact that some of the solids floated had no appreciable effect on the dewatering characteristics.

The broken line average polyelectrolyte dosage curves (milligrams per liter) are of the same general shape for both figures. The solid line average polyelectrolyte required curve (grams per kilogram) for Figure 10 represents only the right hand portion of the parabolic curve obtained in Figure 4, plotting comparable units. This latter condition is attributed to the fact that the sludge solids had an average zeta potential of -2.42 mV at 22.5°C, after the sludge was diluted 100 times in order to observe movement necessary for the electrophoretic mobility measurement. The negative charge is due to the anionic polyelectrolyte which was added during the clarification operation. Initial polyelectrolyte screening tests run on the clarifier blow-down sludge showed a low charge, high-molecular-weight anionic polyacrylamide-based polymer to be most effective in obtaining the flocculation characteristics needed for dewatering in the particular type of dewatering unit chosen, a gravity/belt press. The bridging effect afforded by the high molecular weight dominated over any tendency for particle repulsion due to the increased negative surface charge encountered with increased anionic polymer addition. This is frequently the case with systems employing high-molecular-weight anionic polyelectrolytes. Observed space occupancy by the solids at various volume percents and the increasing inability to move the solids, let alone thoroughly mix them with the viscous polymer solution, as the volume percent increased, made poor mixing the prime candidate for poor polyelectrolyte efficiency.

V. OPERATING PERFORMANCE

Usually the prime concern of anyone dealing with dewatering equipment is the operating performance. Any data published, including that which will be included in this chapter, must not be taken as absolute. At best, it is an indication or approximation of what might be expected. Attempting to firm the final product derived from a sequence of variables, many of them unmeasurable and some unknown, is not prudent.

A. Dewatering Devices or Methods
1. Drying Beds
Drying beds are the oldest and still one of the most widely used methods of dewatering, depending upon land availability, climate, and local regulations. Improved sludge removal by mechanical means and the use of polyelectrolytes to improve drying times have brought drying beds into consideration. When properly designed and operated, they still can produce a drier cake and are less sensitive to influent solids concentrations than mechanical devices.

Design criteria for open, conventional drying beds, with and without chemical conditioning are as follows:

Type sludge	Pretreatment	Sludge loading, dry solids (kg/m²/yr)
Primary and activated	None	73
Primary and activated	Chemically conditioned	269

Dosing anaerobically digested primary sludge in-line with a liquid cationic polyelec-

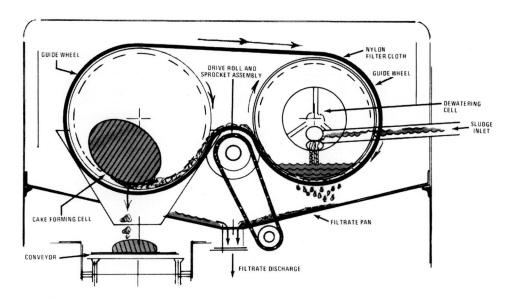

FIGURE 11. Schematic diagram of DCG® Sludge Dewatering Unit gravity drainage and roll cake cells. (Courtesy The Permutit Company, Inc., Paramus, N.J.)

trolyte at a rate of 25 g/kg dry solids ($6.50/ton dry solids) reduces sludge drying time for a liftable cake from 30 days to 8 to 15 days.

The above information was abstracted from a very thorough survey of sludge dewatering methodology and design compiled for the U.S. EPA by Harrison,[3] which also mentions that polyelectrolytes have made mechanical wedgewire drying beds practical.

2. Continuous Mechanical Gravity Dewatering Units

The most recent innovation in dewatering methodology was derived primarily for use in wastewater sludge dewatering. Belt filters (as they have come to be named) are, with rare exception, a practical dewatering method only when used with polyelectrolyte conditioning. Most units employ three stages in varying degrees and configurations: initial gravity drainage, medium pressure solids compaction, and final pressing.

a. Dual Cell Gravity Solids Concentrator (DCG)® Multi-Roll Sludge Dewatering Press (MRP)

This system, one of the first employed in the U.S., divides the three filtration stages between two pieces of equipment. The first unit (DCG®) can accept clarifier sludge, without thickening, and produce a cake that is free of most loose moisture. Figure 11 is a schematic view of the filtration area, the dual cell.

The dual cell is formed by a continuous nylon filter cloth zipper fastened to seals which move along guide wheels. A drive roll assembly located between the two guide wheels forms a hump, dividing the filtration screen into two separate cells. In operation, flocculated sludge, conditioned in an adjacent mixing vessel, is fed by chutes into the first or drainage cell. There the clear, essentially solids-free water rapidly drains through the conditioned floc. The only filtration force employed in this stage is the water head itself, approximately 4 in. maximum. The partially dewatered solids are carried over the hump into the back cell where they form a plug or roll, which increases in size until it extends over the rim of the cake cell guide wheel (out of the plane of

FIGURE 12.   DCG® Sludge Dewatering Unit photograph. (Courtesy The Permutit Company, Inc., Paramus, N.J.)

the paper), and falls onto a conveyor. The roll forms because properly conditioned floc has a greater affinity for itself than it does for the nylon screen. It serves dual purposes of cleaning the screen for its return to the front drainage cell, and further expressing water from the rolling sludge cake due to the weight and motion of the cake itself. The discharged cake can be disposed of directly when applicable, or fed into the Multi-Roll Press (MRP), where the solids undergo continually increasing nip pressure. Roll pressure is pneumatically controlled and adjustable. Continuous or stepwise solids concentration is important in all the gravity/press type units because the closer packing of the solids allows increasing pressure to be applied for further dewatering with optimum solids capture. Figures 12 and 13 show the DCG® and MRP respectively.

Table 9 contains typical operating and performance data for this system operating on a variety of sludges.[48] Gravity type units are usually the least complicated mechanical dewatering devices, requiring the least amount of power. Solids capture is very good and maintenance fairly low. Chemical conditioning costs are the major operating cost.

b. Single Unit, Multiple Stage Systems

This type of machine represents the latest innovation in continuous gravity belt filter

FIGURE 13. Photograph of MRP Multi-Roller Press which can be used to further dewater DCG® discharge cake. (Courtesy The Permutit Company, Inc., Paramus, N.J.)

presses. It was introduced into the U.S. from West Germany approximately 6 to 7 years ago and has gained considerable popularity in a relatively short time period. There are at least 10 to 12 versions of the basic machine concept, differing primarily in the way they perform each of the three basic steps for applying progressive pressure for water removal.

Figure 14 is a schematic diagram of one of the more popular German-developed belt or "bandfilters", the Ecopress®, which is now sold in the U.S. It differs slightly from most of the other units in that it employs a low vacuum zone following the gravity straining zone and preceding the two successively increasing pressure zones.

Harrison describes several other versions of this popular basic design in detail, including operating data for most of the units.[3] Since the purpose of presenting operating data in this chapter is only to give some indication of operating characteristics for units employing predominately polyelectrolyte conditioning, or citing comparisons of polyelectrolyte conditioning vs. none or other forms of conditioning, delineation between performance of various units of the same gender will not be attempted. Table 10 is presented as being typical performance data for "beltfilters,"[49] and Figure 15 shows an installation.

3. Vacuum Filtration

In the past, polyelectrolytes have been used only sparingly with vacuum filters, es-

Table 9
TYPICAL DUAL CELL GRAVITY SOLIDS
CONCENTRATOR AND MULTI-ROLL SLUDGE
DEWATERING PRESS DATA

Performance and Capacity

Type sludge	Influent	% Dry solids DCG®	MRP	Approx. capacity (kg/hr)[a] DCG® 200
Sanitary				
Aerobic digested	0.5	10—12	16—20	56
Aerobic digested	3	10—12	16—20	181
Anaerobic digested	4	12—15	18—20	136
Waste activated	1.9	9—13	18—23	136
Waste activated	3	9—13	18—23	181
Raw primary	3	12—17	20—23	136
UNOX	3	10	16	102
Industrial				
Alum clarification	3.5	9.2	16	109
Iron clarification	6	19	34	272
Flue scrubber waste	5	20—38	33—50	113
Hydroxide, aluminum	1.2	8—13	15—19	24
Hydroxide, ferric/ferrous	3.5	13—15	25—30	68
Hydroxide, zinc	1	7—10	18—22	40
Hydroxide, iron oxide	5.5	18.1	41	175
Lime softening, cold	8	48—53	60—64	544
Hot lime softening	4	16—17	37—38	318
Lead oxide	7.5	54	—	340
Calcium phosphate	2	13	25	109
Iron sulfide	2	9—10	18	95
Iron hydroxide/oil	6	26—36	—	159
Paper mill waste	2.4	5—10	14—25	127
Waste activated (phenolic)	4	8—10	15—20	181
Tanning waste	3	14—15	20—25	113
Latex	4	12—14	—	181
Meat packing	1.5	9.5	15	51

[a]   Dry solids.

Courtesy The Permutit Company, Inc., Paramus, N.J.

pecially those using cloth filtration media. Ferric chloride and lime are usually the preferred flocculant combination for direct addition to the filtration process. Rotary vacuum filters employing a filtration media comprised of a multiplicity of coiled springs, such as the Coilfilter®, manufactured by Komline-Sanderson Engineering Company, have obtained improved performance when operating on a primary and secondary sludge blend, where a polyelectrolyte conditioning was used to improve secondary solids capture during the dissolved air flotation sludge thickening process.

The use of polyelectrolyte conditioning at least doubles the loading rates for flotation thickeners over those obtained without polymer, according to Komline,[50] in a paper which cites the economic benefits derived in sludge dewatering, from employing polymer conditioning for flotation secondary sludge thickening. Every 22.7 kg (50 lb) of polyelectrolyte added in the flotation operation resulted in a savings of 7030 kg (15,500 lb) of lime, 2495 kg (5500 lb) of ferric chlorice (all dry weight basis), and 42

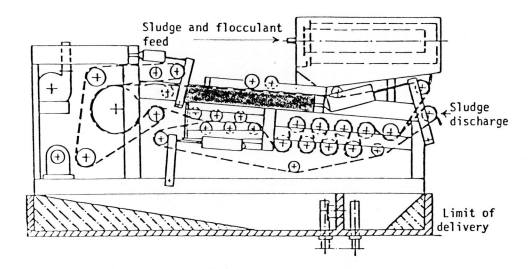

FIGURE 14. Schematic diagram of VN Ecopress belt filter press dewatering unit. (Courtesy Roediger Pittsburgh, Inc., Pittsburgh, Pa. and Euramca Ecosystems, Inc., Addison, Ill.)

Table 10
TYPICAL BELTPRESS PERFORMANCE AND CAPACITY DATA

Dry Solid Content Values   Throughput Values (m³/hr)

Kind/origin of sludge	Volatiles (max % of dry solid)	Dry solid input (%)	Dry solid cake (%)	Belt width (m)		
				0.9	1.5	2.5
Anaerobic digested sludge (domestic)	45	4—6	25—45	2—6	4—10	6—16
Anaerobic digested sludge in combination with sludge from chemical precipitation	50	3—5	25—35	2—5	3—8	9—12
Raw sludge (domestic)	65	3—5	25—40	2—6	4—10	6—12
Raw sludge mixed with waste activated sludge	70	2—4	23—35	3—7	5—11	7—17
Aerobic digested	55	2—4	18—30	2—6	4—10	6—15
Mineral (mining) sludge	20	10—15	55—80	1—5	3—7	5—9
Papermill sludge		3—5	30—45	3—6	5—10	6—15
Bleaching clay sludge	17	25—30	50—60	1—3	2—5	3—7
Flushing water slurry from gravel filtration	35	4—5	35—55	5—8	6—10	8—14
Slurry from cooling water circuit decarbonization	30	9—13	60—70	1—3	2—6	3—8

Courtesy Roediger Pittsburgh Inc., Pittsburgh, Pa., and Euramca Ecosystems, Inc., Addison, Ill.

labor hours in Coilfilter® operation. The yearly savings was $56,225 for a 1420 m³/hr (9 million gal/day) influent flow wastewater treatment plant.

Polyelectrolyte choice can significantly affect the loading rates and solids capture of this type, but appears to have less effect on cake concentration. Data from a plant dewatering digested primary sludge without ferric chloride and lime indicates that a

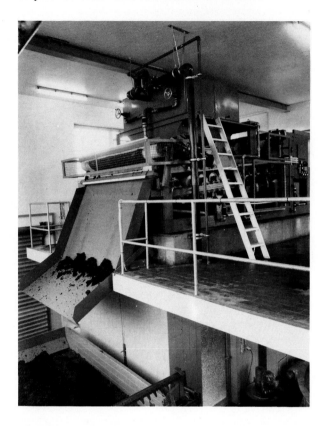

FIGURE 15.   VN Ecopress belt filter press installation. (Courtesy Roediger Pittsburgh, Inc., Pittsburgh, Pa. and Euramca Ecosystems, Inc., Addison, Ill.)

Table 11
OPERATING COSTS AND RESULTS FOR INORGANIC AND POLYELECTROLYTE CONDITIONED SLUDGE[52]

	Unit cost ($/NDT)[a]	
	Lime and ferric chloride	Polyelectrolyte
Chemicals	11.99	5.63
Gas[b]	12.75	9.52
Combined	24.74	15.15
Annual cost ($)	869,400	532,400
Filter cake data		
Yield (net lb/ft²/hr)	7.1	7.8
Percent solids	30	28
Percent volatile	42	56

[a]   NDT = net dry ton.
[b]   Auxiliary gas needed for incineration.

29.3 kg/m²/hr loading rate, a 20% solids filter cake and 93 to 95% solids capture can be obtained using polymer. The same plant could obtain only 19.53 kg/m²/hr loading rate, a 19% solids cake, and 80% solids capture using a different polymer.[51]

Schillinger,[52] in his accounting of the conversion of a vacuum filter sludge condition-

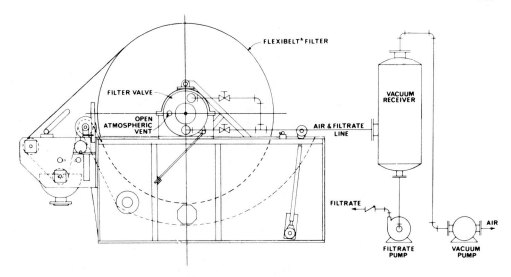

FIGURE 16. Schematic diagram of simple Flexibelt vacuum filter dewatering system. (Courtesy Komline-Sanderson, Peapack, N.J.)

ing system from ferric chloride and lime to polyelectrolyte, tabulated the operating costs and results given in Table 11. Filter cloth blinding was not mentioned as a problem, even though they used a dual polymer system (anionic and cationic polymers).

Figure 16 is a schematic diagram of a Flexibelt® rotary drum vacuum filter which has the basic design of the filters discussed. Figure 17 shows a rotating horizontal drum mixer employed with vacuum filters for both inorganic chemical and polymer conditioning of sludges. This basic rotating drum mixer design, with baffles running the length of the drum, has been adopted by most of the beltfilter manufacturers.

Harrison claims that more of the sludge processed in plants equipped with rotary vacuum filters is conditioned with polymer flocculants than with inorganic conditioners, and that the chemical costs are about the same for the two methods.[3] Polyelectrolytes are preferred because of convenient handling, ease of preparation, fewer corrosion problems, and elimination of significant quantities of inorganic solids in the sludge. Table 12 lists typical rotary vacuum filter results for polymer-conditioned biological sludges.

4. Centrifugals

Polyelectrolyte conditioning is used more with centrifugal dewatering devices than it is with vacuum filters. The primary function of polymer conditioning is to improve solids capture, however, centrifugals are employed without any sludge conditioning when a solids capture range of approximately 70 to 90% for activated sludge is acceptable. As in the case with vacuum filters, the solids capture can, at times, be controlled by the polymer dosage. This differs from gravity beltfilters in that there is a minimum polymer dosage required for gravity units in order for them to operate. Once this minimum dosage is met, little improvement is obtained with additional polymer.

There are a number of types of centrifugals, each generally being best fitted for various specific applications. Bowl and basket type units are usually employed for sludge dewatering. Basket centrifugals operate in a batch mode as opposed to the continuous bowl types. Bowl centrifugals are of two types, imperforate and solid. They are also classified by their rotational speed and gravitational force developed. Low speed (500 to 750 × g) bowl type centrifugals are gaining acceptance because they offer significant improvement in maintenance and operating costs over the high speed units without greatly reducing cake solids. See Figure 18.

FIGURE 17.   Vacuum filter revolving drum sludge conditioning tank. (Courtesy Komline-Sanderson, Peapack, N.J.)

Table 12
TYPICAL ROTARY VACUUM FILTER RESULTS FOR POLYELECTROLYTE CONDITIONED SLUDGES[3]

Type sludge	Chemical costs[a] ($/ton)	Yield (kg/m²/hr)	Cake solids (%)
Raw primary	1—2	39—49	25—38
Anaerobically digested primary	2—5	34—39	25—32
Primary and humus	3—6	20—29	20—30
Primary and air activated	5—12	20—25	16—25
Primary and oxygen activated	5—10	20—29	20—28
Anaerobically digested primary and air activated	6—15	17—29	14—22

[a]   Short and metric tons are considered to be equal because of the significant figures involved.

Centrifugals are sometimes employed as thickeners, operating on wastes with solids contents as low as 0.5 wt%. Table 13 shows the effect various amounts of polymer have on the solids capture when thickening waste activated sludge in a solid-bowl centrifugal.[53]

Polyelectrolyte is usually added to the sludge in the feed line prior to the centrifugals, or in the centrifugal interior; no separate conditioning vessel is usually employed. When a dual polymer system is required both injection methods can be employed. Another method of handling a dual polymer system is found in U.S. Patent #3,228,594, to Amero.[54] In this method, two adjustable conditioning agent pipes are capable of placing the chemicals at various locations inside the bowl, affording optimum usage of the conditioning chemicals in conjunction with the centrifugal action.

Figure 19 shows the effect increased polyelectrolyte dosage has on both the solids

FIGURE 18. Cutaway view of low-speed solid-bowl centrifugal dewatering unit. (Courtesy Bird Machine Company, Inc., South Walpole, Mass.)

Table 13
SOLID-BOWL CENTRIFUGAL THICKENING
PERFORMANCE DATA FOR ACTIVATED
SLUDGE AT VARIOUS POLYELECTROLYTE
DOSAGE LEVELS[48]

Capacity (m³/hr)	Feed solids (%)	Underflow solids (%)	Solids recovery (%)	Polymer requirement (g/kg)
25—36.3	0.5—0.7	5—8	65	None
			85	<2.5
			90	2.5—5
			95	5—7.5

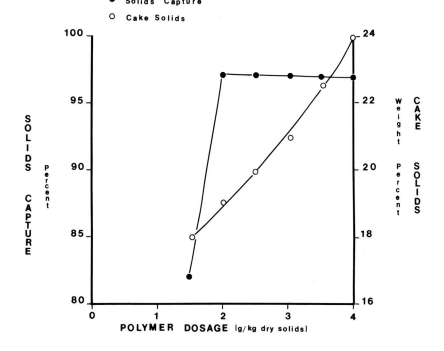

FIGURE 19. Solids capture and cake solids as a function of polyelectrolyte dosage for a 50/50 raw primary/waste activated sludge mixture.

Table 14

FARNHAM POLLUTION CONTROL WORKS, U.K.,
DEWATERING RESULTS

Conditioning agent	Dose (% on dry solids)	Cost ($/tonne dry solids)	Pressing cycle time range (hr)
Aluminum chlorhydrate (batch)	2.5	24.20	6—18
Aluminum chlorhydrate (in-line)	2.5	24.20	6—12
Zetag 63[a] (batch)	0.2—0.3	7.40—11.10	6—9
Zetag 63 (in-line)	0.2—0.3	7.40—11.10	3—6
Ferric chloride and lime (batch)	3 and 25	16.30	3—13
Ferric chloride and lime (in-line)	3 and 25	16.30	3—5

[a]   Manufactured by Allied Colloids Co.

Table 15

THORNBURY, U.K. SLUDGE DEWATERING
PRIMARY AND SECONDARY SLUDGE

Conditioner[a]	Percent dry solids		Conditioner cost ($/tonne dry solids)	Press cycle time (hr)
	Feed	Cake		
Aluminum chlorhydrate	4.6	38	25.75	4.9
Polyelectrolyte (Zetag 94)	4.6	37	5.10	4.9

[a]   Added in-line.

capture and cake solids for a 50/50 raw primary and waste activated sludge, employing a low-speed solid-bowl centrifugal at $780 \times g$. In this instance the solids capture peaked rapidly and then remained constant, whereas the cake solids rose gradually with increasing polyelectrolyte dosage.[55]

5. Filter Presses

Until recently the use of polyelectrolytes with filter presses was not recommended by press manufacturers. As in the case of vacuum filters, inorganic conditioners such as ferric chloride and lime, and other types of body feeds were employed. The two reasons cited against polymer use were floc rupture due to the turbulence and pressure encountered during the cycle and polymer filter media blinding. Again, as in the case of vacuum filters, polymer conditioning in earlier liquid-solids separation operations aided pressure filtration.

Certain polyelectrolytes have been employed with filter presses in Europe and the U.K.[56] Table 14 compares chemical costs and pressing cycle times for polyelectrolyte and inorganic sludge conditioners acting on a combined primary and trickling filter sludge, showing considerable chemical cost reductions and equal or lower cycle times when employing a polyelectrolyte. Table 15 shows a similar comparison for a 45/55 mixture of primary sludge and mixed secondary sludges.[57] Reduced chemical conditioner cost is again the most outstanding difference between the polyelectrolyte and the aluminum chlorhydrate.

VI. CONCLUSIONS

There is no doubt that polyelectrolytes have made a place for themselves in sludge dewatering. The ease of handling and storage, no added solids volume, and lower sludge ash content make them desirable compared to inorganic coagulants. However, until such time as waste treatment moves to the overall process control required for "closing-the-loop" philosophy, they will not reach their full potential. Sludge dewatering is still the last liquid-solids separation step prior to disposal, and as such must contend with the products of compounded problems that can occur any place in the system. Certain excursions in previous treatment steps can have a marked effect on the operability of a polymer. Reports concerning day-to-day operations of units relying on polyelectrolyte sludge conditioning are already hinting at this.

Properly mixing the polyelectrolyte and sludge is almost, if not, as necessary as choosing the correct polymer (electrical charge and molecular weight). Mixing of viscous polyelectrolytes and high solids sludges, required for most cost-effective operation of gravity belt filters and vacuum filters, has had very little, if any, investigation. Better understanding of the type of mixing best suited for polymer flocculation should result in lower operating costs and wider operating latitude.

REFERENCES

1. U.S. EPA, Discussion comments, U.S. EPA Environ. Res. Inf. Center Technol. Trans. Design Seminar Sludge Treatment and Disposal, Philadelphia, March 30 to 31, 1978.
2. Schwoyer, W. L., Dewatering of Lime Softening Sludge, in Trans. of 22nd Annu. Conf. on Sanitary Eng., University of Kansas, Lawrence, February 2, 1972, 41.
3. Harrison, J. R., Review of Developments in Dewatering Wastewater Sludges, prepared for U. S. EPA Environ. Res. Inf. Center Seminar on Sludge Treatment and Disposal, Philadelphia, March 30 to 31, 1978.
4. Karr, P. R. and Keinath, T. M., Influence of particle size on sludge dewaterability, *J. Water Pollut. Control Fed.*, 50, 1911, 1978.
5. *Standard Methods for the Examination of Water and Wastewater*, 14th ed., published jointly by Am. Public Health Assoc., Am. Water Works Assoc., Water Pollut. Control Fed., Washington, D.C., 1975, 61.
6. La Mer, V. K., Coagulation versus the flocculation of colloidal dispersions by high polymers (polyelectrolytes), in *Principles and Applications of Water Chemistry*, Proc. 4th Rudolfs Res. Conf., Faust, S. D. and Hunter, J. V., Eds., John Wiley & Sons, New York, 1967, 246.
7. Vostrcil, J. and Juracka, F., Commercial Organic Flocculants, Noyes Data Corp., Park Ridge, N. J., 1976, 1.
8. O'Brien, J. H., Prediction of Polymer Performance for Water Treatment Sludge Conditioning, presented at national meeting Am. Water Works Assoc., Boston, June 19, 1974.
9. Bradley, R. L., Chemical Conditioning of Sludge Dewatering, Proc. 6th Annu. Ind. Pollut. Conf., Water and Wastewater Equipment Manufacturers Association, St. Louis, April 1978, 353.
10. Handbook for Monitoring Industrial Wastewater, U.S. EPA Technol. Trans., Washington, D.C., August 1973.
11. Glauser, J. and Schwoyer, W., personal correspondence, 1978.
12. Luttinger, L. B. and Schwoyer, W. L., unpublished internal report, 1971.
13. Cassel, A. F. and Johnson, B. P., Evaluation of Dewatering Units to Produce High Sludge Solids Cake, presented at 51st Annu. Conf. Water Pollut. Control Fed., Anaheim, Calif., 1978.
14. Schwoyer, W. L., private notes, 1970.
15. Swanwick, J. D., Theoretical and Practical Aspects of Sludge Dewatering, Proc. 2nd Eur. Sewage and Refuse Symp., Munich, 1972.
16. Vesilind, P. A., *Treatment and Disposal of Wastewater Sludges*, Ann Arbor Science, Ann Arbor, Mich., 1974, 211.
17. Metcalf and Eddy, Inc., *Wastewater Engineering: Treatment, Collection, Disposal*, McGraw-Hill, New York, 1972, 304.

18. **Roberts, K. and Olsson, O.,** The Influence of Colloidal Particles on the Dewatering of Activated Sludge with Polyelectrolytes, presented at Div. of Environ. Chem., American Chemical Society, Atlantic City, September 8 to 13, 1974, 10.

19. **Roberts, K. and Olsson, O.,** The Role of Colloidal Particles in Dewatering of Activated Sludge with Aluminum Hydroxide and Cationic Polyelectrolyte, presented at Div. of Environ. Chem., American Chemical Society, Chicago, September 1975, 5.

20. **Humenick, M. J., Jr.,** *Water and Wastewater Treatment, Calculations for Chemical and Physical Processes,* 1st ed., Marcel Dekker, New York, 1977, 98.

21. **Vesilind, P. A.,** *Treatment and Disposal of Wastewater Sludges,* Ann Arbor Science, Ann Arbor, Mich., 1974, 116.

22. **Coackley, P.,** *Laboratory Scale Filtration Experiments and Their Application to Sewage Sludge Dewatering,* McCabe, J. and Eckenfelder, W. W., Jr., Eds., Van Nostrand Reinhold, New York, 1958, 270.

23. **Gale, R. S. and Baskerville, R. C.,** Capillary suction method for determination of the filtration properties of a solid/liquid suspension, *Chem. Ind. (London),* 9, 355, 1967.

24. **Jones, B. R. S.,** Vacuum sludge filtration. II. Filter performance prediction, *Sewage Ind. Wastes,* 28, 1103, 1956.

25. **Wilhelm, J. H.,** The use of specific resistance data in sizing batch-type pressure filters, *J. Water Pollut. Control Fed.,* 50, 471, 1978.

26. **Nebiker, J. H., Sanders, T. G., and Adrian, D. D.,** An investigation of sludge dewatering rates, *Journal Water Pollut. Control Fed.,* 41, R255, 1969.

27. **Kos, P. and Adrian, D. D.,** Transport phenomena applied to sludge dewatering, *J. Environ. Eng. Div. Am. Soc. Civ. Eng.,* 6, 947, 1975.

28. **Mangravite, F. J., Jr. Leitz, C. R., and Juvan, D. J.,** Application of polyelectrolytes for industrial sludge dewatering, in *Chemistry of Wastewater Technology,* Rubin, A. J., Ed., Ann Arbor Science, Ann Arbor, Mich., 1978, chap. 17.

29. **Dahlstrom, D. A. and Silverblatt, C. E.,** Continuous filters, in *Solid/Liquid Separation Equipment Scale-Up,* Purchas, D. B., Ed., Uplands Press Ltd., Croydon, England, 1977, chap. 11.

30. **Komline-Sanderson Rotary Drum Vacuum Filters,** Bulletin KSB 510-7703, Komline-Sanderson Engineering Corp., Peapack, N.J., 1977, 4.

31. **Vesilind, P. A.,** Estimation of sludge centrifuge performance, *J. Sanit. Eng. Div. Am. Soc. Civ. Eng.,* 137, 95, 1970.

32. **Records, F. A.,** Sedimenting centrifuges, in *Solid/Liquid Separation Equipment Scale-Up,* Purchas, D. B., Ed., Uplands Press Ltd., Croydon, England, 1977, chap. 6.

33. **Vesilind, P. A.,** *Treatment and Disposal of Wastewater Sludges,* Ann Arbor Science, Ann Arbor, Mich., 1974, 226.

34. **AIChE Equipment Testing Procedure — Batch Pressure Filters,** American Institute of Chemical Engineers, New York, 1967.

35. **Tiller, F. M. and Shirato, M.,** The role of porosity in filtration. VI. New definition of filtration resistance, *Am. Inst. Chem. Eng. J.,* 10(1), 61, 1964.

36. **Tiller, F. M., Haynes, S., Jr., and Lu, W.-M.,** The role of porosity in filtration. VII. Effect of sidewall friction in compression-permeability cells, *Am. Inst. Chem. Eng. J.,* 18(1), 13, 1972.

37. **Tiller, F. M. and Lu, W.-M.,** The role of porosity in filtration. VIII. Cake nonuniformity in compression-permeability cells, *Am. Inst. Chem. Eng. J.,* 18(3), 569, 1972.

38. **Shirato, M., Hayashi, N., and Fukushima, T.,** Constant pressure expression of solid-liquid mixtures with medium resistance, *J. Chem. Eng. Jpn.,* 10(2), 154, 1977.

39. **Shirato, M., Aragaki, T., Ichimura, K., and Ootsuji, N.,** Porosity variation in filter cake under constant-pressure filtration, *J. Chem. Eng. Jpn.,* 4(2), 172, 1971.

40. **Shirato, M., Murase, T., Negawa, M., and Senda, T.,** Fundamental studies of expression under variable pressure, *J. Chem. Eng. Jpn.,* 3(1), 105, 1970.

41. **Shirato, M., Aragaki, T., Mori, R., and Sawamoto, K.,** Predictions of constant pressure and constant rate filtrations based on an approximate correction for side wall friction in compression permeability cell data, *J. Chem. Eng. Jpn.,* 10(1), 86, 1968.

42. **Shirato, M., Sambuichi, M., and Okamura, S.,** Filtration behavior of a mixture of two slurries, *Am. Inst. Chem. Eng. J.,* 9(5), 599, 1963.

43. **Schwartzenberg, H. G., Rosenau, J. R., and Richardson, G.,** The removal of water by expression, in *Am. Inst. Chem. Eng. Symp. Series, Water Removal Processes,* 73(163), American Institute of Chemical Engineers, New York, 1977, 177.

44. **Thomas, C. M. and Purdy, J. A.,** Batch filters, in *Solids/Liquid Separation Equipment Scale-Up,* Purchas, D. B., Ed., Uplands Press Ltd., Croydon, England, 1977, chap. 10.

45. **Purchas, D. B., Ed.,** *Solids/Liquid Separation Equipment Scale-Up,* Uplands Press Ltd., Croydon, England, 1977, 1.

46. **Hahn, H. H. and Klute, R.,** Effect of Turbulent Flow Conditions Upon the Process of Coagulation, presented at Div. of Environ. Chem., American Chemical Society, Chicago, August 24 to 29, 1975.
47. **Keys, R. O. and Hogg, R.,** Mixing Problems in Polymer Flocculation, presented 71st Annual Meeting Am. Inst. Chem. Eng., Miami Beach, November 16, 1978.
48. DCG® for Gravity Sludge Dewatering, Bulletin H5161 11775M, The Permutit Co., Inc., Paramus, N.J., 1972.
49. Sludge Dewatering with the ROEDIGER-ECOPRESS® Beltfilter Press with Suction Zone, General Data Sheet 9.04, Euramca "Ecosystems" Inc., Addison, Ill., 1979.
50. **Komline, T. R.,** Sludge Thickening by Dissolved Air Flotation in the U.S.A., presented at Flotation for Water and Wastewater Treatment, A Water Research Center Conference, Felixstowe, England, June 8 to 10, 1976.
51. **Kovacs, K., personal communications,** Komline-Sanderson Engineering Co., Peapack, N.J., December 12, 1978.
52. **Schillinger, G. R.,** Plant Scale Conversion of Sludge Conditioning Chemicals for Vacuum Filters, presented 51st Annu. Conf. Water Pollut. Control Fed., Anaheim, Calif., October 3, 1978.
53. Sludge Thickening, in Process Design Manual For Upgrading Existing Wastewater Treatment Plants, U.S. EPA Technol. Transf., Washington, D.C., 1974, chap. 10.
54. **Amero, C. L.,** Centrifugal Separator, U.S. Patent 3,228,594, 1966.
55. **Guidi, E. J. and Schwoyer, W.,** personal communications, Bird Machine Co., 1979.
56. Sludge Treatment and Disposal, Sludge Treatment, U.S. EPA Technol. Trans., Vol. 1, EPA-625/4-78-012, Washington, D.C., 1978, chap. 6.
57. **White, M. J. D. and Baskerville, R. C.,** Full scale trials of polyelectrolytes for conditioning of sewage sludges for filter pressing, *Water Pollut. Control,* 73(5), 1974, p. 486

Chapter 7

THE USE OF POLYELECTROLYTES IN FILTRATION PROCESSES

L. B. Luttinger

TABLE OF CONTENTS

I. INTRODUCTION

A. Filters for Potable Water and Wastewater Treatment

1. Single, Dual, and Multimedia Granular Filters

Up until about 1950, the only type of filter in common use for most large scale water and dilute waste applications was the single media filter. Sand was by far the most commonly used filter medium, although crushed coal was sometimes used where it was more readily available than sand. Two types of sand filters were employed. The earlier type, or "slow" sand filter, generally consisted of an approximately 1-m deep bed of upgraded sand. Filtration rate was of the order of 0.07 gal/min/ft² or about 0.048 ℓ/m²/sec of filter surface area. Filtration effect was limited to the top few inches of the deep bed, where a colloidal/biological "Schmeutzdecke" formed. No backwashing was practiced. When the flow rate through the bed fell to unacceptable levels, the top few inches of bed were removed by scraping, and filtration could then proceed.

In the later, rapid sand filter, approximately the same depth of sand was employed but filtration was at much more rapid rates, e.g., approximately 2 gal/min/ft² or 1.4 ℓ/m²/sec. Graded sand (effective size 0.4 to 0.5 mm) was used in these beds. The upper portion of such filter beds tended to plug with time as did the earlier slow sand filters, however, the rapid sand filter could be backwashed to remove foulants and reestablish the original filtration rate.

All downflow single media filters are either ungraded with respect to media size (slow type) or tend to grade fine to coarse in the direction of flow, as a result of backwashing (rapid type). The result, in either case, is a filter which is relatively impermeable at the top of the filter bed, where the influent first contacts the medium. In consequence, filtration here is virtually a two-dimensional effect. The suspended solids in the influent tend to deposit at the top of the medium bed, resulting in rapid plugging, high and rapid pressure drop increases through the bed, and rapid suspended solids breakthrough. A more effective means of filtration would comprise coarse to fine filtration rather than fine to coarse filtration.

Two methods of achieving coarse to fine filtration have been developed. The first consists essentially in filtering in the *upflow* direction, rather than in the conventional *downflow* direction. While this arrangement constitutes a real advantage over downflow single media filtration, a more common solution to the problem has been to retain the downflow filtration feature, but to substitute a second, larger, less dense medium for a certain portion of the single medium.

For example, a 3-ft deep bed of, say, 0.5-mm effective size* sand is substituted by a 1-ft depth of the same sand, over which is placed a 2-ft depth of, e.g., 1.0-mm effective size anthracite coal. The coal effective size and uniformity coefficient** are selected so that in backwashing, the coal layer will remain above the sand layer, with perhaps a small zone of intermixing. The result is coarse to fine filtration, as in the upflow filter. Filter runs in coarse-to-fine filtration are longer, and headloss development is slower, since the coarse-to-fine arrangement results in usage of much of the depth of the bed for dirt storage (rather than restricting dirt removal to the top few inches of the bed, as in single media, downflow filtration). Removal of large particles by the upper coal also permits more depth as against more surface filtration by the underlying sand. Higher filter rates also become feasible.

In an attempt to produce even better results along the same lines as dual media filtration, tri, tetra, and even larger numbers of media layers have been proposed and used. As in dual media filters, the media are so selected with respect to size, uniformity coefficient, and density, that the individual media layers in such filters remain more or less discrete, with, ideally, little mixing even after backwashing. Longer filter runs, slower headloss development, and better effluent quality have been claimed, and in some applications, established for such multimedia filters. However, filter bed upset problems are often aggravated with such complex arrangements.

A schematic of a typical "gravity" filter is given in Figure 1. A typical pressurized filter is shown in Figure 2.

2. Diatomaceous Earth (DE)-Type Filters

A quite different type of media filter in common use for potable water and waste treatment is the DE-type filter (Figure 3). In this type of filter, fine (5 to 30 μm) DE media is circulated against a porous support so as to deposit a precoat. Uusually, fine DE media is also fed with the filter influent ("body feed") to further enhance filtration. Typical flow rates of such filters in potable water applications are of the order of 1 gal/min/ft² (0.68 ℓ/m²/sec) and run lengths are often of the order of 40 + hr.

The media most commonly employed in this type of filtration is DE, a calcined amorphous silica powder which is composed of minute fossil diatom skeletons; however, other media are frequently employed. Excellent effluent clarity may be obtained through the use of such filters.

3. Other Types of Filters

The use of polyelectrolytes with other types of filters, for example, in the filtration of heavy suspensions such as sludges, is discussed elsewhere in this volume.

B. Polyelectrolytes for Potable Water and Wastewater Filtration

1. Polyelectrolytes (Flocculants) and "Coagulants" — Basic Mechanism

Two general types of reactions are used to agglomerate colloids and particles in water. These are coagulation and flocculation.

In coagulation processes, the electrical double layer surrounding the colloid particles is discharged by addition of suitable reagents. This discharge can be measured by investigation of the zeta potential (ZP) of the colloidal particles in such systems. (For a discussion of such measurements, see for example, Riddick[1].) This discharge results from the addition of ions of opposite electrical sign to the colloid-containing system. Since most colloidal particles are negatively charged in aqueous systems at most pH

---

* Effective size (e.s.) = that size that 10% by weight is smaller than.

** Uniformity coefficient (u.c.) = the ratio of sizes, such that 60% by weight is smaller, and the effective size, or the 60%/10% size.

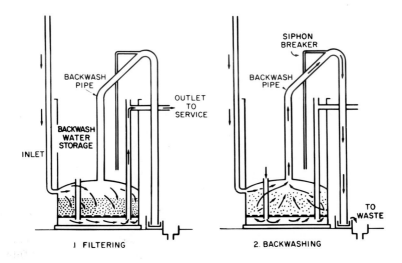

FIGURE 1.   Automatic valveless type gravity filter. (Courtesy The Permutit Company, Paramus, N.J.)

FIGURE 2.   Cutaway view of a typical vertical pressure filter. (Courtesy The Permutit Company, Paramus, N.J.)

levels of interest, the significant ions for discharge are cations, i.e., sodium or calcium ions, rather than chloride or sulfate ions.

A profound effect is exercised by the magnitude of the ionic charge of the coagulating ions. On a molar basis, divalent cations such as calcium ions are roughly 16 to 40 times as effective as monovalent cations such as sodium. Trivalent ions such as aluminum and ferric ions are roughly 30 times as effective as divalent ions, under conditions where the trivalent ions exist as such. However, due to hydrolysis reactions at the pH

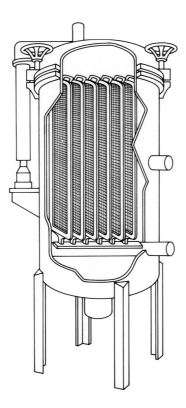

FIGURE 3.   Cutaway view of diatomaceous earth vertical leaf pressure filter. (Courtesy Johns-Manville, Denver, Colo.)

levels most commonly encountered in water and waste treatment, "trivalent" ions are often even more effective than this,[2] and the situation is considerably more complex, as will be discussed below.

The electrical "double layer" surrounding a colloid particle is shown in Figure 4.

In pure coagulation reactions, the double layer is largely discharged by the cationic "counterions", and even more significantly, multivalent cations are sorbed on the anionic colloid surface or in the "Stern Layer" (see Figure 4).[3] This state of affairs leads to a drastic reduction in the repulsive forces between colloidal particles. The latter can then approach and collide with one another more effectively, so that the very short range, attractive Van der Waal's forces, which are always present between particles, can come into play, resulting in agglomeration or "coagulation".

The resulting coagulum constitutes a dense phase which usually settles rapidly. The inidividual colloid particles tend to pack together closely in this phase, since charge repulsion has been drastically reduced. The coagulum tends to be relatively impermeable to water drainage, as a result of the close packing of the individual particles. This relative impermeability can give rise to problems in water retention and filtration.

The basic flocculation mechanism is quite different. In flocculation, the basic mechanism is the bridging of colloid particles by long-chain polymer molecules.[4]. Discharge of double layers is often completely absent, as when negatively charged colloids are flocculated by negatively charged (q.v.) polymers. Flocculation is usually more effective, the higher the molecular weight of the polymer, i.e., the greater its geometric extension in the system, and the greater the number of attachments it can form (Figure 5).

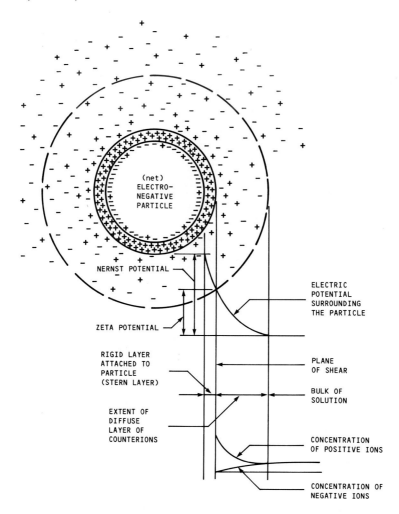

FIGURE 4.   Electric "double layer" surrounding a colloid particle. (Courtesy Zeta-Meters, New York, N.Y.)

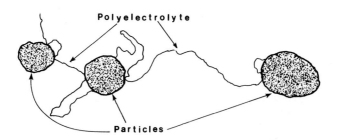

FIGURE 5.   Polyelectrolyte configuration on colloid particles.

Flocculation results typically in the formation of loosely packed flocs, in the sense of their permeability to water, although the individual particles constituting the floc are usually toughened by the presence of the polyelectrolyte.

Since the electrical double layer has often not been discharged in flocculation, electrostatic repulsion is many times still operative, preventing the type of close packing of the reacted colloids which is characteristic of coagulation. The loose packing often results in better water drainage through such a flocculated system than through a coagulated system. Another consequence, of particular importance in filtration, is the very strong attachment of the polymer chains to filter media (sand, anthracite, DE) and to the water/waste colloids and particles. This strong attachment is brought about by the very large number of hydrogen bonding sites on a long-chain polymeric flocculant molecule.[5,6]

Hydrogen bonds, as well as other attractive forces, hold the flocculated colloid particles fast to the filter media. No such mechanism is generally available via simple coagulation, i.e., with calcium ions.

Another advantage of flocculants over coagulants in filtration processes is a consequence of the extremely small dosages of polymers, as compared with coagulants, required for effective filtration. When a "pure" coagulant, such as calcium ion (lime), or a "mixed" coagulant (alum, ferric hydroxide) is used as a filtration aid, the solids volume contributed by the coagulant is often of the same order as the solids volume due directly to removed colloids. On the other hand, when polymeric flocculants are used as filter aids, the volume contributed by these materials is often minute (i.e., in water filtration, a feed level of the order of parts per billion of flocculant is often effective, where parts per million of coagulant or alum would be required. The same ratio holds in many waste applications, although the absolute magnitude of the alum or polymer dosage required is often much greater.) This results in considerably longer filter runs between backwashings.

The tougher floc often formed when polymers are used affords further advantage in that it also contributes to longer filter runs, or permits more rapid filtration, i.e., the tougher floc is better retained on the filter at higher pressure drops. Effluent clarity is often greatly enhanced, as well.

A note on alum and ferric hydroxide — Alum and ferric salts are usually referred to as coagulants, however, their mode of action is undoubtedly more complex. These materials form inorganic polymers in the pH ranges of greatest utility, so that some form of bridging may be operative. In addition, these materials are usually somewhat positively charged in the pH ranges of interest, so that sorption to the colloid surface with a degree of surface discharge of negatively charged colloids must occur. The presence of such effects is demonstrated by ZP measurements on alum and iron-coagulated systems.[7] Also, a highly effective, gelatinous "sweep floc" is formed in the initial stages of the reactions of these reagents with colloids.

2. Types of Polyelectrolytes

The term "polyelectrolyte" is something of a misnomer when applied to many organic, high-molecular-weight flocculants, since for many of these compounds, electrolyte properties are minimal under conditions of use. We will nonetheless follow this commonly used terminology.

Polyelectrolytes are most conveniently classified by their charge type, i.e., as either anionic, cationic, or nonionic materials. Other types of classification are also useful, for example, molecular weight, structure, and extent of electrical charge. These will be discussed below.

a. Anionic Polyelectrolytes

The charge on these polyelectrolytes is negative. A wide range of charge levels is possible, from very small to very large (q.v.).

The most common type of anionically charged group in these materials is the carboxylate moiety, $-COO^{(-)}$. In most contemporary anionic flocculants, a few percent of carboxylate groups are distributed along the polyelectrolyte chain. The structural repeating unit most commonly encountered among anionic polyelectrolytes is the acrylamide unit, or "mer," i.e.

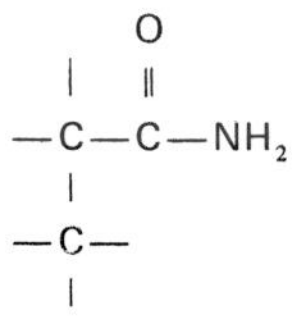

$$
\begin{array}{c}
O \\
\| \\
-C-C-NH_2 \\
| \\
-C- \\
|
\end{array}
$$

among which "mers" the carboxylate groups are interspersed, usually as a result of hydrolysis of some of the amide groups. The polyelectrolyte therefore has the schematic structure shown below, containing a rather random distribution of a relatively few carboxylate groups along the predominantly polyacrylamide chain.

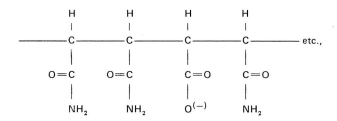

Moieties such as $-COO(^-)$ are relatively weakly ionized, i.e., only a fraction of the $-COO(^-)$ groups are actually in the carboxylate

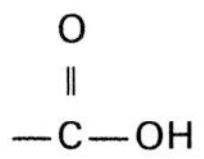

$$
\begin{array}{c}
O \\
\| \\
-C-OH
\end{array}
$$

form, and make a contribution to the molecular charge. This class of anionic polyelectrolytes is sometimes referred to as "weak" polyelectrolytes, in the same sense as the common carboxylic acids, such as acetic acid, are referred to as "weak acids" or "weak electrolytes". A close relationship also exists between these polyelectrolytes and the "weak acid" cation exchange resins of commerce.

Some other types of anionic polyelectrolytes are much more highly ionized than these polycarboxylates. A well-known example is polystyrene sulfonic acid:

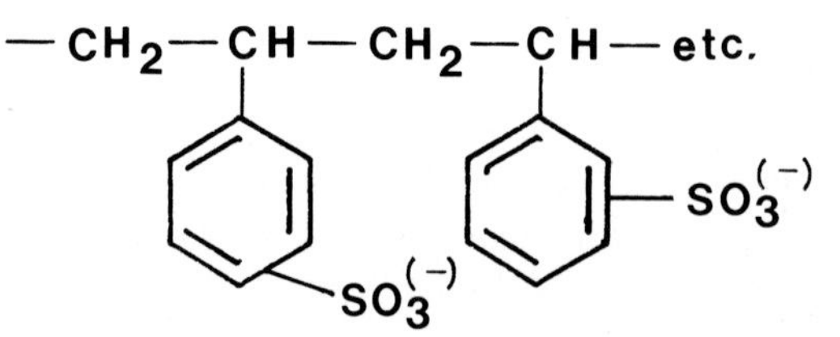

$$
-CH_2-CH-CH_2-CH-etc.
$$

These materials are extensively ionized down to very low pH levels. In analogy with sulfuric and sulfonic acids, as well as with the strong acid cation exchange resins which they closely resemble, these materials are often referred to as "strong" anionic polyelectrolytes. These strong polyelectrolytes therefore possess a high level of charge, which is, moreover, little influenced by the pH of the solution, in contradistinction to the "weak" anionics containing carboxylate groups. Intermediate types also occur, such as polyelectrolytes possessing phosphonic acid groups with acid strengths intermediate between the weak and strong anionics.

b. Nonionic Polyelectrolytes

The term "nonionic" polyelectrolyte is both a misnomer and, actually, a contradiction in terms, since electrolytes are by definition ionic. What is really meant here is that the charge level of the polyelectrolyte is very low, and generally has not been deliberately placed in the polymer by hydrolysis or copolymerization, but rather was created by (1) a very low extent of unavoidable hydrolysis, i.e., typically, in the attempt to prepare a pure or nonionic polyacrylamide, a small degree of perhaps 0.2 to 1% of hydrolysis of the amide groups to carboxylate is virtually unavoidable or (2) even where a true nonionic can be prepared, i.e., with polymers having no possibility of hydrolysis, the polymer will sorb a small quantity of negative charge in water* and so become slightly anionic.

The most typical examples of nonionic polyelectrolytes are polyacrylamides containing a small fraction, about 0.5 to 1% of carboxylate groups. As could be expected from such entities, the behavior of nonionics is often not far different from that of those anionics containing a still low but somewhat higher level of charge. There are some significant differences, however, especially in applications to filtration. The low charge level of the nonionics precludes any significant extension of the polymer chain due to mutual charge repulsion along that chain. The nonionics are therefore less extended in solution than are the more highly charged anionics. Also, the low charge on these materials may cause less interference in the collection of anionic colloids by the anionic (q.v.) filter media, i.e., one is less likely to overload the filter with negative charges, when collecting anionic particles, when a nonionic polyelectrolyte is used than when a more anionic polyelectrolyte is used.

c. Cationic Polyelectrolytes

These flocculants contain positively charged groups, such as amino ($-NH_3^+$), imino ($-CH_2-NH_2^+-CH_2-$), or quaternary amino ($-^+NR_4$). The first two are weak bases in the same sense as the carboxyl group is a weak acid, while the last named is a strong base, i.e., analogous to the strong acid sulfonate group. For the same reasons, the charge levels of the amino- and imino-containing materials are strongly dependent on pH, while the charge level of the quaternary ammonium group is relatively independent of pH. Again, various proportions of these functional, chargeable groups may be contained on different grades of polyelectrolytes.

A close analogy exists between these cationic polyelectrolytes and the weak and strong anion exchange resins of commerce.

Other cationic groups, such as complexes

$$(e.g., \quad -COOCa^+ \quad or \quad -CH_2-NH-CH_2-CH_2-NH-)$$
$$\diagdown_{Cu^{++}}$$

are also encountered. The complexes may be significant in certain biological and wastewater systems. These cationic complexes are rarely sold as such, but form in solutions containing polyelectrolytes and polyvalent cations such as Ca^{++} and Fe^{+++}.[8,9]

Cationics usually function by a combination of charge-neutralization and bridging mechanisms, i.e., these are not purely flocculating reagents but, as a result of their charge, actually tend to neutralize the colloid particles as well. There is evidence that many of the cationic polyelectrolytes used in filtration function to a large extent as charge-neutralizers, i.e., in large measure as coagulators as well as flocculators. Cer-

---

* Via the same mechanism which tends to impart a negative charge to most colloids and solids in water, i.e., negative ions in water are relatively less hydrated than positive ions, and so may be in part "squeezed out" of the aqueous phase and therefore onto solids and polymers.

tain very high-molecular-weight cationics that have recently come into use probably function primarily as flocculators, however.

The use of certain cationics, such as American Cyanamid Magnifloc 573C, has had a major impact on ultrafine filtration (see below). Protein polyelectrolytes, of which glues are typical, are usually of relatively low molecular weight (20 to 60,000), are somewhat hydrophobic (due to their hydrocarbon side chains), and in some cases form rather cationic complexes with certain metals:

$$
\begin{array}{c}
 M^{++} \\
R \quad COO^{(-)}\cdots\quad\cdots NH_2 \\
/ \qquad | \qquad\qquad | \\
-NH-CH \quad CH_2 \qquad\qquad | \\
| \qquad | \qquad\qquad (CH_2)_4 \\
CO \quad CH_2 \qquad\qquad \backslash \\
| \qquad \backslash \\
NH-C-CO-NH-CH-CO-NH- \\
| \\
H
\end{array}
$$

where R stands for a hydrocarbon residue such as CH_2, C_2H_5, C_6H_5, etc. These materials often have an isoelectric point (sum of positive charge−sum of negative charge = 0) at pH 4.6 to 5.0 in the absence of polyvalent metal ions, but in the presence of the latter they may form cationic metal complexes well into the alkaline pH range. The function of such cationics is in part to electrically discharge the negatively charged colloid particles, just as with other low-molecular-weight cationic polyelectrolytes such as polyethylene imine.

The hydrophobic groups on proteins and other polyelectrolytes such as copolymers of styrene and maleic acid (Lytron 820)* can surely contribute special properties to the corresponding polyelectrolytes, which special properties have not, as yet, been elucidated. One can speculate that such reagents might be of value in the filtration of oils and other hydrophobic materials.

Glues and other low-molecular-weight cationics have long been noted as removers of fine turbidity, of types that are often not responsive to high-molecular-weight anionics and nonionics. This fact might make them seem ideal for use in ultrafine filter applications. However, except when used on very finely divided media, such as diatomaceous earth, glues have not usually proved very successful as filtration aids. On the other hand, synthetic cationic polyelectrolytes such as the polyethylene imines are often extremely effective in granular filtration applications.

Typical polyelectrolyte flocculants used in filtration are shown in Table 1. Molecular weights and structures, etc. are indicated.

3. Influence of Charge and Molecular Weight on Polyelectrolyte Behavior in Filtration

In filtration, as well as in settling/flocculation, there is a rather clear-cut relationship between molecular weight and effectiveness, with the higher-molecular-weight polymers of a given type (among typical linear polyacrylamides of a given charge density) being the most effective[10-12] This is the rule for both anionic and nonionic polyacrylamides, and very likely holds for cationic polymers as well. In the latter case, the relationship between molecular weight and effectiveness is less clear because (1) a clear-cut homologous series of cationic polyelectrolytes, differing only in molecular weight, has not yet been studied in this connection, and (2) at the much higher dosage levels usually used with cationic polymers, other factors, such as pressure drop development,

---

*   See Table 1.

Table 1

TYPICAL POLYELECTROLYTES USED IN FILTRATION

Name	Structure	Charge type	Molecular weight	Manufacturer and location
Magnifloc 835A	Polyacrylamide $+CH_2-CH +_x$ $\quad$ $CONH_2$ (with some $-COO-$groups)	Anionic, high charge	15×10^6	American Cyanamid Co. Wayne, New Jersey
Magnifloc 820A	As above	Anionic intermediate charge	6×10^6	American Cyanamid Co. Wayne, New Jersey
Magnifloc 837A	As above	Anionic, low charge	15×10^6	American Cyanamid Co. Wayne, New Jersey
Purifloc A—21	Polystyrene sulfonic acid, $+CH_2-CH+_x$ $\quad \phi-SO_3^{(-)}$	Anionic very high charge		Dow Chemical Co. Midland, Michigan
Wisprofloc 20	Causticized potato starch, crosslinked with borate	Anionic	Very high	Gamlen Chemical Co. South San Francisco, California
Magnifloc 905N	Polyacrylamide, very few $-COO-$ groups	Nonionic	15×10^6	American Cyanamid Co. Wayne, New Jersey
Purifloc N—17	As above	Nonionic	Very high, approx. $5-10 \times 10^6$	Dow Chemical Co. Midland, Michigan
Betz 1260	Polyacrylamide copolymer	Cationic	Very high 10×10^6?	Betz Laboratories, Inc. Trevose, Pennsylvania
Catfloc T	Poly (diallyldimethyl ammonium chloride) (quarternary amine)	Cationic, very high charge	0.5×10^6?	Calgon Corp. Pittsburgh, Pennsylvania
Catfloc B	Amine?	Cationic, moderate charge		Calgon Corp. Pittsburgh, Pennsylvania

Table 1 (continued)
TYPICAL POLYELECTROLYTES USED IN FILTRATION

Name	Structure	Charge type	Molecular weight	Manufacturer and location
Magnifloc 507C	Tertiary amine	Cationic, moderate charge	10^4—10^5	American Cyanamid Co. Wayne, New Jersey
Magnifloc 573C	Quarternary amine	Cationic, very high charge	5×10^4—10^5	American Cyanamid Co. Wayne, New Jersey
Magnifloc 2535	Tertiary amine/ polyacrylamide copolymer	Cationic, moderate charge	1—3×10^6	American Cyanamid Co. Wayne, New Jersey
Primafloc C—7	Polyamine—HSO_4	Cationic, fairly high charge	1—2×10^6	Rohm and Haas Co. Philadelphia, Pennsylvania
Alcolac DV530	Polyethylene imine $[(CH_2-NH-CH_2)_x]$	Cationic	Approx. 5×10^4	Alcolac Inc. Baltimore, Maryland
Tydex 12	As above	Cationic	Approx. 6×10^4	Dow Chemical Co. Midland, Michigan

Note: While other equally effective polymers could have been substituted for those selected for this table, the availability of further information, such as molecular weight and structure, in many cases determined the polymers chosen here.

strongly influence the filter effectiveness of the higher-molecular-weight materials. It is evident, however, that with regard to cationic polyelectrolytes of a variety of types, intrinsic effectiveness, i.e., effectiveness as colloid collectors, is generally proportional to molecular weight.

This general relationship is a result of the greater extension into the liquid phase of the higher-molecular-weight materials, especially with regard to polyelectrolyte-aided filtration (but also in polymer-assisted settling). The larger number of binding hydrogen or other bonds (capable of being formed on filter media and colloids by the longer chains) makes for stronger binding to both filter media and colloids, and hence enhances both the collection mechanism and the resistance of the polymer to removal from the filter media at high flow rates. More colloids can also be collected by the longer chain. These, in essence, are the reasons why the higher the molecular weight of the polyelectrolyte,

1. The clearer the filter effluent (more colloids collected)
2. The longer the filter run (the polymer floc is more difficult to break off the filter)
3. The greater the amount of solids removed (see 1)
4. The greater the ability to operate at high trans-filter pressure drops and at high flow rates (bonding strength of the polymer to filter media and colloids is proportional to the number of bonds formed with the latter and hence to molecular weight).

In certain cases involving sedimentation/settling, the optimum in effectiveness is sometimes achieved at intermediate molecular weights. Schwoyer and Luttinger[13] have noted this in some work involving magnesium hydroxide sludges at high pH. The reason for this behavior is obscure.

In polyelectrolyte-assisted filtration, particularly when using finely divided media such as diatomaceous earth or even fine filter sand (e.s. 0.5 mm) and anthracite (e.s. 0.7 to 0.9 mm), lower-molecular-weight cationic polymers have often been preferred. This is largely a consequence of the lower pressure drop encountered with such polymers in comparison with materials of higher molecular weight, as mentioned above.

a. Effect of Polyelectrolyte Charge Density on Filtration

Polyelectrolyte charge density often exerts a profound effect on polyelectrolyte-assisted filtration. When using anionic polymers, cf. polyacrylamides, very low polyelectrolyte charge densities are usually preferred in "in-line" (q.v.) coagulation processes. The nonionic polymers of very high molecular weight are often preferred to any more highly charged types. The reason would appear to be that due to like-charge repulsion, the more anionic the polymer, the less firmly it would bond with filter media and colloidal particles, both of which are usually anionic, unless they have been "over" pretreated with an excess of alum or iron coagulants.

The effect of polymer charge density on the behavior of cationic polyelectrolytes has been investigated less. It would appear that low-charge-density, rather than high-charge-density polyelectrolytes, might be favored, in that the more highly charged materials would tend to convert both filter media and colloids to cationic particles, which would in turn tend to repel one another as well as repelling the cationic polyelectrolyte remaining in solution at higher levels of surface coverage by the polyelectrolyte. However, there is an important advantage in the use of high-charge quaternary cationics in that these materials are much more resistant to chlorine (often present in the system) than are other cationics such as low-charge-level amine and imine polyelectrolytes.

4. Polyelectrolytes and/or Alum, Iron Salts, Etc.

In filtration applications, as well as in settling, polyelectrolytes are often used together with, as well as in place of, hydrous oxides such as alum and iron salts. Commonly, the addition of relatively very small quantities of polyelectrolytes to such systems will result in a very pronounced reduction in the alum/iron requirements, i.e., in a system which would otherwise require the addition of 10 ppm alum for satisfactory treatment, the addition of, say, 0.5 ppm polyelectrolyte may reduce the alum requirement to perhaps 1 ppm. Often, all of the hydrous salt may be replaced. Considerably smaller sludge volumes (resulting in smaller sludge disposal problems, and longer filter runs) can therefore be achieved through replacement of some or all of the alum/iron with polyelectrolyte. Very importantly in waste applications, polyelectrolyte addition also frequently results in a substantial reduction or virtual elimination of surge upsets in filtration, i.e., the often drastic changes in wastewater flow rates to filters (which in the past have so frequently resulted in filter breakthroughs) are many times obviated by the employment of polyelectrolytes. This effect probably reflects the aforementioned stronger floc binding to the filter media due to the polyelectrolyte. This effect will be further discussed below.

It is interesting that nonionic and anionic polyelectrolytes, which function in most cases by a purely bridging (flocculation) mechanism can replace the hydrous alum/iron salts which function at least in part by a coagulation/charge neutralization mechanism. There do exist, however, qualitative and quantitative differences between the behavior of these polyelectrolytes and hydrous salts, as indicated above. In Section VII following, the mode of action of these various materials will be further discussed.

It must not be supposed that the substitution of organic polyelectrolytes for alum/iron is not without disadvantages. In some applications, a major disadvantage of polyelectrolyte use is cost, the greater cost per pound of the organic materials in some cases offsetting their generally greater effectiveness and lower dosage requirements. Another problem is the effort sometimes required to find the most effective polyelectrolyte. Unlike the hydrous salts, of which there are very few varieties (practically speaking, about four), hundreds of different polyelectrolytes are available from various manufacturers. The manufacturers can often be helpful in a general way, in assisting the user to determine which of the various materials in his product line are most apt to be beneficial (and often at roughly what dosage levels); however, the user himself must determine which polyelectrolyte from all the various product lines is best for a particular application. In waste treatment, "best" is usually interpreted as "most cost effective".

Other potential drawbacks include the following: (1) while a very large number of polyelectrolytes have EPA approval for potable water use, some do not. This can create problems, not only in potable water treatment, but where waste filter effluents must be discharged to a stream or lake or even to ground wells, although the polyelectrolytes themselves are nontoxic materials, and the quantity of polyelectrolyte passing a properly operated filter is extremely minute, (2) more rapid headloss development, and (3) greater difficulty in backwashing when polyelectrolytes are used.

Both (2) and (3) are problems not unique to polyelectrolyte usage in filters, but are often encountered with iron and alum usage as well. Means are available to drastically reduce such problems. These include: avoidance of overfeeding the polymer (overfeeding involves a waste of money in any case), the use of larger media (especially "top" media, i.e., anthracite in a sand/anthracite filter), and the use of auxiliary backwashing devices, such as air scour, rotary surface cleaning, etc. to assist in backwashing. Of these, the use of air scour is probably the most effective.[14,15]

II. METHODS OF USING POLYELECTROLYTES IN FILTRATION

A. Conventional Methods

By "conventional" methods, we refer to those means of polyelectrolyte addition to a filter influent which are not specifically designed along the lines to be discussed in Sections II.B and II.C. Such methods include incidental or even accidental addition of polyelectrolyte to the filter, resulting from (1) overfeeding of the polyelectrolyte to equipment upstream of the filter such as a clarifier or settling basin, (2) the normal filtration of those polyelectrolyte-treated solids that escape from a clarifier sludge blanket and reach the filter, or (3) the more or less haphazard addition of polyelectrolyte before a filter, without particular attention being given to point or method of addition, as set forth below.

B. Direct Filtration

Direct filtration is a rather specialized technique which was developed as a means of economizing in water and wastewater treatment. In direct filtration, one or more steps upstream of the filter is eliminated. Most commonly, a settling basin is eliminated, but in some cases a clarifier may be eliminated. A larger load of suspended solids must therefore be handled by the filter. Whereas, up until the recent past, a conventional filter was normally considered overloaded if suspended solids loadings above approximately 10 ppm were continuously fed to it, in direct filtration, suspended solids loadings of well over 100 ppm are often handled, particularly in waste applications.

To handle such heavy loads, the filters must be suitably modified. To achieve filter runs of reasonable length, larger media, deeper beds, and tough floc are required. A flash mix step (q.v.) of suitable magnitude and duration is highly desirable. Addition of coagulants/flocculants such as alum, iron salts, and/or polyelectrolytes is generally made before the flash mix. The flash mix often results in the formation of a fine, strong pinpoint floc which penetrates deeply into the filter, thus minimizing the rapid headloss development which often results when larger, conventional floc (such as that produced in a sludge blanket clarifier) reaches a filter. Larger media in the upper part of the bed also minimize headloss and permit longer filter runs. Filtration at rates of the order of 10 gal/min/ft² (6.8 ℓ/m² sec) are attainable by such means.

Following European practice, such filters may be constructed with very deep (7 to 9 ft or 2.3 to 3 m) filter beds to give greater floc storage capacity, i.e., longer runs and greater insurance against filter breakthrough. In such deep filters operating on waste streams, advanced backwashing techniques such as air scour are highly desirable. However, the use of air scour requires that special attention be paid to the design and installation of the backwash system, to selection of media and media size and relative size, to the underdrain arrangement, etc. In general, the design of such filters appears to be reasonably well understood by a number of filter manufacturers.

C. In-line Coagulation

The technique of in-line coagulation resembles that of direct filtration, however, the key step in in-line coagulation is the addition of polyelectrolyte directly ahead of the filter. Flash mix is generally avoided here, and therefore the energy considerations attendant with the latter (q.v.) are not relevant. While flash mix is avoided, thorough mixing of the polyelectrolyte with the filter influent stream is very important. Most often, polyelectrolyte is added just a few feet ahead of the filter. Polyelectrolyte injection is to the center of the feed stream. A dependable positive displacement feed pump is used. In filters of only moderate size, mixing is accomplished via mixing elbows, or with the aid of an in-line mixer such as the Kenics mixer.

The term in-line coagulation is really a misnomer. Polyelectrolytes, rather than alum or iron salts, are most frequently employed, so that in-line flocculation would be a better term. Small quantities of alum or iron salts may also be added, usually upstream of the polyelectrolyte.

In-line coagulation is presently employed more in water than in waste treatment. Filter media of conventional size and depth are most often used. The aim is usually to produce a very high quality effluent (very low turbidity) at rather high filter rates. Some of the advantages (and disadvantages) which apply to direct filtration also apply to the in-line mode. The advantages (besides high filter rates and high effluent quality) include long filter runs and protection from effects of rate surges on filter effluent quality. Disadvantages include the need for somewhat larger quantities of backwash water and more effective backwash techniques, such as air scour with the attendant requirements.

Polyelectrolyte dosages in in-line systems may be quite small, from a few parts per billion when nonionic and anionic, very high-molecular-weight polyelectrolytes are used on low suspended solids influents to a few parts per million when cationic polymers are used and filter effluent turbidities well below 0.1 FTU are required.[16]

Although in-line coagulation and direct filtration are formally quite different, a variety of what amounts to intermediate techniques have been described (see Section III.A). Applegate[17] has noted that when the filter influent suspended solids are very low, longer contact times (before the filter) are required for reaction between water colloids and polyelectrolytes. The use of two or three different polyelectrolytes fed at different points upstream of the filter has also been described.[18]

D. Polyelectrolyte Addition to Filter Backwash

Polyelectrolyte has been added to the filter backwash as an alternative to the various modes of polyelectrolyte addition ahead of the filter. The pioneering work in this area is apparently due to Harris[19,20] and has been confirmed by others.[11,21] In this mode of treatment, polyelectrolyte is added for a short period of time to the backwash water. Polyelectrolyte addition is usually at a 0.1 ppm level. Very high-molecular-weight polyelectrolytes, most commonly nonionics, are used. Filter rates, in potable water treatment, of 10 gal/min/ft² (6.8 ℓ/m² sec) can be obtained. The filter effluent is of very high quality, and the filter effluent quality is resistant to rate surge effects. Filter runs are very long. Most remarkably, the filter effluent attains an extremely high quality almost immediately upon the start of a run.

The polyelectrolyte is added to the filter as follows. The bed is first greatly expanded, then lowered slightly from this level. Addition of polyelectrolyte and also of air to the backwash occurs after this slight lowering. Total lift (air plus water) may be 50+ in. Polyelectrolyte addition is relatively brief, but the air lift is maintained to the end of the wash period.

While confirming these results of Harris, Luttinger and Beach[21] successfully extended this work to the treatment of high suspended solids influents, corresponding to certain wastewaters in suspended solids level. We found further that combinations of polyelectrolytes, i.e., a cationic plus a nonionic or anionic, often gave better results (longer runs of good effluent clarity) than did single nonionic or anionic polymers. Best results were obtained when a nonionic polyelectrolyte was added after a cationic (low-molecular-weight) polymer. The reverse order of addition did not provide results as good. We believe this may be due to the higher-molecular-weight nonionic polymer being deposited on top of the low-molecular-weight cationic, and therefore "holding (the cationic) down" in filter service operation. Very high-molecular-weight cationics also gave good results when added in the backwash mode.

No particular problems with backwashing or high headloss development were observed in this work, in spite of the conventionally small sized dual and tri media used. This may reflect relatively uniform coating of the entire filter bed by the polymer(s) when they are added in the backwash mode.

III. APPLICATIONS

A. Potable Water

Potable water treatment presently constitutes by far the largest area of application of polyelectrolytes to filtration. In addition, high rate and direct filtration, as well as filtration theory, etc., were developed largely with references to potable water treatment.

Regulatory emphasis today is on the production of ever purer filter effluents. In the near future, 0.1 JTU will be the accepted standard. Polyelectrolyte treatment combined possibly with the use of more effective media and/or further pre- and post-filtration treatment will probably be necessary to meet these requirements. At the same time, economic factors are pushing in the direction of simplifications in water treatment practices such as the use of higher filtration rates and direct filtration. Experience indicates that no deterioration of effluent quality need result from the use of a properly designed and operated direct filtration plant in place of full (i.e., with retention of a sedimentation step) treatment. Proper mixing is crucial for both direct filtration and in-line coagulation. This subject is fairly well understood with respect to direct filtration due to the work of Camp and Kreske,[22] and others.[23,12]

DE as well as granular filter potable water plants can be expected to make much greater utilization of polyelectrolytes to obtain better effluents. A very large number of polyelectrolytes have obtained U.S. Public Health Service approval for potable water use. A brief review of the role of polyelectrolyte in filtration has recently been presented by Shuster and Wang.[24]

The role of polyelectrolytes in filtration has also been discussed by O'Melia.[25] Conclusions, as to mechanisms of polyelectrolyte action in filtration and results that may be obtained using polyelectrolytes, are discussed. There is evidence here confirming the earlier work of Luttinger[11] and Luttinger and Beach[21] on the efficacy of cationic plus anionic polymer addition to a filter. This and other reports on polyelectrolyte addition to filters are very limited as to the effects of polymer structure on filtration. This writer believes polymer structure to be of great importance in filtration.

Loganathan and Maier[26] have discussed the effects of pH, ionic strength, and polyelectrolyte dosage on sand filtration. The effects of cationic, anionic, and nonionic polymers were also assessed. It was shown in this work that enhancement of filter efficiency by nonionic and anionic polyelectrolytes was greatly improved by the presence of divalent cations. The effect of such ions was much smaller when cationic polyelectrolytes were used.

Eunpu[27] found that in cold weather operation, in-line coagulation with polyelectrolytes improved filtration. This improvement may reflect the floc-strengthening effects of these polymers. Ives (quoted by Norris[28]) noted that in-line coagulation using polyelectrolytes constitutes one of the major areas for improvement in filtration. Conley[29] discussed the use of polyelectrolytes at the Hanford, Washington water plant. Robeck and associates[30] have mentioned the use of cationic polyelectrolytes in high rate filtration of Little Miami River water.

Conley and Pitman[31] discussed high rate (up to 10 gal/min/ft²) filtration of Columbia River water, using 5 to 50 ppb of polyelectrolyte. Conley[32] has also referred to the use of polyelectrolytes as floc-strengthening agents in high rate filtration. Rice[33] also discusses the use of polyelectrolytes in high rate, potable water filtration as does Riddle.[34]

Camp and Kreske[22] reported the use of high speed mixing to condition floc for filtration and to eliminate the sedimentation step. Proper treatment is said to reduce the size of the flocculation unit and to cut chemical costs, while giving faster throughput. When the mean velocity gradient* is in the range 150 to 1000 sec⁻¹, flocculation can be completed in a very short time in small flocculation units and floc particle size maintained small enough to permit direct filtration (with deep floc penetration into the filter) of highly colored and turbid waters without prematurely clogging the filter.

Hutchison and Foley[35] have presented operational and experimental results of direct filtration as studied in several large Canadian water plants. In their experience with direct filtration, these workers found that the filters tended to clog with diatoms, shortening filter runs. The diatom problem was overcome in several ways. This is a valuable contribution to the direct filtration literature, containing careful, exhaustive work carried out in a number of large plants over long periods of time. The authors conclude:

1. Further research into floc-volume reduction and floc-strengthening aids is required (these latter would usually be polyelectrolytes).
2. For successful operation at rates above 6 gal/min/ft², the quantity and size of the floc must be reduced (probably with the help of polyelectrolytes).
3. The pretreatment aspect of direct filtration is just as important as with systems using sedimentation.
4. Flocculation (velocity) gradient and flocculation time are important determinates of filter performance.
5. The diatom filter-clogging problem can be overcome by the use of coarser coal in the dual media filters; however, use of coarser media calls for the exercise of greater operational care (floc breakthrough).

In a more recent paper, Hutchison[36] reports further on studies of high rate direct filtration in Ontario. Variables include raw water turbidity, diatom concentration, and mixing. Results indicate that direct filtration can be a viable alternative to conventional sedimentation-filtration when the coagulant dosage is below 15 ppm and the diatom concentration below 1000 ASU/mℓ.

Pilot plant work showed that direct filtration was viable at raw water turbidities up to 175 FTU if nonionic polymers were used as filter aids. However, both underdosing and overdosing must be avoided. An influent of 150 FTU could be treated to give a filter effluent of 1 FTU or less.

The pre-filter treatment here contained in-line as well as direct filtration features such as mixing at a G value of about 20 sec⁻¹ for about 4.5 min.

Anthracite of effective sizes of 1.4 to 2.0 mm could be successfully used with polyelectrolyte. The polymer demand for the prevention of breakthrough increased as the (coal) filter media size increased. Nonionic polymers used in-line (just ahead of the filters) prevented breakthrough on all media up to 1.5 mm e.s. under all conditions.

Filtration rates as high as 7.2 gal/min/ft² (4.9 ℓ/m² sec) did not affect the filter

---

* G value = "velocity gradient" =

$$\sqrt{\frac{2\pi sT}{\mu V}}$$

where s = rotor speed, rps, T = mixing torque of the rotor, μ = viscosity, and V = fluid volume (in chamber) effluent quality. Higher rates than 7.2 gal/min/ft² may be practical in "direct" filtration if means of floc strengthening and floc volume reduction (i.e., by polymer addition or by use of different mixing speeds) can be achieved. Dual media filters were employed here.

At present, seven direct filtration plants exist in Ontario and six others are under construction. Tredgett[37] has presented a related discussion, based on direct filtration studies.

Spink and Monscvitz[38] have reported on direct filtration at the 200-million gal/day water treatment plant at Lake Mead. As at other direct filtration installations, filtration rate is high, up to 5 gal/min/ft² (3.4 ℓ/m² sec). Alum and sometimes polyelectrolytes are fed. In addition, the filter is also treated in-line with polyelectrolyte so that this plant also utilizes a combination of direct filtration and in-line coagulation (Hutchison[36]). Zeta potential measurements are used to determine chemical dosages. Filter runs are of the order of 50 hr. Spink and Monscvitz in a later paper[39] further discuss the design and operation of the same plant.

Adin and Rebhun[40] have also discussed what they refer to as "direct high rate contact flocculation filtration". The better economics of such systems for low turbidity waters are stressed. Successful direct high rate filtration is said to depend on the use of large diameter filter media to decrease headloss, and the use of in-line coagulation to bring about strong attachment of the suspended particles to the filter and to fully utilize the storage capacity of the bed. The name "contact flocculation-filtration" results from this feature.

Adin and Rebhun found that cationic polymers were superior to alum in such applications. The system they discuss is again a crossbreed between direct filtration and in-line coagulation, in that no special mixing of flocculant and influent water is provided for, except what occurs in the feed pipe to the filter (as in in-line coagulation); yet the contact time of the flocculant with the influent lasts for a few minutes (closer to direct filtration than to in-line coagulation in this regard, due to the position of the flocculant feed point).

In this experimental study, flow rates up to 7 gal/min/ft² (5 ℓ/m² sec) were achieved with the aid of very small quantities of strongly cationic polyelectrolytes.

This and some of the papers mentioned earlier suggest that in a practical sense the distinction between in-line coagulation and direct filtration is less clear than is usually assumed, and that possibly, in some cases, the exact point of addition or mixing time of the polymer may not be so critical. If this is so, it would introduce a comfortable latitude into practical operations which, it was believed, was not present.

Robinson et al.[41] have discussed direct filtration of Lake Superior waters for asbestiform-solids removal. Tests performed using media filtration and diatomaceous earth filtration were compared. Alum, ferric chloride, and polyelectrolytes were used with the media filters. The authors recommended that provisions be made to feed polyelectrolyte to the flash mixers, to the flocculation chambers, and to the filters, or to any combination of these. Once again, here is a case where the distinction between in-line and direct filtration tends to be blurred.

The authors' point is well taken. In practice, the ability to add polymer at several points should provide better overall control of any "direct" filtration process.

Harbert[42] has also discussed direct filtration (he refers to the process as in-line filtration). In the work reported by Harbert, cationic polyelectrolyte was added at a variety of points covering the range of direct filtration to in-line coagulation. It was found that the direct filtration mode of polymer addition gave superior results, but this may well have been due to the admittedly poor mixing experienced when the in-line mode was used.

Ainsworth and Mosley[43] of the Santa Barbara (California) Water Department have discussed the great advantage they observed in going to direct filtration utilizing 1 ppb cationic polymer per 3 ppm alum in the dosage range of a few ppm alum. The small quantity of polymer permitted an enormous reduction in the alum requirement. The filters of this large installation are dual media.

Brailey and associates[23] have presented a detailed paper on the effectiveness of direct filtration in treating New York City metropolitan area waters. These authors point out that for these surface waters, alum is required as a coagulant. Polymers are useful, however, in reducing the alum dosage requirement. Because of the alum requirement and the requirement for approximately 15-hr filter runs, only relatively low filter rates (2 to 3 gal/min/ft²) are possible in direct filtration of such waters. The authors also call attention to the fact that the mixing and flocculation steps conventionally used in prefiltration treatment actually interfere with direct filtration, by causing large size floc to blind the filter surface and hence prevent indepth penetration by smaller floc. A high energy mix, which tends to produce small, tough floc particles, is more compatible with direct filtration. According to these authors, in-line coagulation using polymers only is unreliable in the treatment of these waters. It was found in this study that direct filtration produced water quality as good as conventional treatment. Alum requirements were similar.

Letterman and Logsdon[44] have presented an excellent survey of direct filtration practice. This survey consists of information from about 15 questionnaires filed by direct filtration plant operators. Five of these plants reported alum addition at the rapid mix step, six added alum plus polymer, three added only cationic polymer. Seven plants utilized in-line coagulation. There was clear indication that the use of polymer greatly increased the quantity of water needed per backwash. Twelve of fifteen plants averaged filter runs of 24 hr or more.

Influent turbidities and other parameters are also reported. One trend noted was the use of a short (less than 1 min) flash mix time. Flocculation, when used, was accomplished in a series of steps rather than in one step. Raw waters with turbidities up to 300 "TU" could be treated to give filter effluents of 1 "TU" or less.

Stump[12] has recently discussed selection of polyelectrolytes for direct filtration.

The purpose of this study was to investigate methods to determine an optimal polymer for specific water treatment applications and to propose a laboratory procedure to help make this selection. A synthetic water with an initial turbidity of 80 FTU and a suspended solids concentration of 100 ppm was used for all tests. The effects of rapid mix time and intensity along with flocculation were studied.

The results of this investigation are summarized as follows:

1. Polyelectrolyte molecular weight and charge appear to be the most dominant characteristics to indicate suitability of a polymer for use in direct filtration.
2. Anionic polymers provided unsatisfactory results and in most cases should not be used in direct filtration applications.
3. Cationic polymers in the 20,000 to 100,000 molecular weight range provided the best performance.
4. High-molecular-weight cationic polymers generate high filter headlosses along with excellent turbidity removal, while low-molecular-weight polymers generate low headloss with less effective turbidity removal.
5. Flocculation time makes a significant difference in polymer performance characteristics.
6. Rapid mix time has minor importance while energy input makes a significant difference.

7. Energy input was critical in determining the performance of a polymer. The optimal G value for most polymers was 300 to 600 sec^{-1}.

While Stump's work is important, a number of unsatisfactory features are present. These include: (1) apparent failure to take into account variation of optimum dosage with molecular weight for anionic polymers (anionics are usually more effective at lower dosages than cationics), (2) failure to consider charge characteristics of cationics (strong base vs. weak base) as well as of anionics, and (3) failure to vary the filter media size.

Thompson[45] has described the new direct filtration plant at Springfield, Mass. The direct filtration feature resulted in a savings of $2.5 million. The plant has been operating using a cationic polyelectrolyte at a dosage of about 1.4 to 1.5 ppm at filtration rates on the order of 2 to 3 gal/min/ft^2. The filters have been routinely operated for one week between backwashes. No filter has yet shown excessive headloss or breakthorugh of turbidity and all continue to deliver water of excellent quality.

The long filter runs observed here are noteworthy, but have been observed elsewhere when polyelectrolytes have been used on low-to-moderate turbidity influents being filtered at moderate rates.

Another important contribution in the field of direct filtration was recently made by Letterman et al.[46] who studied direct filtration of high turbidity (32 JTU) water using dual media filtration. Turbidity was derived from clay. Flocculation was affected by addition of cationic polyelectrolyte, but was not in-line, i.e., the polyelectrolyte was added well ahead of, rather than at, the filter. Mixing intensity was systematically varied in the flocculation step. It was found that direct filtration using cationic polyelectrolytes and dual media filters could be used to effectively treat high turbidity water and that effective operation could be achieved by pretreatment control. The operation of a dual media filter using a cationic polyelectrolyte as the sole coagulant is characterized by the formation of a distinct working layer or clogging front which moves down through the bed at a rate which is a function of the filter media size distribution and the pretreatment and filter operating conditions. The rate of clogging front advancement is increased by:

1. Increasing or decreasing the polyelectrolyte concentration above or below a concentration which results in a colloid particle zeta potential of approximately zero
2. Increasing the filtration rate
3. Increasing the pretreatment mixing intensity and to a certain extent
4. Increasing the effective size of the anthracite media

Adin and Rebhun[47] have presented a basic, unconventional model for evaluation and prediction of performance of deep bed filters in high rate in-line coagulation (here called contact flocculation). This work should prove valuable for the evaluation of polyelectrolytes in filtration, particularly in in-line coagulation.

Geise and associates[48] have discussed the role of polyelectrolytes added in-line. Alum was also fed. The filters are dual media.

B. Wastes
1. Industrial Wastes
The use of polyelectrolytes in the filtration of industrial wastes has not advanced to the same extent as their usage in potable water applications; however, use in waste filtration is growing, and part of the apparent lack of wide-scale usage here simply

reflects the generally less complete reporting of waste filtration as compared with potable water filtration. Also, filter manufacturers often claim that "no chemicals are needed with this filter" (a desirable feature, if true). However, the lenient discharge regulations of the past are disappearing, and the use of polyelectrolyte will usually more than pay for itself in the more stringent regulatory climate of today.

Recent developments in the area of industrial waste filtration, particularly with regard to steel mill wastes, often call for the use of very deep, (4 to 8 ft) filter beds composed of coarse, usually dual media (up to 3 to 5 mm), and operating at very high rates (8 to 40 gal/min/ft²). Often, there is provision for air scour, to clean the entire depth of such beds.

Yao and coworkers[49] have discussed the use of cationic polyelectrolytes in waste filtration, and the relation between flocculation and filtration using polyelectrolytes.

a. Steel Mill Wastes

Donovan[50] has discussed the high rate filtration of industrial wastes. He conducted lab scale tests on hot steel rolling mill discharge water using filter tubes containing 90 in. of sand, anthracite, or mixtures. Headloss, effluent quality, filtration rate, etc. were studied and various polyelectrolytes were evaluated. A rough cost estimate was included.

High rate filtration of steel mill wastes was also discussed in another article by Donovan.[51] Some interesting results obtained with polyelectrolytes at filtration rates up to 16 gal/min/ft¹ (11 ℓ/m² sec) are presented. Polycationics performed the best of the various polyelectrolyte types.

The role of metal hydroxide floc (iron and alum) in filtration is discussed by Stevenson.[52] The effects of shear, concentration, and time on floc growth rate are discussed. The role of polyelectrolytes is also discussed in general terms.

The work of Hutchison[36] referred to Section III.A, concerns the place of direct filtration in the treatment of highly turbid potable water. Such water may approximate certain wastewaters. Direct filtration could be useful here.

2. Sewage Treatment

There have been relatively few published reports on in-line coagulation in sewage filtration. More often, polyelectrolytes are added to the clarifier.

Tschobanoglous[53] has shown that the use of polyelectrolytes can increase suspended solids removal in the filtration of secondary effluents. The distribution of suspended solids removed by a filter could be controlled by varying the quantity and point of addition of the polyelectrolyte. Synthetic polyelectrolytes of high molecular weight have been found to be more effective than natural polyelectrolytes, and are partly responsible for the renewed interest in tertiary filtration.

Bell et al.[54] discuss the use of anionic polyelectrolytes plus alum for filtration of phosphorus from municipal wastewater. Bernhardt et al.[55] have discussed phosphate and turbidity removal from a water supply via polyelectrolyte-aided filtration. Lee[56] showed that sewage mixed with related wastes could be successfully treated in a secondary-tertiary system, via alum and polyelectrolyte addition followed by flocculation, sedimentation, and granular media filtration. Culp and Hansen[57] have discussed polyelectrolyte-assisted filtration of sewage effluent.

O'Melia[25] has presented an indepth discussion of the application of polyelectrolytes to the filtration of sewage and related effluents. Results are presented on filtration of trickling filter effluent, on calcium phosphate suspensions, and on suspensions of latex. Conventional treatment and direct filtration are compared, and the use of cationic and anionic polyelectrolytes (the latter in the absence of, and in conjunction with, alum) are compared. Habibian and O'Melia[58] have also discussed the use of cationic polyelectrolytes in the filtration of trickling filter effluent.

Polyelectrolytes are used extensively in the filtration of sludges such as sewage sludge (see Chapter 6). Shireman[59] has discussed the use of all three different types of polyelectrolytes, as well as alum, etc., in the mixed-media filtration (tertiary treatment) of sewage. Operation of such systems in a large-scale plant is described. Excellent results, e.g., effluent turbidities as low as 0.05 JTU, are claimed. An earlier report on polyelectrolyte-aided filtration at this plant was presented by McDonald.[60]

Kavanaugh et al.[61] showed that the use of nonionic polyelectrolytes afforded a sixfold solids removal via filtration of effluent from an activated sludge system. Other polyelectrolyte types were less effective. Polyelectrolyte feed was added, via rapid mix, ahead of the filter. This mode of treatment for phosphorus removal is considered by the authors as an option to conventional coagulation/sedimentation or flotation.

Goodman and Mikkelson[62] have shown that phosphorus and suspended solids removal are significantly enhanced when polyelectrolyte is added to the primary clarifier effluent, and this treatment is followed by dualmedia filtration. Rice and Conley[63] have patented a sewage treatment process in which a filter conditioner (polyelectrolyte) is added more or less in-line to each of two similar dual media filters in series. The final filter effluent is passed through an activated carbon bed.

Related to filtration of sewage effluent is the filtration of a number of wastes which are biochemically rather similar to sewage. One such waste is poultry process water. Sand filtration and activated carbon treatment of poultry process water is discussed by Berry et al.[64] The effluent was pretreated via alum and an anionic polyelectrolyte.

Another related application is filtration of combined sewer overflows. Nebolsine et al.[65] used a deep bed of very large media (4 mm e.s. anthracite over 2 mm e.s. sand) for the filtration of such overflows. High-molecular-weight anionic polyelectrolytes fed more or less in-line greatly increased solids removals by these filters.

As in other applications, judicious selection of polyelectrolytes in sewage filtration will afford clearer effluents, permit higher filter rates (and therefore more economical operation), protect against flow rate surges, and permit longer filter runs.

C. Miscellaneous
1. Filtration as Pretreatment for Reverse Osmosis

Reverse osmosis (RO) is at present the most desirable means of desalination of brackish water and seawater. The purified water produced is used as potable water or since it is a pure 95 to 99% deionized product, it is desirable as make-up and wash water in the electronics, photographic and other industries, and elsewhere.

In addition, certain waste streams can be treated by RO to permit the recovery of up to 90% of the stream as pure water, fit for discharge or reuse, plus a waste concentrate stream which may be further treated to recover individual metals such as copper, zinc, or chrome.

Some of the most effective RO devices are modules consisting, in essence, of many thousands of fine hollow fibers. The latter constitute the RO membranes. Influent water is pressurized against the outside of these fibers. The purified product water passes through the fiber wall and is collected via headers.

While such hollow fiber RO devices are in most respects extremely efficient as to space requirements (the large membrane surface area which can be packed into a small module volume), they are restricted to feedwaters of very low turbidities, of the order of 0.05 to 0.07 FTU or better.* To attain this quality, most waters, exclusive of some well waters, require some type of pretreatment. Of the various types of pretreatment

---

* A question exists as to the exact requirements regarding turbidity here. A far more reliable indicator of required influent quality for hollow fiber RO is the Silt Density Index (SDI). See Reference 16.

available, granular media and DE filtration are by far the most economical. The most effective means of conditioning such filters to obtain filter effluents of the requisite quality for hollow fiber RO is by use of polyelectrolytes, fed in-line.[16,17]

In most instances, cationic polyelectrolytes, possessing rather high charge densities and molecular weights in the range 50,000 to 100,000 are used in such applications, at dosages of the order of 1 to 3 ppm neat. Thorough mixing of the polyelectrolyte with the filter influent is required. The point of addition of the polyelectrolyte may be just ahead of the filter or, in some cases (very low influent solids), up to 30 ft upstream. Conventional dual and mixed media beds, as well as certain single media such as Filter AG or in some cases fine garnet, are employed in such filters. Good filter rates, of the order of 2 to 4 gal/min/ft² (1.4 to 2.8 ℓ/m² sec), as well as ultra-high quality effluents, are produced by such means.

2. Production of Other Ultra-High Quality Filter Effluents

The most practical means of producing large quantities of ultra-high quality liquids is by use of suitably designed filters employing polyelectrolytes, mainly in the in-line or direct filtration modes. Geise et al.[48] have used dual media filters with in-line coagulation to produce effluents of 0.04 to 0.08 JTU. Terrell[18] used a combination of polymers (mainly cationics of molecular weight 50,000 to 100,000) to produce 0.02 to 0.06 NTU effluents from a conventional filtration plant. Letterman et al.[46] have obtained similar results utilizing direct filtration and a strongly cationic polyelectrolyte of relatively high molecular weight. A high turbidity (32 JTU) water gave a filter effluent of 0.05 FTU by this technique, using dual media filtration at the high filter rate of 5 gal/min/ft² (3.4 ℓ/m²/sec).

Tate et al.[66] in pilot testing on coagulated water showed that direct filtration could produce an effluent of 0.04 to 0.15 TU from an influent of approximately 1.8 TU. Alum (0.3 ppm) plus a cationic polyelectrolyte, Catfloc T (0.25 ppm) were the chemicals used. The mixed and partially flocculated water was fed to dual media filters.

Bertsche[67] has discussed the operation of a large dual media (sand/anthracite) filter at rates up to 9.4 gal/min/ft² to produce an average effluent turbidity to 0.03 JTU. 0.2 ppm of a cationic polyelectrolyte was used here.

Spink and Monscvitz[39] have described a 200-million gal/day direct filtration plant producing an effluent with turbidity below 0.1 JTU (average less than 0.07 JTU). The raw water turbidities are below 1 JTU. Conventional dual media filters are used here. Alum and polyelectrolyte, as needed, are fed to the filters in the in-line mode.

In their study of asbestos fiber removal from Lake Superior water, Robinson et al.[41] found that effluent turbidities below 0.1 FTU could be regularly achieved on this low turbidity influent water. The most effective chemicals used were alum plus very high-molecular-weight anionic or nonionic polymers.

Other work in which very high-molecular-weight nonionic polyelectrolytes were used to produce filter effluents of low JTU has been reported by Kirchner and Joines[68] and by Conley and Pitman.[31] The latter workers consistently obtained effluents of 0.04 to 0.08 JTU and sometimes better. High-molecular-weight polyacrylamide was fed, more or less in-line (the filters were preceded by a conventional alum coagulation step).

Adin and Rebhun[40] in their study of in-line coagulation at very high filter rates obtained striking turbidity removals, often greater than 95%, combined with long filter runs. They also demonstrated the superiority of a quaternary cationic polyelectrolyte over alum, and the advantages of employing large media when polyelectrolytes are used in this mode at high filtration rates. The optimum polyelectrolyte dosage used (0.05 ppm) is low for cationic polyelectrolytes used alone.

At the direct filtration plants discussed by Ainsworth and Mosley[43] an effluent averaging below 0.08 JTU is produced under high flow rate conditions. Chemical treatment consists of 1 ppb cationic polyelectrolyte per ppm alum. Harris[19] reported that the addition of polyelectrolyte via backwash produced a filter effluent of excellent quality (<0.1 JTU) and at very high rates.

a. Ultra-High Quality Filtration Using Diatomaceous Earth (DE)

DE filtration is also able to produce ultra-high quality effluents. Thus, Baumann[69] used alum-coated DE as a filter precoat and another grade of DE as body feed together with ~0.4 ppm Cat-Floc B (cationic polyelectrolyte) added in-line to produce an effluent of 0.05 to 0.06 FTU. There can be little doubt, therefore, that polyelectrolyte-assisted DE filters can produce effluents of the same quality as the best dual and multimedia filters.

3. Ultra-High Rate Filtration

Filter rates over approximately 9 gal/min/ft² are often referred to as ultra-high rates. In this chapter, ultra-high rate filtration refers to rates over 9 gal/min/ft² (6.1 ℓ/m²/sec).

It has been established that much higher filter rates than heretofore used are compatible with long filter runs and low turbidity effluents, suitable for potable water. For example, dual and multimedia filters may, in some cases, be operated at rates up to 10 gal/min/ft² and produce fully satisfactory effluent clarity, even though effluent standards are constantly being revised upward.

Ultra-high rate filtration requires the presence of such factors as (1) adequate pretreatment, such as a Precipitator and/or a settling basin, (2) a high quality influent, or (3) in-line coagulation usually with polyelectrolytes and/or proper selection of media sizes (usually dual or multimedia).

Polyelectrolytes have been highly significant in the development of ultra-high rate filtration. By adding polyelectrolyte to the backwash, Harris[19] was able to operate dual media filters at rates of 10 gal/min/ft² (6.8 ℓ/m²/sec). Bertsche[67] using a cationic polyelectrolyte was able to operate at rates up to 9.4 gal/min/ft². Both investigators used dual media filters.

Robeck et al.[30] also achieved high filter rates using cationic polyelectrolytes with dual media filters. In a similar application, Conley and Pitman[31] used polyelectrolytes to operate dual media filters at rates up to 10 gal/min/ft² (6.8 ℓ/m² sec). Conley[32] has reported similar results. Riddle[34] has reviewed this subject. Nebolsine et al.[65] have operated polyelectrolyte-aided dual/large media filters at rates up to 24 g/min/ft² (16 ℓ/m² sec).

4. Production of Long Filter Runs Between Backwashes

The use of polyelectrolytes has permitted substantial increases in filter run lengths to be attained, especially when dual and multi media filters are employed.

The increase in filter run lengths is due to interaction between a number of factors. These include operation at higher rates, allowing deeper penetration of suspended solids into the bed, and the use of larger media. Also made practical by the use of polyelectrolytes, higher filtration rates can be obtained and also increased floc storage capacity. In addition, the use of polyelectrolytes generally strengthens the floc and also the bond between floc and filter media, thus permitting higher headlosses to develop before floc migration into the filter effluent becomes a problem.

5. Protection of Filtrate Quality From the Effects of Rate Surges

In the absence of polyelectrolyte, the rate surges inevitable during the operation of a full-scale filter result in pulses of turbidity passing through filters. This effect is of

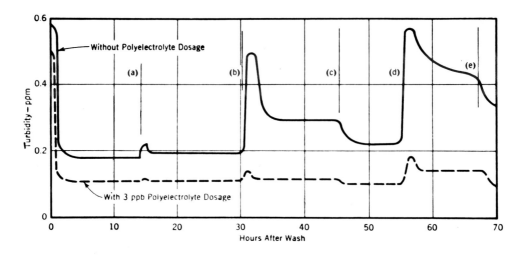

FIGURE 6.   Effluent turbidity with rapid rate changes. At (a) the rate was increased to 2.5 gal/min/ft within 10 sec. At (b) the rate was changed to 3.5 gal/min/ft² within 10 sec. At (c) the rate was reduced to 2.5 gal/min/ft.² At (d) the rate was again increased to 3.5 gal/min/ft² within 10 sec. At (e) the rate was decreased to 2.5 gal/min/ft² (Reprinted with permission of the American Water Works Association, from *J. Am. Water Works Assoc.*, 60, 1380, copyrighted 1968.)

particular importance in waste filtration.

In Figures 6 and 7, due to Tuepker and Buescher[70] this effect is shown. In Figure 6, the upper curve shows the effect of filter rate surge on effluent turbidity in the absence of polyelectrolyte. The striking reduction in this effect, produced by the presence of 3 ppb polyelectrolyte is shown in the lower curve. Figure 7 shows the same effect, but under conditions of more gradual change in the surge rate. Such effects have been demonstrated by other investigators.

IV. SOME SPECIAL APPLICATIONS OF POLYELECTROLYTE-ASSISTED FILTRATION

A. Removal of Color and Other Fine Colloids

A number of published reports,[71,72] as well as work carried out at the Permutit Company,[11] indicate that the very minute colloids which give rise to color in certain waters may be largely removed by in-line or direct filtration techniques with the aid of certain cationic polyelectrolytes. Thus, Duff[71] used polyelectrolytes in-line at high filter rates to remove color. The patent literature in this area, as exemplified by a patent covering the use of a cationic polyelectrolyte on DE for color as well as turbidity removal, is illustrated by Reference 73.

Culp[74] has recently discussed the use of polyelectrolytes for color and turbidity removal in direct filtration. The removal of such very small colloids as color in water illustrates the extremely fine filtration achievable through the use of polyelectrolyte-assisted filters. Other very small colloidal particles in fluids are important in various applications, and could be removed by similar means.

B. Removal of Bacteria, Viruses, Algae, Etc.

1. Removal of Bacteria and Viruses

Better removal of these microorganisms from water is a significant public health goal. Polyelectrolytes are very useful here. As discussed above, better removal of such organisms results from the removal of water turbidity to very low levels, i.e., much

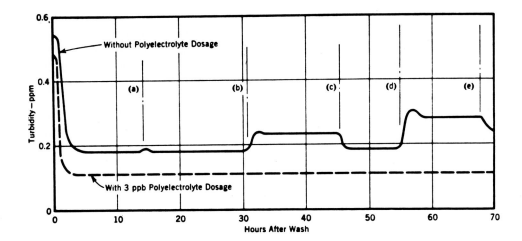

FIGURE 7. Effluent turbidity with gradual rate changes. The rate changes were the same as in Figure 6, but were gradually accomplished over a 10 min period (Reprinted with permission of the American Water Works Association, from *J. Am. Water Works Assoc.*, 60, 1380, copyright 1968.)

fewer organisms are associated with, say 0.1 FTU turbidity passing a filter than with 1 FTU passing a filter.[75,19]* Since polyelectrolytes will reduce filter effluent turbidities, they also reduce the passage of these life forms. In addition, polyelectrolytes will also directly flocculate and remove organisms in the filters, particularly if the latter have already been killed, e.g., by chlorine. Virus removal by polyelectrolyte-aided DE filtration was discussed by Amirhor and Engelbrecht.[76]

2. Removal of Algae and Diatoms

These larger organisms frequently pose problems in filtration by blinding the filter and thus drastically reducing flow, and by passing through the filter. In some cases, a filter will remove only as little as 10% of these organisms.

Proper coagulation and flocculation of these forms, preferably following killing by chlorination, offers the best approach to the filtration of such organisms. The use of polyelectrolytes in these applications can be expected to optimize filter performance.

C. Removal of Oil, Asbestos, Etc.

Oil has been successfully removed from wastewater by feeding alum to the filter. Polyelectrolytes should prove even more effective in such applications.

Asbestos fibers have been successfully removed from Lake Superior water by filtration.[77] Since different types of asbestos fibers may be either positively or negatively charged, it was not surprising to find that polyelectrolytes of different charges were required to afford best removal of the differently charged asbestos fibers. For example, Baumann[77] reported 98 to 100% removal of amphibole asbestos fibers (negatively charged) using a cationic polyelectrolyte fed in-line to a filter containing alum-coated DE media. However, removal of the positively charged chrysotile asbestos was disappointing when the same polyelectrolyte was used. Robinson et al.[44] also studied the removal of asbestos fibers in Lake Superior water by polyelectrolyte-assisted filtration. DE and granular filter media waere investigated.

Lawrence and Zimmerman[78] found that alum-coated diatomaceous earth greatly en-

* Many microorganisms in water are sorbed on "turbidity" particulates such as clays and silica. Such sorbed organisms passing the filter are also more difficult to kill, e.g., with chlorine, ozone, or UV light than are "unsorbed" organisms.

hanced the removal of asbestos fibers from wastewater. There is little doubt that polyelectrolyte-treated DE or granular media filters would be even more effective than alum-coated filters in asbestos removal from waste, since increased effectiveness, using a nonionic, very high-molecular-weight polyelectrolyte, was demonstrated for water filtration by Logston and Symons,[79] and also by Schmitt et al.[80]

V. CONCLUSIONS — THE IMPORTANCE OF POLYELECTROLYTES IN FILTRATION PROCESSES TODAY

We have seen that polyelectrolytes can be used to improve filtration processes in a number of ways. These include:

1. Improvement in filter effluent quality
2. Increase in filter flow rate
3. Increase in filter run length between backwashings
4. Reduction of the deleterious effects of flow rate surges on filter effluent quality
5. In direct filtration applications, the use of expensive primary equipment, such as sedimentation basins and in some cases clarifiers, may be avoided, with concomitant economic benefits

Altogether, the combination of the above effects has produced a highly significant increase in filtration efficiency. For example, a typical modern dual media filter operating without polyelectrolytes might operate at 2 gal/min/ft² and produce a 0.5 JTU effluent during an 8-hr filter run. The filtrate quality would be effected by rate surges, i.e., when one filter in a battery of four were backwashed, or if water demand or wastewater flow increased.

A similar filter, utilizing polyelectrolytes and containing somewhat larger media, might operate at 8 gal/min/ft² and produce a 0.1 JTU effluent during a 24-hr run. The filtrate quality would be little effected by surges in flow rate, and, if the direct filtration technique were utilized, a sedimentation basin would not be required and a savings of as much as 30% in plant capital costs might be realized.

The use of polyelectrolytes in DE filtration can also result in some of benefits (1) to (5) listed above.

Disadvantages of the use of polyelectrolytes include (1) cost, in some cases, (2) greater difficulty in backwashing (more backwash water required, and a more complex backwash arrangement, including air scour, is highly desirable), and (3) somewhat more operator attention is required to control polymer dosage levels and prevent over (or under) dosing with attendant clogging of the filter beds, backwash problems, possible passage of polymer into the filter effluent, etc.

VI. PROJECTIONS FOR THE FUTURE

Some probable new areas for polyelectrolytes in filtration are

1. Development of even higher rate filters, e.g., for waste as well as water treatment, at rates perhaps above 30 gal/min/ft² (20 ℓ/m² sec). Pressure filters as well as larger filter media would often be utilized to attain these high rates. Ultra-high rate filtration offers considerable economic advantages for water and particularly for waste filtration.
2. Further (and improved) utilization of the direct filtration technique will be made, at considerable savings in plant first costs.

3. Higher quality filtrates in particular for potable water treatment are attainable via the use of polyelectrolytes. The production of filter effluents with turbidities consistently below 0.1 FTU (the present water quality goal) will, in most cases, require the employment of polyelectrolytes as well as suitable filter media, etc.

4. The demand for higher clarity products, i.e., soap solutions, beverages, etc., will be met in part by greater use of polyelectrolytes in conjunction with granular media and DE filters.

5. The filtration of very fine colloids, such as color bodies in water, will be required. Certain polyelectrolytes are very effective in such applications.

6. Special polyelectrolytes, such as oil-attracting polyelectrolytes, should prove useful in, for example, oil separations.

7. It should be practical to combine in-line coagulation using, say, a nonionic, high-molecular-weight polymer, with addition of a highly charged cationic to the backwash, to produce even more effective filtration. As usual, a balance must always be sought between polymer type and dosage, the nature of the filter influent stream, media size and type, backwashing, underdrain selection, etc.

8. Particularly in waste treatment, the use of polyelectrolytes plus selected large e.s. dual media in deep beds, with provision for air scour in backwash, will permit filter operation on very high suspended solids level feeds, at very high rates, affording good quality filter effluents.

VII. A NOTE ON THE THEORY OF POLYELECTROLYTE BEHAVIOR IN FILTRATION

A. In-Line Coagulation

For most applications of in-line coagulation, as well as specifically for the production of ultra-high quality filter effluents, the most effective polyelectrolytes appear to be highly charged cationics of moderate (50,000 to 100,000) molecular weight. The objection to higher-molecular-weight materials appears to be largely due to the more rapid headloss development encountered with such materials, although the employment of larger media or better-quality filter influents might be expected to obviate this problem. Another factor which might be affected by employment of higher-molecular-weight cationics is ease of backwashing.

The fact that cationic polymers often prove most effective in such applications indicates that charge neutralization effects are of great importance here. This is also clear from ZP measurements, which indicate that the optimum polyelectrolyte dosage is that dosage which reduces the ZP to ±2 to 3 mV, i.e., precisely the ZP which makes for maximum coagulation in sedimentation applications (alum addition), although optimum dosages to coagulation equipment may be different than optimum dosages to filters. In other words, the charge-neutralizing properties of these polyelectrolytes seems more important here than any bridging mechanisms (if the latter predominated, optimum polyelectrolyte dosage would not occur at the minimum ZP of ± 2 to 3 mV). Obviously, particle bridging is also of importance in in-line coagulation with polyelectrolytes. When anionic or nonionic polyelectrolytes are used, bridging is the only possible mechanism (no charge neutralization is possible), and, as is to be expected, the higher the polyelectrolyte molecular weight, the greater the bridging potential and, usually, the greater the effectiveness. With cationic polyelectrolytes, bridging must also play some role in in-line coagulation. One indication of this is that as the molecular weight of the polycationic increases, the effectiveness (viz. effluent clarity) increases.

The mode of action of a polyelectrolyte in a filter system in in-line coagulation can be described as follows: the polyelectrolyte adsorbs on both the (negatively charged) filter media and the (negatively charged) colloids/particles in the influent stream, by virtue of hydrogen bonding if nonionic or anionic polymers are used, or hydrogen bonding plus electrostatic attraction if cationic polymers are used. If overdosage is avoided, these processes have two results: (1) with cationic polymers, charges of both media and colloids/particulates are reduced to low levels, enabling coagulation phenomena in the usual sense, as in addition of lime or alum, to occur*, and (2) more or less bridging between colloids/particles with one another and with the filter media will also occur, depending on the molecular weight of the polymer. In either case, coagulation/flocculation is the result, leading to enlargement of the particles and colloids in the water and therefore, to their easier capture — particularly by media which bear a much reduced charge, are "sticky" (hydrogen bonds) with adsorbed polymer, and which may project polymer "loops" deeper into the intergranular interstices, i.e., which develop, by this means, a large effective diameter into such interstices for particle/colloid capture.

The binding strength of the polyelectrolyte (with many binding sites) to the water particulates/colloids, and to the filter media, is much greater than the binding strength attainable in the absence of polyelectrolytes. This results, in effect, in a tough floc usually stronger than the flocs or coagula produced using alum. Such a tough floc resists breakthrough forces such as are produced by high flow rates or high headloss.

Certain special features of polyelectrolytes may also play a role in the mechanism of in-line coagulation. Among these are the drag reduction phenomena often encountered in solutions of polyelectrolytes.[81] Drag reduction may result in decreased pressure drop, turbulence, and shearing. Flocculation/coagulation in the conventional sense may also occur in the intergranular interstices during in-line coagulation.

B. Direct Filtration

Effective direct filtration appears to commonly require the formation of a fine, pinpoint-type floc. The pinpoint floc is formed in a high energy mix step. This floc is immediately fed to the filter where, typically, it penetrates deeply into the larger upper layer of the dual or multimedia bed, utilizing much of the latter as floc storage space. This storage space is sufficient to permit operation of the treatment plant without a sedimentation basin before the filter.

The polyelectrolyte probably serves here to toughen the floc (permitting filter operation at higher solids loadings and shear rates), and permits stronger binding of the floc to the filters to produce results similar to those described under in-line coagulation above.

---

*   Such coagulation may be additionally enhanced in a filter, simply because more collisions (particles/ colloids vs. media) may occur in a filter than in other types of coagulation equipment.

REFERENCES

1. Riddick, T. M., Control of Colloid Stability Through Zeta Potential, Zeta-Meter, Inc., New York, 1968.
2. Stumm, W. and O'Melia, C. R., *J. Am. Water Works Assoc.*, 60, 514, 1968.
3. Adamson, A. W., *The Physical Chemistry of Surfaces,* Interscience, New York, 1967, 231; Sparnaay, M. J., *Rec. Trav. Chim.,* 77, 872, 1958.
4. La Mer, V. K., in *Principles and Applications of Water Chemistry,* Faust, S. D. and Hunter, J. V., Eds., John Wiley & Sons, New York, 1967, 246.
5. Michaels, A. S., Aggregation of suspensions by polyelectrolytes, *Ind. Eng. Chem.,* 46, 1485, 1954.
6. Michaels, A. S. and Morelos, O., Polyelectrolyte adsorption by kaolinite, *Ind. Eng. Chem.,* 47, 1801, 1955.
7. Black, A. P., in *Principles and Applications of Water Chemistry,* Faust, S. D. and Hunter, J. V., Eds., John Wiley & Sons, New York, 1967, 274.
8. Luttinger, L. B., Metal-Polyelectrolyte Complexes, Ph.D. thesis, Polytechnic Institute of Brooklyn, New York, 1954.
9. Summerauer, A., Sussman, D. L., and Stumm, W., The role of complex formation in the flocculation of negatively charged sols with anionic polyelectrolyte, *Kolloid Z. Z. Polym.,* 225(2), 147, 1968.
10. Booth, R. B., personal communication with American Cyanamid Company, Wayne, N.J., 1956.
11. Luttinger, L. B., unpublished work, Permutit R&D Center, Princeton, N.J., 1971.
12. Stump, V. L., Proc. 96th Annu. Conf. Am. Water Works Assoc., Vol. 2, Paper No. 17-4, New Orleans, June 20 to 25, 1976.
13. Schwoyer, W. and Luttinger, L. B., unpublished work, Permutit R&D Center, Princeton, N.J., 1971.
14. Cleasby, J. L., Arboleda, J., Burns, D. E., Prendiville, P. W., and Savage, E. S., Proc. 96th Am. Conf. Am. Water Works Assoc., Vol. 2, Paper No. 29-5, New Orleans, June 20 to 25, 1976.
15. Cleasby, J. L., Malik, A. M., and Stangl, E. W., paper presented Annu. Conf. of Water Pollution Control Fed., Cleveland, October 1, 1973.
16. Luttinger, L. B., In-Line Coagulation for the Production of Ultra-High Quality Filter Effluents, presented Am. Inst. Chem. Eng. Meeting, Philadelphia, June 6, 1978.
17. Applegate, L. E., personal communication to L. B. Luttinger, 1977.
18. Terrell, D. L., Organic polymers replace alum and improve water quality in Ithaca, *J. Am. Water Works Assoc.,* 69, 263, 1977.
19. Harris, W. L., *J. Am. Water Works Assoc.,* 62, 515, 1970.
20. Harris, W. L., U. S. Patent No. 3,478,880, 1969.
21. Luttinger, L. B. and Beach, W., unpublished work, Permutit R&D Center, Princeton, N.J., 1970.
22. Camp, J. R. and Kreske, W. J., *Water Works Wastes Eng.,* p.27 1974.
23. Brailey, D., Shanmugam, A., and Fulton, G., paper presented Am. meeting N.Y. State Section Am. Water Works Assoc., September 16, 1976.
24. Shuster, W. W. and Wang, L. K., NTIS Report No. AD-A027-330, June 1976.
25. O'Melia, C. R., NTIS Report No. PB-233-271, April 1974.
26. Loganathan, P. and Maier, W. J., *J. Am. Water Works Assoc.,* 67, 336, 1975.
27. Eunpu, F. C., *J. Am. Water Works Assoc.,* 62, 340, 1970.
28. Norris, W. G., *Filtr. Sep.,* 12, 702, 1975.
29. Conley, W. R., *J. Am. Water Works Assoc.,* 53, 1473, 1961.
30. Robeck, G. G., Dostal, K. A., and Woodword, R. L., *J. Am. Water Works Assoc.,* 56, 198, 1964.
31. Conley, W. R. and Pitman, R. W., *J. Am. Water Works Assoc.,* 52, 1319, 1960.
32. Conley, W. R., *J. Am. Water Works Assoc.,* 64, 203, 1972.
33. Rice, A. H., J. Am. Water Works Assoc., 66, 258, 1974.
34. Riddle, W. G., *Public Works,* 105, 86, 1974.
35. Hutchison, W. and Foley, P. D., *J. Am. Water Works Assoc.,* 66, 79, 1974.
36. Hutchison, W. R., *J. Am. Water Works Assoc.,* 68, 292, 1976.
37. Tredgett, R. G., *J. Am. Water Works Assoc.,* 66, 103, 1974.
38. Spink, C. M. and Monscvitz, J. T., *Am. City,* p.90 1973.
39. Spink, C. M. and Monscvitz, J. T., *J. Am. Water Works Assoc.,* 66, 127, 1974.
40. Adin, A. and Rebhun, M., High-rate contact flocculation-filtration with cationic polyelectrolytes, *J. Am. Water Works Assoc.,* 66, 109, 1974.
41. Robinson, J. H., Schmidt, O. J., Stukenberg, J. R., Jacob, K. M., and Bollier, G. H., *J. Am. Water Works Assoc.,* 68, 531, 1976.
42. Harbert, R. H., *Public Works,* 60.
43. Ainsworth, L. D. and Mosley, J. A., *Water & Sewage Works,* 64, 1972.
44. Letterman, R. and Logsdon, G. S., paper presented 9th Annu. Am. Water Works Assoc. Conf., New Orleans, June 20 to 25, 1976.

45. Thompson, J. G., *Water & Sewage Works,* 122, 78, 1975.
46. Letterman, R. D., Sama, R. R., and Domenico, E. J., *Proc. 96th Annu. Conf. Am. Water Works Assoc.* Vol. 2, Paper No. 15-6, New Orleans, June 20 to 25, 1976.
47. Adin, A. and Rebhun, M., *Proc. 96th Annu. Conf. Am. Water Works Assoc.,* Vol. 2, Paper No. 17-5, New Orleans, June 20 to 25, 1976.
48. Geise, G. D., Pitman, R. W., and Wells, G. W., *J. Am. Water Works Assoc.,* 59, 1303, 1967.
49. Yao, K. M., Habibian, M. T., and O'Melia, C. R., Water and wastewater filtration: concepts and applications, *Environ. Sci. Technol.,* 5, 1105, 1971.
50. Donovan, E. J., Jr., *Water Resour. Symp.,* 3, 167, 1970.
51. Donovan, E. J., Jr., Water Qual. Improvement by Physical and Chemical Processes, Vol. 3, Gloyna, E. F. and Eckenfelder, W. W., Eds., U. of Texas Press, Austin, 1970, 167.
52. Stevenson, D. G., *Filtr. Sep.,* 9, 75, 1972.
53. Tschobanoglous, G., Filtration techniques in tertiary treatment, *J. Water Pollut. Control Fed.,* 42, 604, 1970.
54. Bell, G. R., Libby, D. V., and Lordi, D. T., AN-Pb-196, 734, 1970.
55. Bernhardt, H., Clasen, J., and Schell, H., *J. Am. Water Works Assoc.,* 63, 355, 1971.
56. Lee, J. W. J., AN-Pb-222 251/1, 1973.
57. Culp, G. L. and Hansen, S. P., *Water & Sewage Works,* 46, 114, 1967.
58. Habibian, M. R. and O'Melia, C. R., *J. of Env. Eng. Div. Proc. Am. Soc. Civ. Eng.,* 101, EE4, 567, 1967.
59. Shireman, H. C., *Water Works Wastes Eng.,* 9, 34, 1972.
60. McDonald, K. R., *Water & Sewage Works,* 367, 1965.
61. Kavanaugh, M., Engster, J., Weber, A., and Boller, M., *J. Water Pollut. Control Fed.,* 49, 2157, 1977.
62. Goodman, B. L. and Mikkelson, K. A., *Chem. Eng.,* 75, 1970.
63. Rice, A. H. and Conley, W. R., U.S. Pat. 3,171,802, 1965.
64. Berry, L. S., Lafayette, P. F., and Woodard, F. E., *J. Water Pollut. Control Fed.,* 48, 2394, 1976.
65. Nebolsine, R., Harvey, P. J., and Fan, C. Y., EPA Report on Project No. 11023 EYI, 1972.
66. Tate, C. H., Lang, J. S., and Hutchinson, H. L., *J. Am. Water Works Assoc.,* 69, 379, 1977.
67. Bertsche, E. C., *Water Works Wastes Eng.,* B-9, 1973.
68. Kirchner, W. B. and Jones, W. H., *J. Am. Water Works Assoc.,* 64, 157, 1972.
69. Baumann, E. R., personal communication to L. B. Luttinger, 1978.
70. Tuepker, J. L. and Beuscher, C. A., Jr., *J. Am. Water Works Assoc.,* 60, 1377, 1968.
71. Duff, J. H., *Pap. Trade J.,* 1968.
72. Chamberlain, R., personal communication to L. B. Luttinger, 1978.
73. U.S. Pat. 3,227,650, 1966.
74. Culp, R. L., *J. Am. Water Works Assoc.,* 69, 375, 1977.
75. Hudson, H. E., Jr., *J. Am. Water Works Assoc.,* 54, 1265, 1962.
76. Amirhor, P. and Engelbrecht, R. S., *J. Am. Water Works Assoc.,* 67, 187, 1975.
77. Baumann, E. R., Precoat Filtration, report from Department of Civil Engineers, U. of Iowa, Ames, 1977, p. 10. 47.
78. Lawrence, J. and Zimmermann, H. W., *J. Water Pollut. Control Fed.,* 49, 156, 1977.
79. Logston, G. L. and Symons, J. M., *J. Am. Water Works Assoc.,* 69, 499, 1977.
80. Schmitt, R. P., Lindsten, D. C., and Shannon, T. F., *Environ., Sci. Technol.,* 11, 464, 1977.
81. Savins, J. G. and Virk, P. S., Drag Reduction, *Chem. Eng. Prog. Symp. Ser.,* 11, 67, 1971.

Chapter 8

POLYELECTROLYTE MAKEUP AND HANDLING

R. J. Chamberlain

TABLE OF CONTENTS

I. INTRODUCTION

Great care should be exercised in the designing or choosing of a flocculant feed system in order to get the most out of the product and to avoid problems associated with improper flocculant application such as:

1.   Incomplete solution — wastes flocculant
2.   Undissolved particles (fish eyes) — cause equipment plugging and harm end product
3.   Overmixing — degrades flocculant
4.   Overfeeding — wastes flocculant and can upset treatment

Polyelectrolytes are sold in four forms: dry powders, solutions, emulsions (aqueous solutions dispersed in oil), and gels. Frequently, specific products are available in more than one form. In selecting the form to be used, consideration should be given to cost-performance factors including transportation costs, degree of feed automation desired, and equipment costs.

The degree of automation used will be governed by the size of the operation and the manpower available for flocculant handling. While a number of the largest flocculant users are successfully handling dry products in bulk with strong silos and pneumatic transfer equipment, many others are using completely liquid treatments. Ideally, polyelectrolyte storage and feed systems should be designed with the maximum flexibility to facilitate changing from one product to another as more efficient or more economical flocculants are developed.

II. POLYELECTROLYTE FORMS — PHYSICAL CHARACTERISTICS

Numerical ranges of polymer physical properties listed in this chapter will cover most of the products commonly used today. Properties of specific products should be obtained directly from the manufacturer.

A. Dry Powders
Appearance — Flocculants in the solid form are white to off-white materials which are sold as granular powders, flakes, or beads.

Bulk density — Bulk density can range from 50 lb/ft³ for fine granular products to 18 lb/ft³ for flakes or beads. Bulk density is governed principally by the method of manufacture.

Hygroscopicity — Polyelectrolyte powders are hygroscopic and can pick up moisture

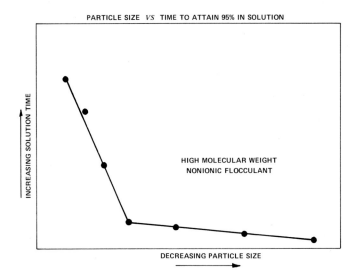

PARTICLE SIZE *VS* TIME TO ATTAIN 95% IN SOLUTION

INCREASING SOLUTION TIME

HIGH MOLECULAR WEIGHT
NONIONIC FLOCCULANT

DECREASING PARTICLE SIZE

FIGURE 1. Effect of particle size on solution rate.

from the air causing them to cake and stick to the walls of containers and feed equipment. Therefore, they must be protected from humidity. Heat lamps are sometimes used in areas where flocculant powders must be exposed to the atmosphere. Air used for conveying dry flocculants must be dry.

Solution rate — The three main factors governing the solution rate of dry polyelectrolytes are the chemical nature of the product, particle size, and the method of manufacture. Nonionic flocculants, usually homopolymers of acrylamides, are the slowest to dissolve. For complete solution, 30 min to 1 hr must generally be allowed. The effect of particle size on solution rate is illustrated in Figure 1. Anionic products dissolve more rapidly than the nonionic materials of the same particle size and form. Solution rate increases with increasing anionic charge on the polymer. For dissolving dry anionic flocculants, 15 to 45 min should be allowed. The use of warm water, up to 120°F, hastens the dissolving process. Water above 120°F should not be used, since it can lead to degradation of the polymer. Cationic products behave much like their anionic counterparts.

Stability — Dry flocculants, when protected from extreme heat or humidity, will lose little of their activity when stored for periods up to one year. For optimum activity, polymer solutions should be prepared as needed. Storage of flocculant solutions for more than 48 hr is not recommended and some cationic products should be used within 24 hr to assure full activity. The effective life of some cationic polymer solutions can be prolonged by adjusting the pH to 6.0. Fresh solutions are always prepared for laboratory evaluation.

Viscosity — Dry organic flocculants, when dissolved in water, can give solutions with considerable viscosity. Some typical examples are

Viscosity in centipoise at

Flocculant type	0.1%	0.5%	1.0%	2.0%
Nonionic	10	50	320	—
Strongly anionic	90	650	2000	—
Cationic	—	140	640	3800

The above figures are apparent viscosities as measured by Brookfield Viscometer at

20°C. Viscosity at infinite shear (which is called effective viscosity) can be two thirds to one half of the above figures. Effective viscosity figures would be used in determining pumping characteristics.

B. Solution Products

Appearance — Polyelectrolytes in solution form vary in color from water white to pale blue to dark amber.

Solids — The active ingredients in solution flocculants can vary from 3 to over 50%.

Viscosity — Apparent viscosities, as measured by Brookfield Viscometer, range from 1.0 cP (like water) to 24,000 cP or higher (like heavy molasses). Effective or pumping viscosity may be one half of the apparent viscosity for the higher viscosity materials.

pH — Product pH can vary widely, some being as low as 2 while others may be as high as 12.

Specific gravity — Specific gravity ranges from essentially 1.0 to 1.2 depending on the solids concentration in the product.

Solubility (dilutability) — Since all liquid flocculants, other than emulsions, are aqueous solutions, they can be further diluted with water with a minimum of mixing. Even the most viscous products can be diluted in-line with the help of static mixers (Figure 6). Sometimes, to facilitate pumping, the more viscous products (i.e., >15,000 cP apparent viscosity) are diluted 1:1 or 2:1 with water prior to metering and final dilution to application strength.

Stability — Before dilution, most liquid products can be stored for 6 to 12 months. A few should be used within 90 days. Specific product stability data is available from individual manufacturers.

As with dry polymers, dilutions of liquid products should be prepared as needed to assure maximum effectiveness. In the absence of extremes in temperature or prolonged agitation, most solutions are stable for 48 hr. Again, solutions of some cationic materials will have a shorter effective life; therefore, manufacturers' specifications must be checked.

C. Emulsion Products

Composition — A relatively recent development in flocculant technology has led to the marketing of very high-molecular-weight polymers in emulsion form. The emulsions are dispersions of high-solids polymer gels in hydrocarbon oil. Stabilizers (surface active agents) are incorporated to prevent separation of the oil and water phases. There are some emulsion-type flocculants on the market which require the separate addition of a second surface active agent (demulsifier) to water before the emulsion will dissolve in it. These so-called "two package" systems can be used in automatic feed equipment but a separate metering pump is required to add the surface active agent, sometimes called an "activator". Products in the emulsion form are free from the presence of insoluble polymer particles which sometimes develop during the manufacture of dry flocculants.

Appearance — Emulsions vary from almost clear, through translucent, to opaque white liquids.

Solids — Generally, flocculants in emulsion form contain from 25 to 30% active ingredients.

Viscosity — Emulsion viscosities range from 300 to 5000 cps.

pH — Since the hydrocarbon oil is the continuous phase in polymer emulsions, pH measurements are not meaningful and usually not reported. In solution, at concentrations of 0.5 to 2.0%, many of the emulsion products give slightly acid pHs, sometimes as low as 4.5. Solution pH will vary with the type of flocculant and also from one manufacturer to another.

Specific gravity — Since emulsions are mixtures of high solids, aqueous polymer particles, oil, and water, the resulting specific gravities are very close to 1.0 (range 0.96 to 1.02).

Solubility — The dissolving of single component emulsion flocculants takes place in two steps. First, high shear forces provided by strong mixing, disperse the oil phase in the dissolving water, freeing the polymer gel particles. Then the gel particles dissolve in the water. The initial step of "breaking" or "inverting" the emulsion requires only a few seconds of high shear mixing. Following the breaking of the emulsion, a few minutes (3 to 15) of quiescent standing or very gentle mixing are required to complete dissolution of the product. Specific flocculants will dissolve much faster as emulsions than they will as dry powders. However, as with powders, nonionic emulsions dissolve most slowly (15 min) while highly anionic products dissolve almost immediately (1 to 5 min). A few emulsion-type flocculants require that an activator (surface active agent) be added to water before the emulsion will dissolve in it. The activator usually amounts to 5 to 10% of the weight of emulsion used. The stability of a flocculant solution is usually the same whether the solution was prepared from a dry powder or an emulsion. Solutions of emulsion-type flocculants are usually prepared at 2% concentration or less because of the relatively high viscosities generated.

D. Gel Products

Appearance — Gels are shipped as individual logs which are approximately 18 in. long by 9 in. in diameter, and weigh about 46 lb. They have a tough, rubbery consistency and are water-white to pale buff in color. Gels comprise high-molecular-weight, polyacrylamide-based flocculants in solution in water.

Solids — The active ingredient (polymer) in the gels is about 30 to 35%.

Solubility — Since the gel products are sold in large solid blocks, they must be reduced to granules to achieve reasonable solution rates. The granulating step is described in the section on gel feeding, VI.D. Once granulated, the gel products require about 1 hr of mixing with water to achieve complete solution.

III. PACKAGING AND SHIPPING

A. Dry Powder Products

Dry flocculants are usually shipped in 25- or 50-lb paper bags and 50- or 100-lb, plastic-lined fiber drums. Bins and cubitainers can be used for handling larger volumes of dry polymer when equipment is available for transferring the product from the container to the solution preparation device. While bulk handling of dry polymers is difficult because of its hygroscopicity, a number of large users are doing it.

B. Liquids (Solutions and Emulsions)

Polymers in liquid form are supplied in 5-gal pails, 55-gal lined drums (450 to 500 lb), 5000-gal tank trucks, and 10,000- to 30,000-gal railroad tank cars. Tank trucks are usually equipped with transfer pumps and hoses for unloading over reasonable distances. Rail cars do not have pumps but are unloaded by gravity, low air pressure on the car, or with the end user's transfer pump. Normal hose fittings are 3-in. diameter camlocks on both tank trucks and tank cars.

C. Gel Products

Gels are shipped in individual sealed plastic bags. They are packed either six to a fiberboard drum, with a net weight of about 250 lb, or in boxes containing 1 ton. No protection from freezing is necessary, and the gels may be stored in any location compatible with their shipping container.

IV. STORAGE

A. Solids

Flocculants in solid form should be stored in a cool dry area away from direct sunlight.

B. Liquids

Drums of either solution or emulsion polymer should be protected from extremes of temperature. While most polymer solutions are not generally harmed by freezing, thawing them out is not always convenient. Some emulsion-type products have excellent freeze-thaw stability but since not all do, it is wise to avoid freezing and/or to check with the manufacturer.

Bulk storage of liquid flocculants is becoming widespread. Most commonly, fiberglass tanks are used; however, stainless steel, rubber-lined, glass-lined, and epoxy-coated tanks are excellent. If iron, mild steel, or aluminum tanks are to be used, they would have to be well painted on the inside with a polyester-type coating.

As a rule, solution-type liquid flocculants do not require mixing during storage but emulsion-type products can separate on long standing. For this reason, agitation in bulk storage tanks should be provided. Slow speed, blade-type agitators are commonly used and it has been found that side mounting of the mixers near the bottom of the tank is most effective. As an alternative, the pump used to transfer the emulsion to the feed or mixing equipment can be used to recirculate the emulsion within the tank. It is recommended for effective mixing that the tank contents be turned over at least once a day.

In climates where freezing temperatures are encountered, storage tanks should be placed indoors, insulated, and heat traced, or buried below the frost line. Storage tanks of large capacity have been successfully built by placing vinyl plastic liners inside cement block walls.

V. TRANSFER

A. Pumps

Pumps for transferring liquid flocculants from bulk storage tanks to makeup stations are usually of the gear, lobe, or progressive cavity type. Because of the viscosities sometimes encountered, pump suction lines are normally 2 to 8 in. in diameter while discharge lines are in the 1- to 6-in. range depending on the flocculant used and the distance to be pumped. Pumps are always installed with flooded suctions. If suction strainers are used, two strainers must be installed in parallel to allow periodic cleaning without interruption of service.

Transfer pump controls should be placed at the receiving vessel to allow for visual control of the volume transferred or a reliable meter must be installed. As detailed later, the transfer pump can also connect to automatic controls so that a measured amount of flocculant is transferred from the bulk tank on demand.

Stainless steel, plastic, or other noncorroding materials should be used in pump construction. Since all emulsion-type flocculants contain mineral oil, wetted parts of rubber (Buna-N, Hypalon®, etc.) must be avoided when these products are pumped. Viton® parts have generally been acceptable. Brass, bronze, and aluminum should also be avoided because of the extremes in pH encountered.

B. Pipes and Hoses

In new polymer feed installations, extensive use of rigid PVC pipe is being made. Many older plants which formerly used ferric chloride or alum and lime for flocculation are now using organic polymers in their existing systems without equipment changes.

For temporary or experimental use, plastic tubing and hose of any material is satisfactory with one exception — rubber hose should not be used with undissolved emulsion-type flocculants.

VI. APPLICATION METHODS

A. Dry Powder Products

1. Batch Method for Small Applications

In applications where flocculant usage is less than one bag (50 lb) per day, it may be satisfactory to manually prepare batches of flocculant solution as needed.

Equipment required for this type of operation is

1. Small scale for weighing polymer
2. Dissolving/storage tank
3. Low speed mixer (400 to 500 rpm)
4. Flocculant disperser
5. Solution metering pump

To prepare flocculant solution using a batch system one would first determine the solution concentration to be prepared. (Some lower molecular weight products can be used at 1 to 2% concentration but most high-molecular-weight materials can be used only at 0.1 to 5% concentration because of their very high solution viscosity.)

Next, weigh out the desired quantity of flocculant. After the initial weighing, the volume of flocculant may be marked on the container so that thereafter the flocculant may be measured out by volume.

Add water to the solution tank until the blades of the mixer are covered. Ideally, potable water should be used to prepare flocculant solutions. However, if "process" or recycle water is available which is free from suspended solids, it may be satisfactory for polymer makeup. If other than potable water is to be used, it is essential to make sure that the pH is acceptable for the specific polymer being used. This information is available from the flocculant supplier. Water containing high levels of dissolved solids should also be avoided. If such water must be used, check its compatibility with the polymer. In some cases, plant effluent containing suspended solids has been used for dissolving polymer but some flocculant effectiveness was lost.

Then, start the mixer. Add the flocculant to the stirred water in the tank through a flocculant disperser of the type illustrated in Figure 2. Do not attempt to add flocculant directly to the water without a disperser as lumps or "fish-eyes" will form, and these lumps are very difficult to dissolve. Continue adding water until the tank is filled to the desired level. Continuing mixing for 1 hr or as recommended by the flocculant manufacturer. The arrangement of equipment for a single tank system is illustrated in Figure 3.

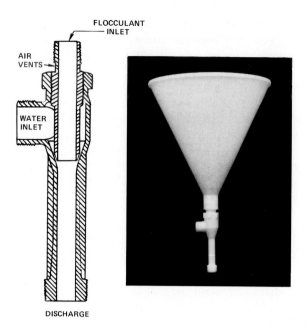

FIGURE 2. Dry flocculant disperser. (Courtesy Pentech Houdaille, Cedar Falls, Iowa.)

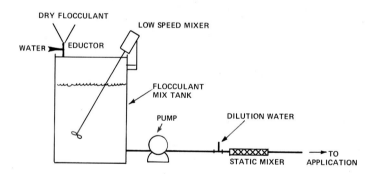

FIGURE 3. Dry flocculant batch system — single tank.

When dissolving flocculant in the same tank from which it is fed, some undissolved particles will be carried into the water being treated. In many cases, this does no harm but represents a waste of flocculant. However, it can be avoided by transferring the prepared solution from the mix tank to a separate feed tank by pumping or by gravity as illustrated in Figure 4.

Prepared flocculant solution is pumped to the point of application using variable delivery metering pumps. Many different types of pumps have been used satisfactorily including piston, diaphragm, peristaltic, gear, lobe, and progressive cavity. Feed pumps with mechanical, hydraulic, or electrical (SCR) speed controls may be used. When selecting a flocculant feed pump, be certain to take into consideration the viscosity of the solution to be pumped as well as the length and diameter of the feed lines.

Feed rate control by flocculant solution recirculation is *not* recommended because of the possiblity of flocculant degradation. All flocculant feed pumps must be calibrated when a system is put into service. It is dangerous, however, to rely on these

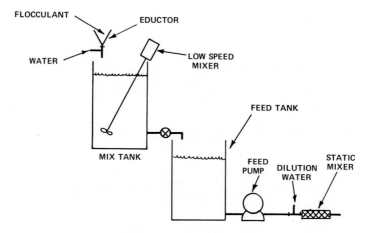

FIGURE 4.   Dry flocculant batch system — double tank.

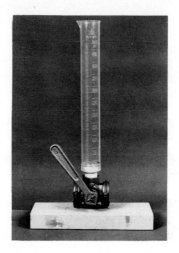

A

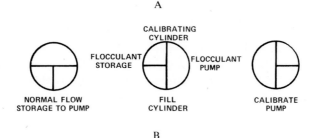

B

FIGURE 5.   (A) Pump calibrating cylinder; (B) three-way
valve positions.

calibrations over a long period of time because check valves become obstructed and
pump parts wear. Recalibrating through the use of a petcock on the feed pump discharge is not completely reliable because under these conditions the pump delivery is
measured at atmospheric pressure and not at the pressure of the closed feed system.
Very accurate measurements of pump deliveries can be made by installing a pump
calibrating cylinder in the line between the flocculant storage tank and the pump (see
Figure 5).

FIGURE 6.   Static mixer. (Courtesy Koch Engineering Co., Wichita, Kan.)

In almost all water clarification operations by settling or flotation, it is desirable to add the flocculant in the most dilute form practical. Since high dilution in the solution tank would require very large tanks or the preparation of many batches each day, it is preferable to prepare a 0.1 to 1.0% solution in the tank and post dilute the solution to 0.01 to 0.1% after it leaves the metering pump. In-line mixing of the flocculant solution and dilution water can be accomplished through the use of "static" or "motionless" mixers of the type illustrated in Figure 6. Water "ejectors" like that illustrated in Figure 7 can efficiently mix flocculant solutions and dilution water and have the added benefit of reducing the back pressure against which the flocculant feeder must pump.

An inexpensive system for feeding batches of flocculant solution without the use of any pumps is illustrated in Figure 8. With this gravity feed system, flocculant flow to several different application points can be regulated with simple valves or with flow meters and valves as shown in Figure 9.

When it is necessary to deliver flocculant solution to elevated as well as lower application points, simply allow the flocculant to flow into the suction side of one or more liquid ejectors as shown in Figure 10. The water, which acts as the motive force in the ejector, also serves to dilute the flocculant prior to its addition to the stream being treated. Again, valves and flow meters are used to regulate flocculant and dilution water. Systems of this type have been widely used in coal preparation plants. They can be used to feed solutions of dry, emulsion, liquid, or gel flocculant.

2. Automatic Feed System — From Components

Many plants undertaking the use of organic polyelectrolytes have previously used inorganic flocculants or at least have storage tanks, mixers, and perhaps metering

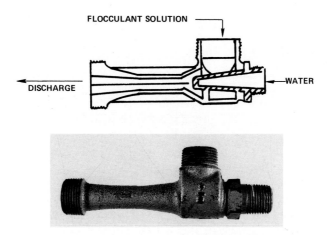

FIGURE 7. Flocculant solution ejector. (Courtesy Pentech Houdaille, Cedar Falls, Iowa.)

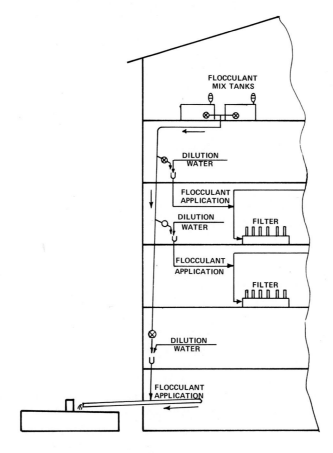

FIGURE 8. Gravity flocculant feed system.

pumps which could be incorporated into a polymer feed system. A typical arrangement is shown in Figure 11.

a. Tanks

Two tanks are required, one for dissolving the dry flocculant and the other for so-

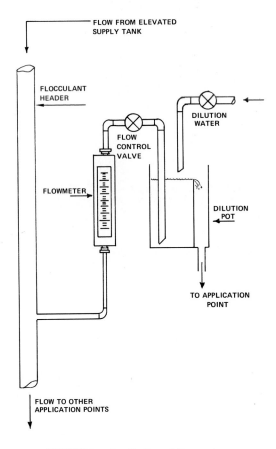

FIGURE 9.   Application of flow meters.

lution storage. Fiberglass or other plastics and stainless steel make ideal tanks. If an iron tank is used, it should be coated inside with a polyester- or epoxy-type finish.

b. Transfer Pump

Fixed speed gear, lobe, or progressive cavity pumps capable of emptying the mix tank in 5 to 10 min are commonly used.

c. Dry Flocculant Feeder

Either a volumetric or gravimetric feeder may be used to meter the dry flocculant, the choice being made on the basis of the accuracy required (see Figure 12). In their publication *Dry Solids Metering: An Equipment Buyer's Guide,* Riccardi and Russo say that a feed rate accuracy of ± 2% to ± 5% can be expected from volumetric equipment. For accuracy of ± 2% and below, gravimetric feeders should be considered. The added accuracy, however, will increase the feeder cost three to five times over that of a similar volumetric unit.

d. Eductor

To prevent the accumulation of undissolved lumps of flocculant in the solution, the dry flocculant powder must pass through a wetting device such as the disperser shown in Figure 2.

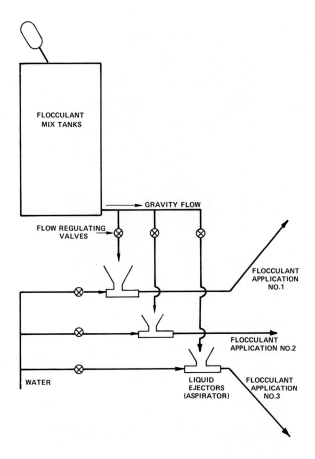

FIGURE 10.   Application of liquid ejectors.

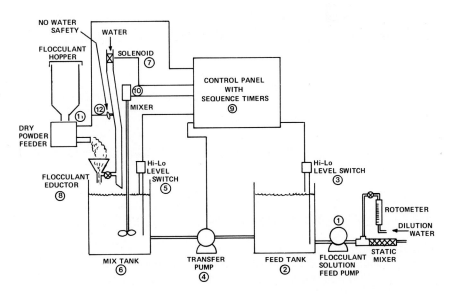

FIGURE 11.   Automatic dry flocculant feed system — from components.

e. Polyelectrolyte Solution Feed and Dilution

Solution metering and diluting equipment would be the same as that used with the nonautomatic system described in the previous section.

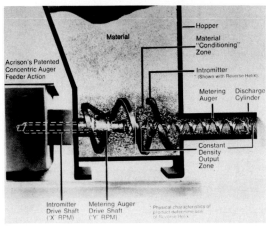

FIGURE 12.   Dry flocculant feeders. (Courtesy of Acrison, Inc., Moonachie, N.J.)

f. Mixer

The mixer should be rugged enough to handle viscous solutions and should have a maximum speed of 400 rpm.

g. Level Controls

Pneumatic, electrical, and mechanical (float type) level controls are commonly used.

h. Automatic Controls

A series of switches, timers, solenoid valves, etc. are used to accomplish the following sequence of events. (Refer to Figure 11.) Starting with all tanks full:

1.   Solution feed pump (1) draws solution down in feed tank (2) until low level control (3) starts transfer pump (4).
2.   When feed tank (2) is refilled, level control (3) shuts off transfer pump (4).

3. Level control (5) in mix tank (6) signals low level which opens solenoid (7) allowing makeup water to enter the mix tank (6) with a side stream flowing through the flocculant eductor (8).

4. A time delay switch in the control panel (9) allows the water in the tank to cover the mixer blades before starting the mixer (10).

5. Shortly after the mixer has started, a second timer in the control panel starts the dry flocculant feeder (11). The feeder timer is set to dispense the amount of powder needed for the solution concentration desired. A pressure activated switch (12) should be located in the water line so that the dry chemical feeder cannot start unless water is flowing through the eductor.

6. Dry feeder shuts off when present amount of powder has been fed.

7. Solenoid shuts off water when activated by high level switch in mix tank.

8. Mixer continues to run until shut off by preset timer in control panel. The mix time set will vary with the product used and could be anywhere from 10 to 90 min.

3. Automatic Feed System — Commercial Units

Dry flocculant feed systems are available that combine into one compact package all the components described in the section above. Once these units have been connected to a water and power supply and calibrated, one needs only occasionally refill the storage bin or hopper with flocculant powder. The better units are equipped with vacuum or air scrubbing devices which prevent the escape of flocculant dust into the atmosphere. All feeders currently on the market dispense the dry flocculant volumetrically, using either an auger screw or star valve arrangement. Various wetting principles are used including a swirling water vortex, a plane of water, or a sprayed water curtain.

Some units have only a flocculant mix tank. These would be used to prepare polymer solutions for transfer to a larger external storage tank. Other feeders include not only the mixing tank but also the feeding tank and even the solution feed pump so that no additional equipment is needed.

One of several good commercial dry feeders is shown in Figure 13. Its operation is shown in Figure 14, and can be easily understood by referring to the following steps.

"Dry polymer is usually manually loaded into the feeder hopper (A) and then accurately metered at a preset rate by feeder (B). The metered polymer is pneumatically dispersed by atomizer (C) and simultaneously conveyed into the cyclone wetting chamber (E). In the wetting chamber, the fluidized particles are brought into contact with a continuously flowing water cyclone. The positive atomizing of the polymer and subsequent contact and mixing of each particle with water prevents clumping or agglomerating. The completely wetted polymer then drops into the mixing tank (F) where a slow speed mechanical mixer (H) facilitates dissolving without damaging the polymer chain. The air scrubber (P) exhaust blower (R) assures a positive downdraft at the wetting chamber inlet and prevents even the smallest particles from escaping. Only clean air is then returned into the atmosphere. Level probe (G) automatically closes solenoid valve (T) and stops feeder (B) at high level; and also opens solenoid valve (T) and starts feeder (B) at low level. Level probe (G) also closes transfer valve (K) when mixing tank (F) is empty. Level probe (M) opens the transfer valve (K) to ensure a continuous supply of polymer solution to the aging tank (L) which is sized to provide sufficient time for full growth of the specific polymer being utilized. Aged polymer is then fed at the desired rate by metering pump (N).

When liquid polymer only is to be used, the feeder (B), the atomizing and conveying system (C), and the scrubber components (P) and (R) are replaced with a positive displacement metering pump and solution line.

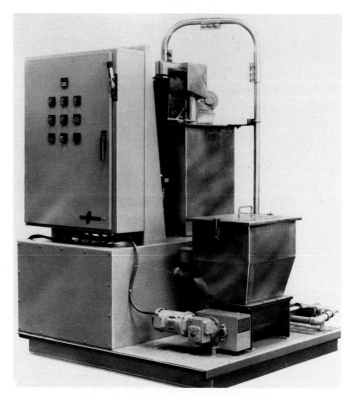

FIGURE 13.   Commercial dry flocculant feed system. (Courtesy Acrison, Inc. Moonachie, N.J.)

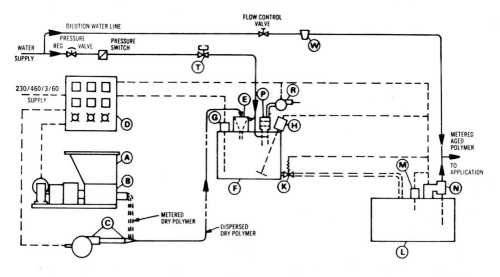

FIGURE 14.   Commercial feeder operations. (Courtesy Acrison, Inc., Moonachie, N.J.)

Should it be desirous to arrange the system to handle either dry or liquid (emulsion) polymers, the same basic system as described is utilized. To accommodate this provision, a switch is provided in the control panel for the selection of dry or liquid polymer which automatically selects the appropriate hardware. Either dry or liquid polymers can be placed in solution through the standard wetting chamber (E) without modifications.'' (From Acrison, Inc., Moonachie, New Jersey. With permission.)

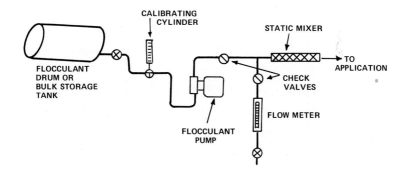

FIGURE 15.   Liquid flocculant feed system.

B. Liquid Products

Liquid flocculants (other than emulsions) are by far the easiest products to use since they require no mix tank, agitator, or disperser. In normal application, a liquid flocculant would be pumped directly from a drum or bulk storage tank into a dilution stream of water and hence to the point of application (Figure 15). The notes on water quality in the dry polymer section of this chapter apply to liquid flocculants also.

The metering pump selected should have a high enough capacity to treat the maximum anticipated plant flow at the highest expected dosage level. At the same time, the pump must also be capable of treating the lowest flow at the lowest dosage. Normally, metering pump deliveries can be varied over a tenfold range. In cases where plant flow and/or dosage levels vary widely, special wide-range pumps or perhaps multiple pumps may be used. It is also important to consider the pressure (dilution water or receiving stream) against which the pump must deliver the flocculant. In addition, some of the most viscous products (15,000 to 30,000 cP) will require pumps fitted with extra-large suction and discharge lines.

In virtually all clarification applications, the flocculant should be added in as dilute a form as possible, certainly no higher than 1% and preferably at 0.1% or lower. In a few rare instances, it has been found that liquid flocculant performs best when fed neat. A flow meter may be used in the dilution water line but is not really necessary as long as the output of the flocculant pump is accurately known, and the dilution water flow is known approximately.

C. Emulsion Products

As mentioned earlier in this chapter, emulsion flocculants are prepared for use in two steps — a short high-energy mix with a bladed mixer, static mixer, or flocculant disperser, followed by a short period of aging with or without gentle mixing. With some types of flocculants, the aging or dissolving time will be so short (1 to 3 min) that it can be built into a continuous feed system.

1. Batch Method

This method of flocculant solution preparation is generally used for small or occasional applications. When an emulsion flocculant is put into solution in batches, it is handled almost exactly as if it were a dry powder.

The solution tank is filled with water and the mixer is started. (Refer to the notes on water quality earlier in this chapter.) The proper amount of emulsion is measured out by weight or volume. Usually, solution viscosity will limit concentrations to 2% or less. Solutions can be made as dilute as desired.

The measured emulsion is poured slowly into the vortex formed by the mixer. Mixing

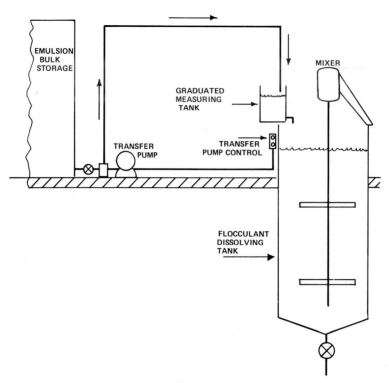

FIGURE 16.   Large batch mixing system for emulsions.

is continued only until all emulsion particles have disappeared, usually about 5 to 20 min. Occasionally, emulsions are encountered which do not disperse readily in the water. In these cases, it is perfectly acceptable to add the emulsion through an aspirator-type polymer disperser such as is used for dry powdered products (see Figure 2).

When very large batches of flocculant are mixed at one time, it is more convenient to pump the emulsion from a drum or bulk storage tank into a small graduated measuring tank mounted over the solution tank (see Figure 16). From the measuring tank, the emulsion is allowed to drain into the stirred mixing tank below.

It is also possible to pump emulsion directly from storage into the mix tank for a given period of time if the delivery rate of the pump is known accurately. However, pumping rates can change with small batch-to-batch variations (in produc t viscosity, with foreign matter in the pump or lines, and with pump wear), therefore, direct measurement of the emulsion is suggested.

2. Automatic Feed Systems — Conversion of Dry Feeders

Plants which have commercial automatic dry flocculant feeders can easily adapt them to feed emulsions without losing the capability of reverting to dry products. The only auxiliary equipment needed for the conversion is an emulsion metering pump which is wired into the timer on the control panel which normally operates the dry powder feeder. The connection can even be made through a selector switch which allows the choice of dry or liquid feed. If emulsion is to be fed from drums, a drum mixer must be provided.

3. Automatic Feed Systems — From Components

An emulsion feed system may be assembled from the components shown in the dry flocculant diagram, Figure 11, by merely substituting a liquid metering pump for the

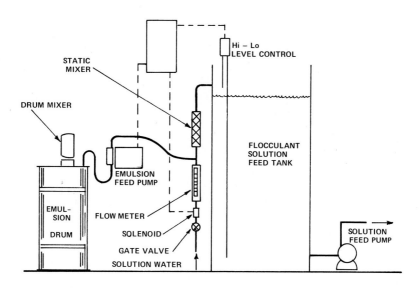

FIGURE 17. Simple emulsion feed system from components.

dry feeder (item no. 11 in the diagram). However, since emulsion flocculants dissolve so rapidly once the emulsion is "broken", it is usually not necessary to use separate mixing and feeding tanks. Elimination of the extra tank also obviates the need for a transfer pump, and one set of level controls. A simplified emulsion feed system is shown in Figure 17. With this system, a low level in the solution tank signals the solenoid valve to open allowing water to flow through the flow meter and mixing device. Simultaneously, the low level control starts the emulsion feed pump. By regulating the water flowing through the flow meter with a valve and by setting the emulsion pump delivery rate, a solution of known concentration is sent to the feed tank. In plants where fluctuating water pressure would cause uneven flow, a constant flow valve or fixed delivery water pump may be substituted for the flow meter and gate valve. It should be noted that any flocculant make-up system which is piped directly to a potable water line must contain an approved positive back-flow prevention device.

A still simpler method of feeding emulsion-type flocculant to an open system is illustrated in Figure 18. Here, the only components are an emulsion metering pump, a static mixer, and a container of large enough size to give the retention time needed for complete activation of the flocculant. The container could be something as simple as an empty drum or several drums in series. Additional mixing in the container is not needed.

When feeding to closed systems, the same technique may be used except that the aging tank must be closed. Obviously, this system will work only where plant water pressure is substantially higher than the pressure in the line into which the flocculant is being pumped. In plants where line water pressure is not high enough, it may be supplemented by a simple booster pump. When using such a system, it is necessary to vary only the flocculant pump to change dosages since the actual solution concentration is not critical as long as there is sufficient water to keep the final concentration at 1% or preferably less. When a change in the polymer pump setting is made, a delay equal to the retention time in the tank is encountered before the new concentration gets to the point of application. However, since the retention time is seldom more than 15 to 30 min, this delay will not affect most uses. A continuous closed system emulsion feeder is illustrated in Figure 19.

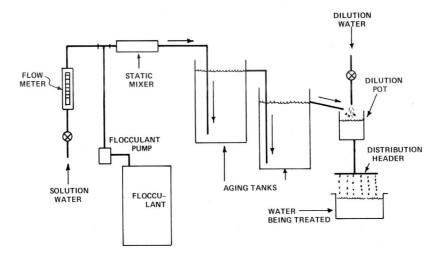

FIGURE 18.   Emulsion feed system for open channels.

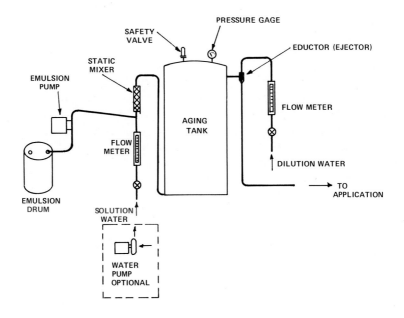

FIGURE 19.   Continuous closed system emulsion feeder.

4. Automatic Feed Systems — Commercial Units

There is equipment now on the market which will continuously and automatically draw flocculant emulsion from a drum or bulk storage tank, prepare a dilute solution, and finally feed the solution from a storage tank to the point of application at the desired rate.

Installation of the commercial unit shown in Figures 20 and 21 takes only a few minutes. First, attach the unit's single power line to a source of 110V, 25 amp service. Attach a plant water line to the pump reservoir box. Connect the flocculant solution pump to the point of polymer application. Attach the emulsion pump suction line to a drum of flocculant or to a bulk storage tank. (Note — the unit illustrated is equipped with a drum mixer which operates automatically off a timer in the control box.)

FIGURE 20.   Commercial emulsion feeder — photograph. (Courtesy Gaco Manufacturing, Bartlesville, Okla.)

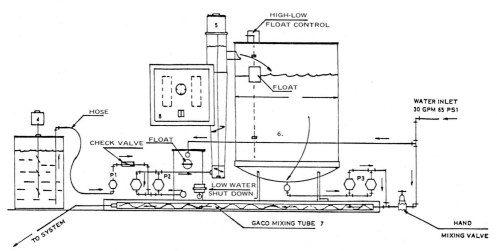

FIGURE 21.   Commercial emulsion feeder — schematic. (Courtesy Gaco Manufacturing, Bartlesville, Okla.)

After installation, the unit illustrated in Figure 21 would be operated by first checking the output of the water pump (P2). This nonvariable speed piston pump delivers about 250 gal/hr. Secondly, the SCR controlled emulsion pump (P1) would be calibrated. Finally, the SCR-controlled flocculant solution pump would be calibrated. With the fixed water flow rate, the emulsion pump would next be set to supply emulsion for the desired solution concentration, usually not over 2%. Hereafter, the water pump and emulsion pump will start and stop simultaneously on command from the float switch in the solution storage tank. As can be seen in Figure 21, the emulsion enters the water stream just before the water pump. Passing through the pump gives an initial mix to the water and emulsion. At this point, the partially mixed stream enters the tall mixing column where the emulsion is "broken" by a high-speed blade at the bottom. The flocculant solution rises in the column and overflows into the solution storage. No further mixing is required in the storage tank since only a short aging period is needed to bring the flocculant solution to full activity.

Also illustrated in Figure 21 is a secondary mixing tube which is an auxiliary piece of equipment used when it is desired to further dilute the solution being fed from the storage tank.

The emulsion feeder pictured in Figure 20 is capable of converting two drums per day (~900 lb) of emulsion into 2% flocculant solution. Larger units are available which will handle up to 12 drums per day. Obviously, these feed devices will also operate off a bulk storage tank for the liquid polymer.

D. Gel Products

To use polyelectrolytes in the gel form, it is necessary to have a grinder or granulator specially designed to chop the 40-lb blocks of gel into particles small enough to give reasonable solution rates. This special machine uses a powerful auger screw to break up the large gel blocks as they are fed down from a storage hopper. The still large gel pieces are extruded by the screw, through a perforated plate where high-speed knife blades chop them into tiny pieces. A stream of water passing through the chamber housing the rotating blades carries the cut-up polymer gel to a storage tank where dissolution is completed with conventional mixing. When the gel has completely dissolved, it is transferred to a holding tank from which it is fed to the point of application. As with the other flocculant forms, gel feeding can be completely automated by the incorporation of the appropriate tank level controls, solenoid valves, and sequence timers. The unit is capable of dispensing gel, as a partially dissolved slurry in water, at rates up to 20 lb/min.

A typical installation for dissolving and feeding gel polymer is shown in Figure 22.

VII. FLOCCULANT FEED RATE MEASUREMENT AND DOSAGE CONTROL

A. Feed Rate Monitoring
1. Chemical Metering Pumps

Chemical metering pumps are the most common method of controlling the flow of flocculant to the point of use. As mentioned earlier in the chapter, a wide variety of pumps have been used successfully for feeding polymer. Most popular are the diaphragm, gear, piston, lobe, progressive cavity, and roto-dip. Each type has its benefits and shortcomings. Care should be taken, therefore, to fit the feed pump to the end use. Materials of solution pump construction would be the same as for the transfer pumps discussed earlier.

Once calibrated by plotting pump delivery against dial settings, a good pump will

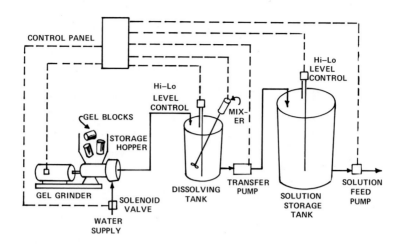

FIGURE 22.   Automatic flocculant gel feeder.

continue to deliver at the set rate over a long period of time. However, things like fouled check valves, changes in flocculant viscosity, and changes in back pressure can affect pump delivery. It is, therefore, good practice to recheck pump delivery rates periodically and also to monitor flocculant usage by measuring drum and storage tank levels.

2. Flow Meters

These are commonly used to measure the output of feed pumps as well as to monitor the flow from gravity and pressurized systems. Increasing in popularity are flow meters with direct digital readout based on both the rotating vane and sonic principle. Like the older tube-type meters, both the vane and sonic meters must be calibrated for each type and concentration of flocculant.

3. Flocculant Concentrations

Flocculant concentration as well as flow rate must sometimes be monitored. This is usually of special importance when applying flocculants to paper machines. Since both the conductivity and refractive index of water are affected by the addition of flocculants, either of these properties may be used to measure flocculant concentration once a standard curve has been established. Naturally corrections in reading will have to be made as changes occur in conductivity and refractive index of the raw water. In an automatic feed system, it would be possible to have the flocculant makeup pump respond to a signal from a conductivity probe located in the flocculant solution stream.

A typical concentration conductivity curve for a flocculant is shown in Figure 23.

B. Feed Regulation

Flocculant dosage (ppm or pounds per dry ton) is governed principally by the amount of solids in the treated stream and the nature of the solids being treated. The flocculant feed rate is determined by the dosage required and the flow rate of the stream being treated. pH also can have a profound effect on flocculant requirements, but usually that property is measured and adjustments made before the point of flocculant addition.

Pumps are on the market, the output of which can be varied in two ways, by stroke length or stroke frequency. Both of these parameters can be changed automatically in response to an electrical signal. It is, therefore, possible to have such a pump change

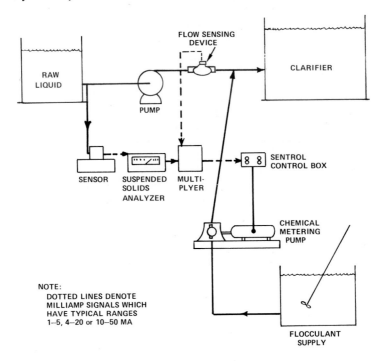

FIGURE 23.   Automatic flocculant dosage control. (Courtesy AMF Cuno, Meriden, Conn.)

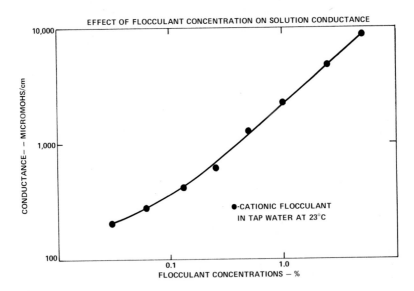

FIGURE 24.   Conductivity/concentration curve.

stroke length in response to a signal from a flow meter located in the stream being treated and to change stroke frequency in response to a turbidity meter also located in the stream. Figure 24 illustrates the system described.

Many plants regulate their flocculant dosage in response to the residual electrical charge (zeta potential) on the solids suspended in the treated water. However, this remains a manual operation since a charge analyzer has not been successfully coupled to flocculant feed pumps to give an automatic system.

INDEX